COMPUTER
VISION

COMPUTER VISION

Dana H. Ballard
Christopher M. Brown

Department of Computer Science
University of Rochester
Rochester, New York

PRENTICE-HALL, INC., Englewood Cliffs, New Jersey 07632

Library of Congress Cataloging in Publication Data

BALLARD, DANA HARRY.
    Computer vision.

    Bibliography: p.
    Includes index.
    1. Image processing.   I. Brown, Christopher M.
II. Title.
TA1632.B34   621.38′0414   81-20974
ISBN   0-13-165316-4   AACR2

Cover design by Robin Breite

© 1982 by Prentice-Hall, Inc.
Englewood Cliffs, New Jersey 07632

Printed in the United States of America

10 9 8 7 6 5

ISBN 0-13-165316-4

PRENTICE-HALL INTERNATIONAL, INC., *London*
PRENTICE-HALL OF AUSTRALIA PTY. LIMITED, *Sydney*
PRENTICE-HALL OF CANADA, LTD., *Toronto*
PRENTICE-HALL OF INDIA PRIVATE LIMITED, *New Delhi*
PRENTICE-HALL OF JAPAN, INC., *Tokyo*
PRENTICE-HALL OF SOUTHEAST ASIA PTE. LTD., *Singapore*
WHITEHALL BOOKS LIMITED, *Wellington, New Zealand*

Contents

Part I
GENERALIZED IMAGES
13

PART II
SEGMENTED IMAGES
115

11 MATCHING   352

12 INFERENCE   383

13 GOAL ACHIEVEMENT 438

APPENDICES
465

A1 SOME MATHEMATICAL TOOLS 465

A2 ADVANCED CONTROL MECHANISMS 497

AUTHOR INDEX 509

SUBJECT INDEX 513

Preface

The dream of intelligent automata goes back to antiquity; its first major articulation in the context of digital computers was by Turing around 1950. Since then, this dream has been pursued primarily by workers in the field of *artificial intelligence,* whose goal is to endow computers with information-processing capabilities comparable to those of biological organisms. From the outset, one of the goals of artificial intelligence has been to equip machines with the capability of dealing with sensory inputs.

Computer vision is the construction of explicit, meaningful descriptions of physical objects from images. Image understanding is very different from image processing, which studies image-to-image transformations, not explicit description building. Descriptions are a prerequisite for recognizing, manipulating, and thinking about objects.

We perceive a world of coherent three-dimensional objects with many invariant properties. Objectively, the incoming visual data do not exhibit corresponding coherence or invariance; they contain much irrelevant or even misleading variation. Somehow our visual system, from the retinal to cognitive levels, understands, or imposes order on, chaotic visual input. It does so by using *intrinsic information* that may reliably be extracted from the input, and also through assumptions and *knowledge* that are applied at various levels in visual processing.

The challenge of computer vision is one of *explicitness*. Exactly what information about scenes can be extracted from an image using only very basic assumptions about physics and optics? Explicitly, what computations must be performed? Then, at what stage must domain-dependent, prior knowledge about the world be incorporated into the understanding process? How are world models and knowledge represented and used? This book is about the representations and mechanisms that allow image information and prior knowledge to interact in image understanding.

Computer vision is a relatively new and fast-growing field. The first experiments were conducted in the late 1950s, and many of the essential concepts have been developed during the last five years. With this rapid growth, crucial ideas have arisen in disparate areas such as artificial intelligence, psychology, computer graphics, and image processing. Our intent is to assemble a selection of this material in a form that will serve both as a senior/graduate-level academic text and as a useful reference to those building vision systems. This book has a strong artificial intelligence flavor, and we hope this will provoke thought. We believe that both the intrinsic image information and the internal model of the world are important in successful vision systems.

The book is organized into four parts, based on descriptions of objects at four different levels of abstraction.

1. Generalized images—images and image-like entities.
2. Segmented images—images organized into subimages that are likely to correspond to "interesting objects."
3. Geometric structures—quantitative models of image and world structures.
4. Relational structures—complex symbolic descriptions of image and world structures.

The parts follow a progression of increasing abstractness. Although the four parts are most naturally studied in succession, they are not tightly interdependent. Part I is a prerequisite for Part II, but Parts III and IV can be read independently.

Parts of the book assume some mathematical and computing background (calculus, linear algebra, data structures, numerical methods). However, throughout the book mathematical rigor takes a backseat to concepts. Our intent is to transmit a set of ideas about a new field to the widest possible audience.

In one book it is impossible to do justice to the scope and depth of prior work in computer vision. Further, we realize that in a fast-developing field, the rapid influx of new ideas will continue. We hope that our readers will be challenged to think, criticize, read further, and quickly go beyond the confines of this volume.

D. H. Ballard
C. M. Brown

Acknowledgments

Jerry Feldman and Herb Voelcker (and through them the University of Rochester) provided many resources for this work. One of the most important was a capable and forgiving staff (secretarial, technical, and administrative). For massive text editing, valuable advice, and good humor we are especially grateful to Rose Peet. Peggy Meeker, Jill Orioli, and Beth Zimmerman all helped at various stages.

Several colleagues made suggestions on early drafts: thanks to James Allen, Norm Badler, Larry Davis, Takeo Kanade, John Kender, Daryl Lawton, Joseph O'Rourke, Ari Requicha, Ed Riseman, Azriel Rosenfeld, Mike Schneier, Ken Sloan, Steve Tanimoto, Marty Tenenbaum, and Steve Zucker. Anil K. Jain, Janet Walz, and the Japanese translation team, led by Teruo Fukumura, pointed out many errors in the first printing.

Graduate students helped in many different ways: thanks especially to Michel Denber, Alan Frisch, Lydia Hrechanyk, Mark Kahrs, Keith Lantz, Joe Maleson, Lee Moore, Mark Peairs, Don Perlis, Rick Rashid, Dan Russell, Dan Sabbah, Bob Schudy, Peter Selfridge, Uri Shani, and Bob Tilove. Bernhard Stuth deserves special mention for much careful and critical reading.

Finally, thanks go to Jane Ballard, mostly for standing steadfast through the cycles of elation and depression and for numerous engineering-to-English translations.

As Pat Winston put it: "A willingness to help is not an implied endorsement." The aid of others was invaluable, but we alone are responsible for the opinions, technical details, and faults of this book.

Funding assistance was provided by the Sloan Foundation under Grant 78-4-15, by the National Institutes of Health under Grant HL21253, and by the Defense Advanced Research Projects Agency under Grant N00014-78-C-0164.

The authors wish to credit the following sources for figures and tables. For complete citations given here in abbreviated form (as "from . . ." or "after . . ."), refer to the appropriate chapter-end references.

Fig. 1.2 from Shani, U., "A 3-D model-driven system for the recognition of abdominal anatomy from CT scans," TR77, Dept. of Computer Science, University of Rochester, May 1980.

Fig. 1.4 courtesy of Allen Hanson and Ed Riseman, COINS Research Project, University of Massachusetts, Amherst, MA.

Fig. 2.4 after Horn and Sjoberg, 1978.

Figs. 2.5, 2.9, 2.10, 3.2, 3.6, and 3.7 courtesy of Bill Lampeter.

Fig. 2.7a painting by Louis Condax; courtesy of Eastman Kodak Company and the Optical Society of America.

Fig. 2.8a courtesy of D. Greenberg and G. Joblove, Cornell Program of Computer Graphics.

Fig. 2.8b courtesy of Tom Check.

Table 2.3 after Gonzalez and Wintz, 1977.

Fig. 2.18 courtesy of EROS Data Center, Sioux Falls, SD.

Figs. 2.19 and 2.20 from Herrick, C.N., *Television Theory and Servicing: Black/White and Color,* 2nd Ed. Reston, VA: Reston, 1976.

Figs. 2.21, 2.22, 2.23, and 2.24 courtesy of Michel Denber.

Fig. 2.25 from Popplestone et al., 1975.

Fig. 2.26 courtesy of Production Automation Project, University of Rochester.

Fig. 2.27 from Waag and Gramiak, 1976.

Fig. 3.1 courtesy of Marty Tenenbaum.

Fig. 3.8 after Horn, 1974.

Figs. 3.14 and 3.15 after Frei and Chen, 1977.

Figs. 3.17 and 3.18 from Zucker, S.W. and R.A. Hummel, "An optimal 3-D edge operator," *IEEE Trans. PAMI 3,* May 1981, pp. 324-331.

Fig. 3.19 curves are based on data in Abdou, 1978.

Figs. 3.20, 3.21, and 3.22 from Prager, J.M., "Extracting and labeling boundary segments in natural scenes," *IEEE Tans. PAMI 12,* 1, January 1980. © 1980 IEEE.

Figs. 3.23, 3.28, 3.29, and 3.30 courtesy of Berthold Horn.

Figs. 3.24 and 3.26 from Marr, D. and T. Poggio, "Cooperative computation of stereo disparity," *Science,* Vol. 194, 1976, pp. 283-287. © 1976 by the American Association for the Advancement of Science.

Fig. 3.31 from Woodham, R.J., "Photometric stereo: A reflectance map technique for determining surface orientation from image intensity," *Proc. SPIE,* Vol. 155, August 1978.

Figs. 3.33 and 3.34 after Horn and Schunck, 1980.

Fig. 3.37 from Tanimoto, S. and T. Pavlidis, "A hierarchical data structure for picture processing," *CGIP 4,* 2, June 1975, pp. 104-119.

Fig. 4.6 from Kimme et al., 1975.

Figs. 4.7 and 4.16 from Ballard and Sklansky, 1976.

Fig. 4.9 courtesy of Dana Ballard and Ken Sloan.

Figs. 4.12 and 4.13 from Ramer, U., "Extraction of line structures from photgraphs of curved objects," *CGIP 4,* 2, June 1975, pp. 81-103.

Fig. 4.14 courtesy of Jim Lester, Tufts/New England Medical Center.

Fig. 4.17 from Chien, Y.P. and K.S. Fu, "A decision function method for boundary detection," *CGIP 3,* 2, June 1974, pp. 125-140.

Fig. 5.3 from Ohlander, R., K. Price, and D.R. Reddy, "Picture segmentation using a recursive region splitting method," *CGIP 8,* 3, December 1979.

Fig. 5.4 courtesy of Sam Kapilivsky.

Figs. 6.1, 11.16, and A1.13 courtesy of Chris Brown.

Fig. 6.3 courtesy of Joe Maleson and John Kender.

Fig. 6.4 from Connors, 1979. Texture images by Phil Brodatz, in Brodatz, *Textures.* New York: Dover, 1966.

Fig. 6.9 texture image by Phil Brodatz, in Brodatz, *Textures.* New York: Dover, 1966.

Figs. 6.11, 6.12, and 6.13 from Lu, S.Y. and K.S. Fu, "A syntactic approach to texture analysis," *CGIP 7,* 3, June 1978, pp. 303-330.

Fig. 6.14 from Jayaramamurthy, S.N., "Multilevel array grammars for generating texture scenes," *Proc. PRIP,* August 1979, pp. 391-398. © 1979 IEEE.

Fig. 6.20 from Laws, 1980.

Figs. 6.21 and 6.22 from Maleson et al., 1977.

Fig. 6.23 courtesy of Joe Maleson.

Figs. 7.1 and 7.3 courtesy of Daryl Lawton.

Fig. 7.2 after Prager, 1979.

Figs. 7.4 and 7.5 from Clocksin, W.F., "Computer prediction of visual thresholds for surface slant and edge detection from optical flow fields," Ph.D. dissertation, University of Edinburgh, 1980.

Fig. 7.7 courtesy of Steve Barnard and Bill Thompson.

Figs. 7.8 and 7.9 from Rashid, 1980.

Fig. 7.10 courtesy of Joseph O'Rourke.

Figs. 7.11 and 7.12 after Aggarwal and Duda, 1975.

Fig. 7.13 courtesy of Hans-Hellmut Nagel.

Fig. 8.1d after Requicha, 1977.

Figs. 8.2, 8.3, 8.21a, 8.22, and 8.26 after Pavlidis, 1977.

Figs. 8.10, 8.11, 9.6, and 9.16 courtesy of Uri Shani.

Figs. 8.12, 8.13, 8.14, 8.15, and 8.16 from Ballard, 1981.

Fig. 8.21 b from Preston, K., Jr., M.J.B. Duff, S. Levialdi, P.E. Norgren, and J-i. Toriwaki, "Basics of cellular logic with some applications in medical image processing," *Proc. IEEE,* Vol. 67, No. 5, May 1979, pp. 826–856.

Figs. 8.25, 9.8, 9.9, 9.10, and 11.3 courtesy of Robert Schudy.

Fig. 8.29 after Bribiesca and Guzman, 1979.

Figs. 9.1, 9.18, 9.19, and 9.27 courtesy of Ari Requicha.

Fig. 9.2 from Requicha, A.A.G., "Representations for rigid solids: theory, methods, systems," *Computer Surveys 12,* 4, December 1980.

Fig. 9.3 courtesy of Lydia Hrechanyk.

Figs. 9.4 and 9.5 after Baumgart, 1972.

Fig. 9.7 courtesy of Peter Selfridge.

Fig. 9.11 after Requicha, 1980.

Figs. 9.14 and 9.15b from Agin, G.J. and T.O. Binford, "Computer description of curved objects," *IEEE Trans. on Computers 25,* 1, April 1976.

Fig. 9.15a courtesy of Gerald Agin.

Fig. 9.17 courtesy of A. Christensen; published as frontispiece of *ACM SIGGRAPH 80 Proceedings.*

Fig. 9.20 from Marr and Nishihara, 1978.

Fig. 9.21 after Tilove, 1980.

Fig. 9.22b courtesy of Gene Hartquist.

Figs. 9.24, 9.25, and 9.26 from Lee and Requicha, 1980.

Figs. 9.28a, 9.29, 9.30, 9.31, 9.32, 9.35, and 9.37 and Table 9.1 from Brown, C. and R. Popplestone, "Cases in scene analysis," in *Pattern Recognition,* ed. B.G. Batchelor. New York: Plenum, 1978.

Fig. 9.28b from Guzman, A., "Decomposition of a visual scene into three-dimensional bodies," in *Automatic Interpretation and Classification of Images,* A. Grasseli, ed., New York: Academic Press, 1969.

Fig. 9.28c from Waltz, D., "Understanding line drawing of scenes with shadows," in *The Psychology of Computer Vision,* ed. P.H. Winston. New York: McGraw-Hill, 1975.

Fig. 9.28d after Turner, 1974.

Figs. 9.33, 9.38, 9.40, 9.42, 9.43, and 9.44 after Mackworth, 1973.

Figs. 9.39, 9.45, 9.46, and 9.47 and Table 9.2 after Kanade, 1978.

Figs. 10.2 and A2.1 courtesy of Dana Ballard.

Figs. 10.16, 10.17, and 10.18 after Russell, 1979.

Fig. 11.5 after Fischler and Elschlager, 1973.

Fig. 11.8 after Ambler et al., 1975.

Fig. 11.10 from Winston, P.H., "Learning structural descriptions from examples," in *The Psychology of Computer Vision,* ed. P.H. Winston. New York: McGraw-Hill, 1975.

Fig. 11.11 from Nevatia, 1974.

Fig. 11.12 after Nevatia, 1974.

Fig. 11.17 after Barrow and Popplestone, 1971.

Fig. 11.18 from Davis, L.S., "Shape matching using relaxation techniques," *IEEE Trans. PAMI 1,* 4, January 1979, pp. 60–72.

Figs. 12.4 and 12.5 from Sloan and Bajcsy, 1979.

Fig. 12.6 after Barrow and Tenenbaum, 1976.

Fig. 12.8 after Freuder, 1978.

Fig. 12.10 from Rosenfeld, A.R., A. Hummel, and S.W. Zucker, "Scene labeling by relaxation operations," *IEEE Trans. SMC 6,* 6, June 1976, p. 420.

Figs. 12.11, 12.12, 12.13, 12.14, and 12.15 after Hinton, 1979.

Fig. 13.3 courtesy of Aaron Sloman.

Figs. 13.6, 13.7, and 13.8 from Garvey, 1976.

Fig. A1.11 after Duda and Hart, 1973.

Figs. A2.2 and A2.3 from Hanson, A.R. and E.M. Riseman, "VISIONS: A computer system for interpreting scenes," in *Computer Vision Systems,* ed. A.R. Hanson and E.M. Riseman. New York: Academic Press, 1978.

Mnemonics
for Proceedings and Special Collections
Cited in the References

CGIP

Computer Graphics and Image Processing

COMPSAC

IEEE Computer Society's 3rd International Computer Software and Applications Conference, Chicago, November 1979.

CVS

Hanson, A. R. and E. M. Riseman (Eds.). *Computer Vision Systems.* New York: Academic Press, 1978.

DARPA IU

Defense Advanced Research Projects Agency Image Understanding Workshop, Minneapolis, MN, April 1977.

Defense Advanced Research Projects Agency Image Understanding Workshop, Palo Alto, CA, October 1977.

Defense Advanced Research Projects Agency Image Understanding Workshop, Cambridge, MA, May 1978.

Defense Advanced Research Projects Agency Image Understanding Workshop, Carnegie-Mellon University, Pittsburgh, PA, November 1978.

Defense Advanced Research Projects Agency Image Understanding Workshop, University of Maryland, College Park, MD, April 1980.

IJCAI

2nd International Joint Conference on Artificial Intelligence, Imperial College, London, September 1971.

4th International Joint Conference on Artificial Intelligence, Tbilisi, Georgia, USSR, September 1975.

5th International Joint Conference on Artificial Intelligence, MIT, Cambridge, MA, August 1977.

6th International Joint Conference on Artificial Intelligence, Tokyo, August 1979.

IJCPR

2nd International Joint Conference on Pattern Recognition, Copenhagen, August 1974.

3rd International Joint Conference on Pattern Recognition, Coronado, CA, November 1976.

4th International Joint Conference on Pattern Recognition, Kyoto, November 1978.

5th International Joint Conference on Pattern Recognition, Miami Beach, FL, December 1980.

MI4

Meltzer, B. and D. Michie (Eds.). *Machine Intelligence 4*. Edinburgh: Edinburgh University Press, 1969.

MI5

Meltzer, B. and D. Michie (Eds.). *Machine Intelligence 5*. Edinburgh: Edinburgh University Press, 1970.

MI6

Meltzer, B. and D. Michie (Eds.). *Machine Intelligence 6*. Edinburgh: Edinburgh University Press, 1971.

M17

Meltzer, B. and D. Michie (Eds.). *Machine Intelligence 7*. Edinburgh: Edinburgh University Press, 1972.

PCV

Winston, P. H. (Ed.). *The Psychology of Computer Vision*. New York: McGraw-Hill, 1975.

PRIP

IEEE Computer Society Conference on Pattern Recognition and Image Processing, Chicago, August 1979.

Computer
Vision

1

Computer Vision Issues

1.1 ACHIEVING SIMPLE VISION GOALS

Suppose that you are given an aerial photo such as that of Fig. 1.1a and asked to locate ships in it. You may never have seen a naval vessel in an aerial photograph before, but you will have no trouble predicting generally how ships will appear. You might reason that you will find no ships inland, and so turn your attention to ocean areas. You might be momentarily distracted by the glare on the water, but realizing that it comes from reflected sunlight, you perceive the ocean as continuous and flat. Ships on the open ocean stand out easily (if you have seen ships from the air, you know to look for their wakes). Near the shore the image is more confusing, but you know that ships close to shore are either moored or docked. If you have a map (Fig. 1.1b), it can help locate the docks (Fig. 1.1c); in a low-quality photograph it can help you identify the shoreline. Thus it might be a good investment of your time to establish the correspondence between the map and the image. A search parallel to the shore in the dock areas reveals several ships (Fig. 1.1d).

Again, suppose that you are presented with a set of computer-aided tomographic (CAT) scans showing "slices" of the human abdomen (Fig. 1.2a). These images are products of high technology, and give us views not normally available even with x-rays. Your job is to reconstruct from these cross sections the three-dimensional shape of the kidneys. This job may well seem harder than finding ships. You first need to know what to look for (Fig. 1.2b), where to find it in CAT scans, and how it looks in such scans. You need to be able to "stack up" the scans mentally and form an internal model of the shape of the kidney as revealed by its slices (Fig. 1.2c and 1.2d).

This book is about *computer vision*. These two example tasks are typical com-

1

puter vision tasks; both were solved by computers using the sorts of knowledge and techniques alluded to in the descriptive paragraphs. Computer vision is the enterprise of automating and integrating a wide range of processes and representations used for vision perception. It includes as parts many techniques that are useful by themselves, such as *image processing* (transforming, encoding, and transmitting images) and *statistical pattern classification* (statistical decision theory applied to general patterns, visual or otherwise). More importantly for us, it includes techniques for geometric modeling and cognitive processing.

1.2 HIGH-LEVEL AND LOW-LEVEL CAPABILITIES

The examples of Section 1.1 illustrate vision that uses *cognitive processes*, *geometric models*, *goals*, and *plans*. These *high-level* processes are very important; our examples only weakly illustrate their power and scope. There surely would be some overall purpose to finding ships; there might be collateral information that there were submarines, barges, or small craft in the harbor, and so forth. CAT scans would be used with several diagnostic goals in mind and an associated medical history available. Goals and knowledge are high-level capabilities that can guide visual activities, and a visual system should be able to take advantage of them.

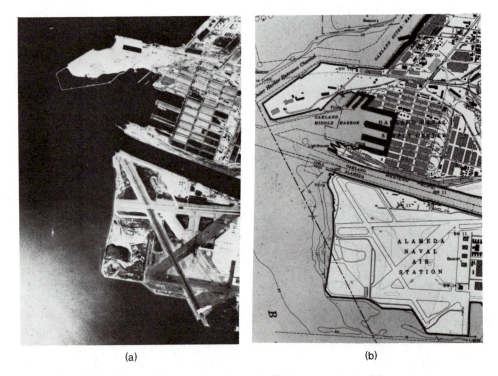

(a)   (b)

Fig. 1.1 Finding ships in an aerial photograph. (a) The photograph; (b) a corresponding map; (c) the dock area of the photograph; (d) registered map and image, with ship location.

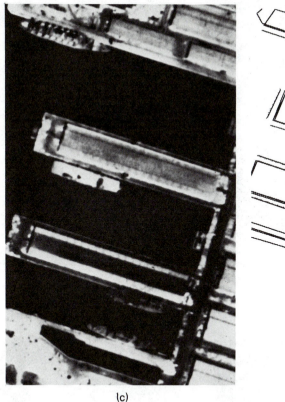

(c)

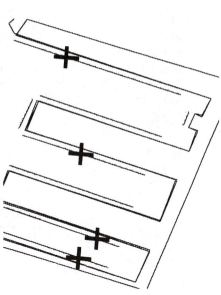

(d)

Fig. 1.1 (cont.)

Even such elaborated tasks are very special ones and in their way easier to think about than the commonplace visual perceptions needed to pick up a baby, cross a busy street, or arrive at a party and quickly "see" who you know, your host's taste in decor, and how long the festivities have been going on. All these tasks require judgment and large amounts of knowledge of objects in the world, how they look, and how they behave. Such high-level powers are so well integrated into "vision" as to be effectively inseparable.

Knowledge and goals are only part of the vision story. Vision requires many *low-level* capabilities we often take for granted; for example, our ability to extract *intrinsic images* of "lightness," "color," and "range." We perceive black as black in a complex scene even when the lighting is such that some black patches are reflecting more light than some white patches. Similarly, perceived colors are not related simply to the wavelengths of reflected light; if they were, we would consciously see colors changing with illumination. Stereo fusion (stereopsis) is a low-level facility basic to short-range three-dimensional perception.

An important low-level capability is *object perception*: for our purposes it does not really matter if this talent is innate, ("hard-wired"), or if it is developmental or even learned ("compiled-in"). The fact remains that mature biological vision systems are specialized and tuned to deal with the relevant objects in their environ-

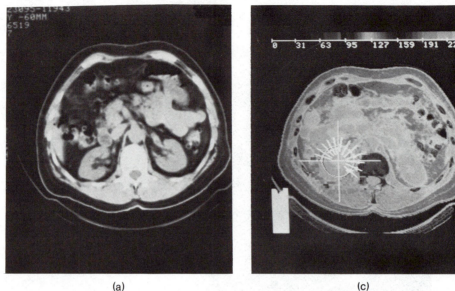

(a) (c)

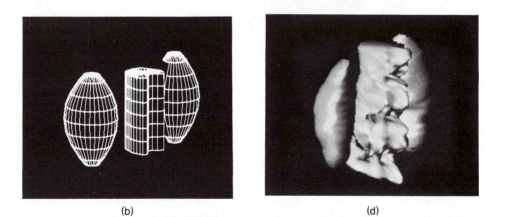

(b) (d)

Fig. 1.2 Finding a kidney in a computer-aided tomographic scan. (a) One slice of scan data;
(b) prototype kidney model; (c) model fitting; (d) resulting kidney and spinal cord instances.

ments. Further specialization can often be learned, but it is built on basic immutable assumptions about the world which underlie the vision system.

A basic sort of object recognition capability is the "figure/ground" discrimination that separates objects from the "background." Other basic organizational predispositions are revealed by the "Gestalt laws" of clustering, which demonstrate rules our vision systems use to form simple arrays of stimuli into more coherent spatial groups. A dramatic example of specialized object perception for human beings is revealed in our "face recognition" capability, which seems to occupy a large volume of brain matter. Geometric visual illusions are more surprising symptoms of nonintuitive processing that is performed by our vision systems, either for some direct purpose or as a side effect of its specialized architecture. Some other illusions clearly reflect the intervention of high-level knowledge. For instance, the familiar "Necker cube reversal" is grounded in our three-dimensional models for cubes.

Low-level processing capabilities are elusive; they are unconscious, and they are not well connected to other systems that allow direct introspection. For instance, our visual memory for images is quite impressive, yet our quantitative verbal descriptions of images are relatively primitive. The biological visual "hardware" has been developed, honed, and specialized over a very long period. However, its organization and functionality is not well understood except at extreme levels of detail and generality—the behavior of small sets of cat or monkey cortical cells and the behavior of human beings in psychophysical experiments.

Computer vision is thus immediately faced with a very difficult problem; it must reinvent, with general digital hardware, the most basic and yet inaccessible talents of specialized, parallel, and partly analog biological visual systems. Figure 1.3 may give a feeling for the problem; it shows two visual renditions of a familiar subject. The inset is a normal image, the rest is a plot of the intensities (gray levels) in the image against the image coordinates. In other words, it displays information

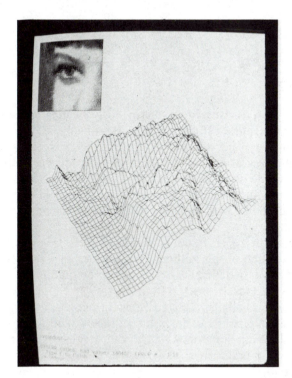

Fig. 1.3 Two representations of an image. One is directly accessible to our low-level processes; the other is not.

with "height" instead of "light." No information is lost, and the display is an image-like object, but we do not immediately see a face in it. The initial representation the computer has to work with is no better; it is typically just an array of numbers from which human beings could extract visual information only very painfully. Skipping the low-level processing we take for granted turns normally effortless perception into a very difficult puzzle.

Computer vision is vitally concerned with both low-level or "early processing" issues and with the high-level and "cognitive" use of knowledge. Where does vision leave off and reasoning and motivation begin? We do not know precisely, but we firmly believe (and hope to show) that powerful, cooperating, rich representations of the world are needed for any advanced vision system. Without them, no system can derive relevant and invariant information from input that is beset with ever-changing lighting and viewpoint, unimportant shape differences, noise, and other large but irrelevant variations. These representations can remove some computational load by predicting or assuming structure for the visual world.

Finally, if a system is to be successful in a variety of tasks, it needs some "meta-level" capabilities: it must be able to model and reason about its own goals and capabilities, and the success of its approaches. These complex and related models must be manipulated by cognitive-like techniques, even though introspectively the perceptual process does not always "feel" to us like cognition.

Computer Vision Systems

1.3 A RANGE OF REPRESENTATIONS

Visual perception is the relation of visual input to previously existing *models* of the world. There is a large representational gap between the image and the models ("ideas," "concepts") which explain, describe, or abstract the image information. To bridge that gap, computer vision systems usually have a (loosely ordered) *range of representations* connecting the input and the "output" (a final description, decision, or interpretation). Computer vision then involves the design of these intermediate representations and the implementation of algorithms to construct them and relate them to one another.

We broadly categorize the representations into four parts (Fig. 1.4) which correspond with the organization of this volume. Within each part there may be several layers of representation, or several cooperating representations. Although the sets of representations are loosely ordered from "early" and "low-level" *signals* to "late" and "*cognitive*" symbols, the actual flow of effort and information between them is not unidirectional. Of course, not all levels need to be used in each computer vision application; some may be skipped, or the processing may start partway up the hierarchy or end partway down it.

Generalized images (Part I) are *iconic* (image-like) and *analogical* representations of the input data. Images may initially arise from several technologies.

(a)

(b)

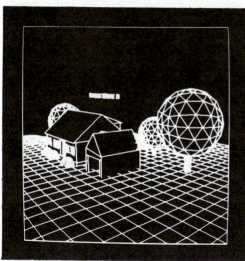

(c)

Fig. 1.4 Examples of the four categories of representation used in computer vision. (a) Iconic; (b) segmented; (c) geometric; (d) relational (next page).

Domain-independent processing can produce other iconic representations more directly useful to later processing, such as arrays of *edge elements* (gray-level discontinuities). *Intrinsic images* can sometimes be produced at this level—they reveal physical properties of the imaged scene (such as surface orientations, range, or surface reflectance). Often *parallel processing* can produce generalized images. More generally, most "low-level" processes can be implemented with parallel computation.

 Segmented images (Part II) are formed from the generalized image by gathering its elements into sets likely to be associated with meaningful *objects* in the scene. For instance, segmenting a scene of planar polyhedra (blocks) might result in a set of *edge segments* corresponding to polyhedral edges, or a set of two-

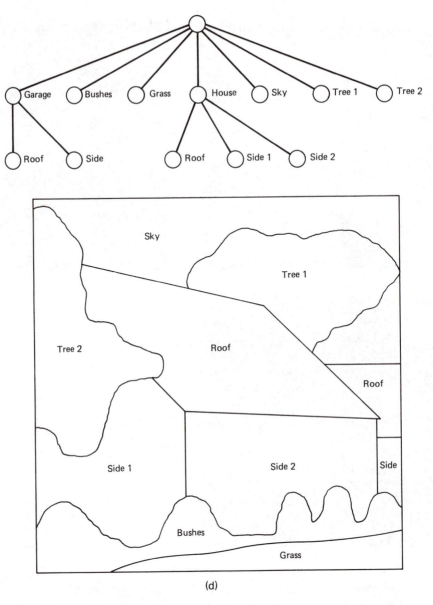

(d)

Fig. 1.4 (cont.)

dimensional *regions* in the image corresponding to polyhedral faces. In producing the segmented image, knowledge about the particular domain at issue begins to be important both to save computation and to overcome problems of *noise* and inadequate data. In the planar polyhedral example, it helps to know beforehand that the line segments must be straight. *Texture* and *motion* are known to be very important in segmentation, and are currently topics of active research; knowledge in these areas is developing very fast.

Geometric representations (Part III) are used to capture the all-important idea of two-dimensional and three-dimensional *shape*. Quantifying shape is as important as it is difficult. These geometric representations must be powerful enough to support complex and general processing, such as "simulation" of the effects of lighting and motion. Geometric structures are as useful for encoding previously acquired knowledge as they are for re-representing current visual input. Computer vision requires some basic mathematics; Appendix 1 has a brief selection of useful techniques.

Relational models (Part IV) are complex assemblages of representations used to support sophisticated high-level processing. An important tool in *knowledge representation* is *semantic nets*, which can be used simply as an organizational convenience or as a formalism in their own right. High-level processing often uses prior knowledge and models acquired prior to a perceptual experience. The basic mode of processing turns from *constructing* representations to *matching* them. At high levels, *propositional* representations become more important. They are made up of assertions that are true or false with respect to a model, and are manipulated by rules of *inference*. Inference-like techniques can also be used for *planning*, which models situations and actions through time, and thus must reason about temporally varying and hypothetical worlds. The higher the level of representation, the more marked is the flow of *control* (direction of attention, allocation of effort) downward to lower levels, and the greater the tendency of algorithms to exhibit *serial processing*. These issues of control are basic to complex information processing in general and computer vision in particular; Appendix 2 outlines some specific control mechanisms.

Figure 1.5 illustrates the loose classification of the four categories into analogical and propositional representations. We consider generalized and segmented images as well as geometric structures to be analogical models. Analogical models capture directly the relevant characteristics of the represented objects, and are manipulated and interrogated by simulation-like processes. Relational models are generally a mix of analogical and propositional representations. We develop this distinction in more detail in Chapter 10.

1.4 THE ROLE OF COMPUTERS

The computer is a congenial tool for research into visual perception.

- Computers are versatile and forgiving experimental subjects. They are easily and ethically reconfigurable, not messy, and their workings can be scrutinized in the finest detail.

- Computers are demanding critics. Imprecision, vagueness, and oversights are not tolerated in the computer implementation of a theory.

- Computers offer new metaphors for perceptual psychology (also neurology, linguistics, and philosophy). Processes and entities from computer science provide powerful and influential conceptual tools for thinking about perception and cognition.

- Computers can give precise measurements of the amount of processing they

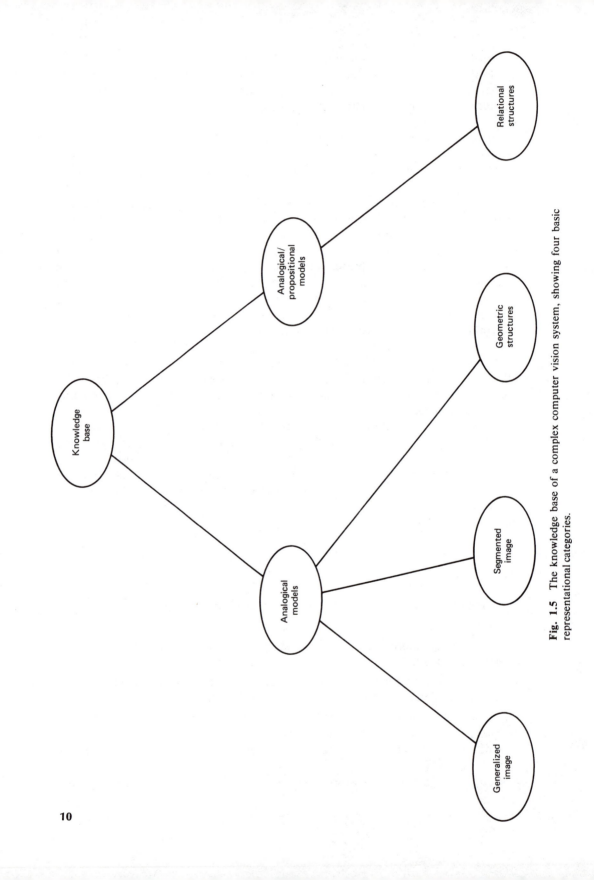

Fig. 1.5 The knowledge base of a complex computer vision system, showing four basic representational categories.

Table 1.1

EXAMPLES OF IMAGE ANALYSIS TASKS

Domain	Objects	Modality	Tasks	Knowledge Sources
Robotics	Three-dimensional outdoor scenes indoor scenes Mechanical parts	Light X-rays Light Structured light	Identify or describe objects in scene Industrial tasks	Models of objects Models of the reflection of light from objects
Aerial images	Terrain Buildings, etc.	Light Infrared Radar	Improved images Resource analyses Weather prediction Spying Missile guidance Tactical analysis	Maps Geometrical models of shapes Models of image formation
Astronomy	Stars Planets	Light	Chemical composition Improved images	Geometrical models of shapes
Medical Macro	Body organs	X-rays Ultrasound Isotopes Heat Electronmicroscopy	Diagnosis of abnor-malities Operative and treatment planning	Anatomical models Models of image formation
Micro	Cells Protein chains Chromosomes	Light	Pathology, cytology Karyotyping	Models of shape
Chemistry	Molecules	Electron densities	Analysis of molecular compositions	Chemical models Structured models
Neuroanatomy	Neurons	Light Electronmicroscopy	Determination of spatial orientation	Neural connectivity
Physics	Particle tracks	Light	Find new particles Identify tracks	Atomic physics

11

do. A computer implementation places an upper limit on the amount of computation necessary for a task.

- Computers may be used either to mimic what we understand about human perceptual architecture and processes, or to strike out in different directions to try to achieve similar ends by different means.
- Computer models may be judged either by their efficacy for applications and on-the-job performance or by their internal organization, processes, and structures—the theory they embody.

1.5 COMPUTER VISION RESEARCH AND APPLICATIONS

''Pure'' computer vision research often deals with relatively domain-independent considerations. The results are useful in a broad range of contexts. Almost always such work is demonstrated in one or more applications areas, and more often than not an initial application problem motivates consideration of the general problem. Applications of computer vision are exciting, and their number is growing as computer vision becomes better understood. Table 1.1 gives a partial list of ''classical'' and current applications areas.

Within the organization outlined above, this book presents many specific ideas and techniques with general applicability. It is meant to provide enough basic knowledge and tools to support attacks on both applications and research topics.

GENERALIZED IMAGES

1

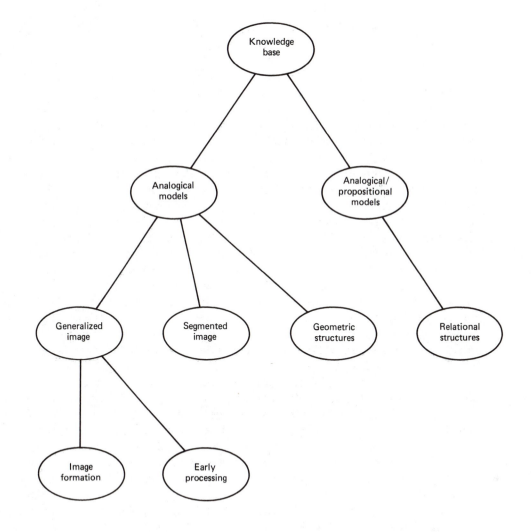

The first step in the vision process is image formation. Images may arise from a variety of technologies. For example, most television-based systems convert reflected light intensity into an electronic signal which is then digitized; other systems use more exotic radiations, such as x-rays, laser light, ultrasound, and heat. The net result is usually an array of samples of some kind of energy.

The vision system may be entirely passive, taking as input a digitized image from a microwave or infrared sensor, satellite scanner, or a planetary probe, but more likely involves some kind of *active imaging*. Automated active imaging systems may control the direction and resolution of sensors, or regulate and direct their own light sources. The light source itself may have special properties and structure designed to reveal the nature of the three-dimensional world; an example is to use a plane of light that falls on the scene in a stripe whose structure is closely related to the structure of opaque objects. Range data for the scene may be provided by stereo (two images), but also by triangulation using light-stripe techniques or by ''spotranging'' using laser light. A single hardware device may deliver range and multispectral reflectivity (''color'') information. The image-forming device may also perform various other operations. For example, it may automatically smooth or enhance the image or vary its resolution.

The *generalized image* is a set of related image-like entities for the scene. This set may include related images from several modalities, but may also include the results of significant processing that can extract *intrinsic images*. An intrinsic image is an ''image,'' or array, of representations of an important physical quantity such as surface orientation, occluding contours, velocity, or range. Object color, which is a different entity from sensed red−green−blue wavelengths, is an intrinsic quality. These intrinsic physical qualities are extremely useful; they can be related to physical objects far more easily than the original input values, which reveal the physical parameters only indirectly. An intrinsic image is a major step toward scene understanding and usually represents significant and interesting computations.

The information necessary to compute an intrinsic image is contained in the input image itself, and is extracted by "inverting" the transformation wrought by the imaging process, the reflection of radiation from the scene, and other physical processes. An example is the fusion of two stereo images to yield an intrinsic range image. Many algorithms to recover intrinsic images can be realized with *parallel* implementations, mirroring computations that may take place in the lower neurological levels of biological image processing.

All of the computations listed above benefit from the idea of *resolution pyramids*. A pyramid is a generalized image data structure consisting of the same image at several successively increasing levels of resolution. As the resolution increases, more samples are required to represent the increased information and hence the successive levels are larger, making the entire structure look like a pyramid. Pyramids allow the introduction of many different coarse-to-fine image-resolution algorithms which are vastly more efficient than their single-level, high-resolution-only counterparts.

Image
Formation   2

2.1 IMAGES

Image formation occurs when a *sensor* registers *radiation* that has interacted with *physical objects*. Section 2.2 deals with mathematical models of images and image formation. Section 2.3 describes several specific image formation technologies.

The mathematical model of imaging has several different components.

1. An *image function* is the fundamental abstraction of an image.
2. A *geometrical model* describes how three dimensions are projected into two.
3. A *radiometrical model* shows how the imaging geometry, light sources, and reflectance properties of objects affect the light measurement at the sensor.
4. A *spatial* frequency model describes how spatial variations of the image may be characterized in a transform domain.
5. A *color model* describes how different spectral measurements are related to image colors.
6. A *digitizing model* describes the process of obtaining discrete samples.

This material forms the basis of much image-processing work and is developed in much more detail elsewhere, e.g., [Rosenfeld and Kak 1976; Pratt 1978]. Our goals are not those of image processing, so we limit our discussion to a summary of the essentials.

The wide range of possible sources of samples and the resulting different implications for later processing motivate our overview of specific imaging techniques. Our goal is not to provide an exhaustive catalog, but rather to give an idea of the range of techniques available. Very different analysis techniques may be needed depending on how the image was formed. Two examples illustrate this point. If the image is formed by reflected light intensity, as in a photograph, the image records both light from primary light sources and (more usually) the light reflected off physical surfaces. We show in Chapter 3 that in certain cases we can use these kinds of images together with knowledge about physics to derive the orientation of the surfaces. If, on the other hand, the image is a computed tomogram of the human body (discussed in Section 2.3.4), the image represents tissue density of internal organs. Here orientation calculations are irrelevant, but general segmentation techniques of Chapters 4 and 5 (the agglomeration of neighboring samples of similar density into units representing organs) are appropriate.

2.2 IMAGE MODEL

Sophisticated image models of a statistical flavor are useful in image processing [Jain 1981]. Here we are concerned with more geometrical considerations.

2.2.1 Image Functions

An *image function* is a mathematical representation of an image. Generally, an image function is a vector-valued function of a small number of arguments. A special case of the image function is the *digital (discrete) image function*, where the arguments to and value of the function are all integers. Different image functions may be used to represent the same image, depending on which of its characteristics are important. For instance, a camera produces an image on black-and-white film which is usually thought of as a real-valued function (whose value could be the density of the photographic negative) of two real-valued arguments, one for each of two spatial dimensions. However, at a very small scale (the order of the film grain) the negative basically has only two densities, "opaque" and "transparent."

Most images are presented by functions of two *spatial* variables $f(\mathbf{x}) = f(x, y)$, where $f(x, y)$ is the brightness of the gray level of the image at a spatial coordinate (x, y). A multispectral image $\mathbf{f}$ is a vector-valued function with components $(f_1 \ldots f_n)$. One special multispectral image is a color image in which, for example, the components measure the brightness values of each of three wavelengths, that is,

$$f(\mathbf{x}) = \left\{ f_{\text{red}}(\mathbf{x}), f_{\text{blue}}(\mathbf{x}), f_{\text{green}}(\mathbf{x}) \right\}$$

Time-varying images $f(\mathbf{x}, t)$ have an added temporal argument. For special three-dimensional images, $\mathbf{x} = (x, y, z)$. Usually, both the domain and range of f are bounded.

An important part of the formation process is the conversion of the image representation from a continuous function to a discrete function; we need some way of describing the images as samples at discrete points. The mathematical tool we shall use is the *delta function.*

Formally, the delta function may be defined by

$$\delta(x) = \begin{cases} 0 & \text{when } x \neq 0 \\ \infty & \text{when } x = 0 \end{cases} \tag{2.1}$$

$$\int_{-\infty}^{\infty} \delta(x)\, dx = 1$$

If some care is exercised, the delta function may be interpreted as the limit of a set of functions:

$$\delta(x) = \lim_{n \to \infty} \delta_n(x)$$

where

$$\delta_n(x) = \begin{cases} n & \text{if } |x| < \dfrac{1}{2n} \\ 0 & \text{otherwise} \end{cases} \tag{2.2}$$

A useful property of the delta function is the *sifting property:*

$$\int_{-\infty}^{\infty} f(x)\, \delta(x - a)\, dx = f(a) \tag{2.3}$$

A continuous image may be multipled by a two-dimensional "comb," or array of delta functions, to extract a finite number of discrete *samples* (one for each delta function). This mathematical model of the sampling process will be useful later.

2.2.2 Imaging Geometry

Monocular Imaging

Point projection is the fundamental model for the transformation wrought by our eye, by cameras, or by numerous other imaging devices. To a first-order approximation, these devices act like a pinhole camera in that the image results from projecting scene points through a single point onto an *image plane* (see Fig. 2.1). In Fig. 2.1, the image plane is behind the point of projection, and the image is reversed. However, it is more intuitive to recompose the geometry so that the point of projection corresponds to a *viewpoint* behind the image plane, and the image occurs right side up (Fig. 2.2). The mathematics is the same, but now the viewpoint is $+f$ on the z axis, with $z = 0$ plane being the image plane upon which the image is projected. (f is sometimes called the *focal length* in this context. The use of f in this section should not be confused with the use of f for image function.) As the imaged object approaches the viewpoint, its projection gets bigger (try moving your hand toward your eye). To specify how its imaged size changes, one needs only the geometry of similar triangles. In Fig. 2.2b y', the projected height of the object, is related to its real height y, its position z, and the focal length f by

$$\frac{y}{f - z} = \frac{y'}{f} \tag{2.4}$$

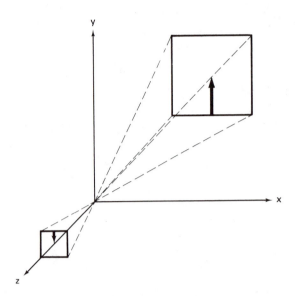

Fig. 2.1 A geometric camera model.

The case for x' is treated similarly:

$$\frac{x}{f-z} = \frac{x'}{f} \tag{2.5}$$

The projected image has $z = 0$ everywhere. However, projecting away the z component is best considered a separate transformation; the projective transform is usually thought to distort the z component just as it does the x and y. *Perspective distortion* thus maps (x, y, z) to

$$(x', y', z') = \left[\frac{fx}{f-z}, \frac{fy}{f-z}, \frac{fz}{f-z} \right] \tag{2.6}$$

The perspective transformation yields *orthographic projection* as a special case when the viewpoint is the *point at infinity* in the z direction. Then all objects are projected onto the viewing plane with no distortion of their x and y coordinates.

The perspective distortion yields a three-dimensional object that has been "pushed out of shape"; it is more shrunken the farther it is from the viewpoint. The z component is not available directly from a two-dimensional image, being identically equal to zero. In our model, however, the distorted z component has information about the distance of imaged points from the viewpoint. When this distorted object is projected orthographically onto the image plane, the result is a perspective picture. Thus, to achieve the effect of railroad tracks appearing to come together in the distance, the perspective distortion transforms the tracks so that they *do* come together (at a point at infinity)! The simple orthographic projection that projects away the z component unsurprisingly preserves this distortion. Several properties of the perspective transform are of interest and are investigated further in Appendix 1.

Binocular Imaging

Basic binocular imaging geometry is shown in Fig. 2.3a. For simplicity, we

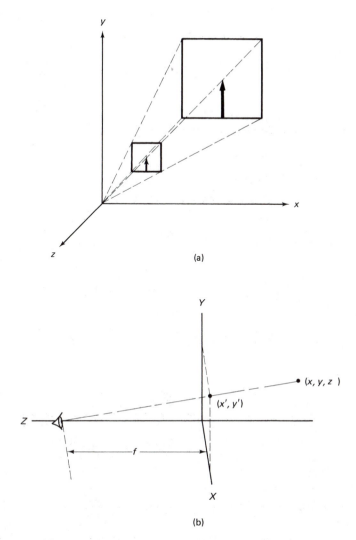

Fig. 2.2 (a) Camera model equivalent to that of Fig. 2.1; (b) definition of terms.

use a system with two viewpoints. In this model the eyes do not *converge*; they are aimed in parallel at the point at infinity in the $-z$ direction. The depth information about a point is then encoded only by its different positions (*disparity*) in the two image planes.

With the stereo arrangement of Fig. 2.3,

$$x' = \frac{(x - d)f}{f - z}$$

$$x'' = \frac{(x + d)f}{f - z}$$

where (x', y') and (x'', y'') are the retinal coordinates for the world point imaged

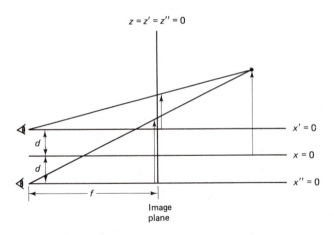

$$z = z' = z'' = 0$$

$$x' = 0$$

$$x = 0$$

$$x'' = 0$$

Image plane

Fig. 2.3 A nonconvergent binocular imaging system.

through each eye. The *baseline* of the binocular system is $2d$. Thus

$$(f - z)x' = (x - d)f \qquad (2.7)$$

$$(f - z)x'' = (x + d)f \qquad (2.8)$$

Subtracting (2.7) from (2.8) gives

$$(f - z)(x'' - x') = 2df$$

or

$$z = f - \frac{2df}{x'' - x'} \qquad (2.9)$$

Thus if points can be matched to determine the disparity $(x'' - x')$ and the baseline and focal length are known, the z coordinate is simple to calculate.

If the system can converge its directions of view to a finite distance, convergence angle may also be used to compute depth. The hardest part of extracting depth information from stereo is the *matching* of points for disparity calculations. "Light striping" is a way to maintain geometric simplicity and also simplify matching (Section 2.3.2).

2.2.3 Reflectance

Terminology

A basic aspect of the imaging process is the physics of the reflectance of objects, which determines how their "brightness" in an image depends on their inherent characteristics and the geometry of the imaging situation. A clear presentation of the mathematics of reflectance is given in [Horn and Sjoberg 1978; Horn 1977]. Light *energy flux* Φ is measured in watts; "brightness" is measured with respect to area and solid angle. The *radiant intensity* I of a source is the exitant flux per unit solid angle:

$$I = \frac{d\Phi}{d\omega} \qquad \text{watts/steradian} \qquad (2.10)$$

Here $d\omega$ is an incremental solid angle. The solid angle of a small area dA measured perpendicular to a radius r is given by

$$d\omega = \frac{dA}{r^2} \qquad (2.11)$$

in units of steradians. (The total solid angle of a sphere is 4π.)

The *irradiance* is flux incident on a surface element dA:

$$E = \frac{d\Phi}{dA} \qquad \text{watts/meter}^2 \qquad (2.12)$$

and the flux exitant from the surface is defined in terms of the *radiance L*, which is the flux emitted per unit foreshortened surface area per unit solid angle:

$$L = \frac{d^2\Phi}{dA\ \cos\theta\, d\omega} \qquad \text{watts/(meter}^2 \text{ steradian)} \qquad (2.13)$$

where θ is the angle between the surface normal and the direction of emission.

Image irradiance f is the "brightness" of the image at a point, and is proportional to scene radiance. A "gray-level" is a quantized measurement of image irradiance. Image irradiance depends on the reflective properties of the imaged surfaces as well as on the illumination characteristics. How a surface reflects light depends on its micro-structure and physical properties. Surfaces may be *matte* (dull, flat), *specular* (mirrorlike), or have more complicated reflectivity characteristics (Section 3.5.1). The *reflectance r* of a surface is given quite generally by its Bidirectional Reflectance Distribution Function (BRDF) [Nicodemus et al. 1977]. The BRDF is the ratio of reflected radiance in the direction towards the viewer to the irradiance in the direction towards a small area of the source.

Effects of Geometry on an Imaging System

Let us now analyze a simple image-forming system shown in Fig. 2.4 with the objective of showing how the gray levels are related to the radiance of imaged objects. Following [Horn and Sjoberg 1978], assume that the imaging device is properly focused; rays originating in the infinitesimal area dA_o on the object's surface are projected into some area dA_p in the image plane and no rays from other portions of the object's surface reach this area of the image. The system is assumed to be an ideal one, obeying the laws of simple geometrical optics.

The energy flux/unit area that impinges on the sensor is defined to be E_p. To show how E_p is related to the scene radiance L, first consider the flux arriving at the lens from a small surface area dA_o. From (2.13) this is given as

$$d\Phi = dA_o \int L \cos\theta\, d\omega \qquad (2.14)$$

This flux is assumed to arrive at an area dA_p in the imaging plane. Hence the irradiance is given by [using Eq. (2.12)]

$$E_p = \frac{d\Phi}{dA_p} \qquad (2.15)$$

Now relate dA_o to dA_p by equating the respective solid angles as seen from the lens; that is [making use of Eq. (2.12)],

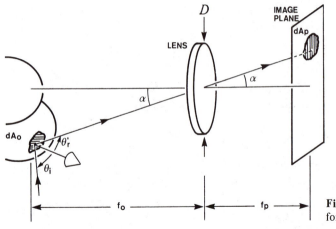

Fig. 2.4 Geometry of an image forming system.

$$dA_o \frac{\cos\theta}{f_o^2} = dA_p \frac{\cos\alpha}{f_p^2} \qquad (2.16)$$

Substituting Eqs. (2.16) and (2.14) into (2.15) gives

$$E = \cos\alpha \left(\frac{f_o}{f_p}\right)^2 \int L\,d\omega \qquad (2.17)$$

The integral is over the solid angle seen by the lens. In most instances we can assume that L is constant over this angle and hence can be removed from the integral. Finally, approximate $d\omega$ by the area of the lens foreshortened by $\cos\alpha$, that is, $(\pi/4)D^2 \cos\alpha$ divided by the distance $f_o/\cos\alpha$ squared:

$$d\omega = \frac{\pi}{4} D^2 \frac{\cos^3\alpha}{f_o^2} \qquad (2.18)$$

so that finally

$$E = \frac{1}{4}\left(\frac{D}{f_p}\right)^2 \cos^4\alpha\,\pi\,L \qquad (2.19)$$

The interesting results here are that (1) the image irradiance is proportional to the scene radiance L, and (2) the factor of proportionality includes the fourth power of the off-axis angle α. Ideally, an imaging device should be calibrated so that the variation in sensitivity as a function of α is removed.

2.2.4 Spatial Properties

The Fourier Transform

An image is a spatially varying function. One way to analyze spatial variations is the decomposition of an image function into a set of orthogonal functions, one such set being the Fourier (sinusoidal) functions. The Fourier transform may be used to transform the intensity image into the domain of *spatial frequency*. For notational convenience and intuition, we shall generally use as an example the continuous one-dimensional Fourier transform. The results can readily be extended to the discrete case and also to higher dimensions [Rosenfeld and Kak 1976]. In two dimensions we shall denote transform domain coordinates by (u, v). The one-dimensional Fourier transform, denoted $\mathcal{F}$, is defined by

$$\mathcal{F}[f(x)] = F(u)$$

where

$$F(u) = \int_{-\infty}^{+\infty} f(x)\exp{(-j2\pi ux)}\,dx \qquad (2.20)$$

where $j = \sqrt{(-1)}$. Intuitively, Fourier analysis expresses a function as a sum of sine waves of different frequency and phase. The Fourier transform has an *inverse* $\mathcal{F}^{-1}[F(u)] = f(x)$. This inverse is given by

$$f(x) = \int_{-\infty}^{\infty} F(u)\exp{(j2\pi ux)}\,du \qquad (2.21)$$

The transform has many useful properties, some of which are summarized in Table 2.1. Common one-dimensional Fourier transform pairs are shown in Table 2.2.

The transform $F(u)$ is simply another representation of the image function. Its meaning can be understood by interpreting Eq. (2.21) for a specific value of x, say x_0:

$$f(x_0) = \int F(u)\exp{(j2\pi ux_0)}\,du \qquad (2.22)$$

This equation states that a particular point in the image can be represented by a weighted sum of complex exponentials (sinusoidal patterns) at different spatial frequencies u. $F(u)$ is thus a *weighting function* for the different frequencies. Low-spatial frequencies account for the "slowly" varying gray levels in an image, such as the variation of intensity over a continuous surface. High-frequency components are associated with "quickly varying" information, such as edges. Figure 2.5 shows the Fourier transform of an image of rectangles, together with the effects of removing low- and high-frequency components.

The Fourier transform is defined above to be a continuous transform. Although it may be performed instantly by optics, a discrete version of it, the "fast Fourier transform," is almost universally used in image processing and computer vision. This is because of the relative versatility of manipulating the transform in the digital domain as compared to the optical domain. Image-processing texts, e.g., [Pratt 1978; Gonzalez and Wintz 1977] discuss the FFT in some detail; we content ourselves with an algorithm for it (Appendix 1).

The Convolution Theorem

Convolution is a very important image-processing operation, and is a basic operation of linear systems theory. The convolution of two functions f and g is a function h of a displacement y defined as

$$h(y) = f*g = \int_{-\infty}^{\infty} f(x)g(y - x)\,dx \qquad (2.23)$$

Table 2.1

PROPERTIES OF THE FOURIER TRANSFORM

	Spatial Domain	Frequency Domain
	$f(x)$	$F(u) = \mathscr{F}[f(x)]$
	$g(x)$	$G(u) = \mathscr{F}[g(x)]$

		Spatial	Frequency
(1)	Linearity	$c_1 f(x) + c_2 g(x)$	$c_1 F(u) + c_2 G(u)$
		c_1, c_2 scalars	
(2)	Scaling	$f(ax)$	$\dfrac{1}{\|a\|} F\left(\dfrac{u}{a}\right)$
(3)	Shifting	$f(x - x_0)$	$e^{-j2\pi x_0 u} F(u)$
(4)	Symmetry	$F(x)$	$f(-u)$
(5)	Conjugation	$f^*(x)$	$F^*(-u)$
(6)	Convolution	$h(x) = f*g = \displaystyle\int_{-\infty}^{\infty} f(x')g(x - x') \, dx'$	$F(u)G(u)$
(7)	Differentiation	$\dfrac{d^n f(x)}{dx^n}$	$(2\pi ju)^n F(u)$

Parseval's theorem:

$$\int_{-\infty}^{\infty} |f(x)|^2 dx = \int_{-\infty}^{\infty} |F(\xi)|^2 d\xi$$

$$\int_{-\infty}^{\infty} f(x)g^*(x) \, dx = \int_{-\infty}^{\infty} F(\xi)G^*(\xi) \, d\xi$$

$f(x)$	$F(\xi)$
Real (R)	Real part even (RE)
	Imaginary part odd (IO)
Imaginary (I)	RO, IE
RE, IO	R
RE, IE	I
RE	RE
RO	IO
IE	IE
IO	RO
Complex even (CE)	CE
CO	CO

Table 2.2

FOURIER TRANSFORM PAIRS

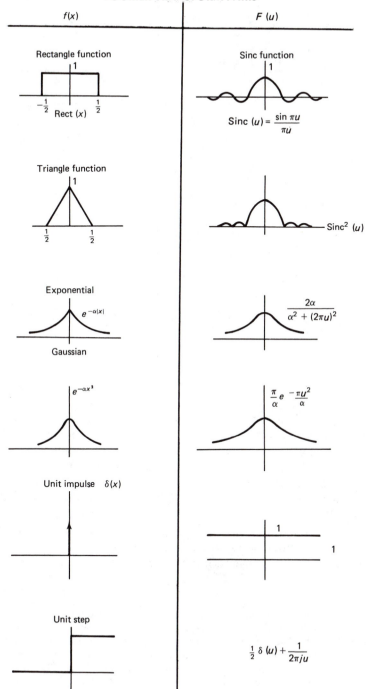

$f(x)$ $F(u)$

Rectangle function

Rect (x)

Sinc function

$\text{Sinc}(u) = \dfrac{\sin \pi u}{\pi u}$

Triangle function

$\text{Sinc}^2(u)$

Exponential

$e^{-\alpha |x|}$

$\dfrac{2\alpha}{\alpha^2 + (2\pi u)^2}$

Gaussian

$e^{-\alpha x^2}$

$\dfrac{\pi}{\alpha} e^{-\frac{\pi u^2}{\alpha}}$

Unit impulse $\delta(x)$

1

Unit step

$\frac{1}{2}\delta(u) + \dfrac{1}{2\pi j u}$

Table 2.2 (cont.)

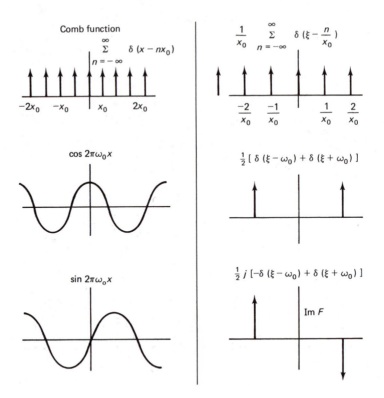

Intuitively, one function is "swept past" (in one dimension) or "rubbed over" (in two dimensions) the other. The value of the convolution at any displacement is the integral of the product of the (relatively displaced) function values. One common phenomenon that is well expressed by a convolution is the formation of an image by an optical system. The system (say a camera) has a "point-spread function," which is the image of a single point. (In linear systems theory, this is the "impulse response," or response to a delta-function input.) The ideal point-spread function is, of course, a point. A typical point-spread function is a two-dimensional Gaussian spatial distribution of intensities, but may include such phenomena as diffraction rings. In any event, if the camera is modeled as a linear system (ignor-

Fig. 2.5 (on facing page) (a) An image, $f(x, y)$. (b) A rotated version of (a), filtered to enhance high spatial frequencies. (c) Similar to (b), but filtered to enhance low spatial frequencies. (d), (e), and (f) show the logarithm of the power spectrum of (a), (b), and (c). The power spectrum is the log square modulus of the Fourier transform $F(u, v)$. Considered in polar coordinates (ρ, θ), points of small ρ correspond to low spatial frequencies ("slowly-varying" intensities), large ρ to high spatial frequencies contributed by "fast" variations such as step edges. The power at (ρ, θ) is determined by the amount of intensity variation at the frequency ρ occurring at the angle θ.

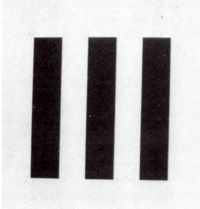

(a)

(b)

(c)

(d)

(e)

(f)

ing the added complexity that the point-spread function usually varies over the field of view), the image is the convolution of the point-spread function and the input signal. The point-spread function is rubbed over the perfect input image, thus blurring it.

Convolution is also a good model for the application of many other linear operators, such as line-detecting templates. It can be used in another guise (called correlation) to perform matching operations (Chapter 3) which detect instances of subimages or features in an image.

In the spatial domain, the obvious implementation of the convolution operation involves a shift–multiply–integrate operation which is hard to do efficiently. However, multiplication and convolution are "transform pairs," so that the calculation of the convolution in one domain (say the spatial) is simplified by first Fourier transforming to the other (the frequency) domain, performing a multiplication, and then transforming back.

The convolution of f and g in the spatial domain is equivalent to the pointwise product of F and G in the frequency domain,

$$\mathcal{F}(f*g) = FG \qquad (2.24)$$

We shall show this in a manner similar to [Duda and Hart 1973]. First we prove the *shift theorem*. If the Fourier transform of $f(x)$ is $F(u)$, defined as

$$F(u) = \int_x f(x) \exp\left[-j2\pi(ux)\right]dx \qquad (2.25)$$

then

$$\mathcal{F}\left[f(x-a)\right] = \int_x f(x-a) \exp\left[-j2\pi(ux)\right]dx \qquad (2.26)$$

changing variables so that $x' = x - a$ and $dx = dx'$

$$= \int_{x'} f(x') \exp\left\{-j2\pi[u(x'+a)]\right\}dx' \qquad (2.27)$$

Now $\exp[-j2\pi u(x'+a)] = \exp(-j2\pi ua)\exp(-j2\pi ux')$, where the first term is a constant. This means that

$$\mathcal{F}\left[f(x-a)\right] = \exp(-j2\pi ua)F(u) \qquad \text{(shift theorem)}$$

Now we are ready to show that $\mathcal{F}[f(x)*g(x)] = F(u)G(u)$.

$$\mathcal{F}(f*g) = \int_y \left\{\int_x f(x)g(y-x)\right\} \exp(-j2\pi uy)\ dx\ dy \qquad (2.28)$$

$$= \int_x f(x)\left\{\int_y g(y-x) \exp(-j2\pi uy)\ dy\right\} dx \qquad (2.29)$$

Recognizing that the terms in braces represent $\mathcal{F}[g(y-x)]$ and applying the shift theorem, we obtain

$$\mathcal{F}(f*g) = \int_x f(x)\exp(-j2\pi ux)G(u)\ dx \qquad (2.30)$$

$$= F(u)G(u) \qquad (2.31)$$

2.2.5 Color

Not all images are monochromatic; in fact, applications using multispectral images are becoming increasingly common (Section 2.3.2). Further, human beings intuitively feel that color is an important part of their visual experience, and is useful or even necessary for powerful visual processing in the real world. Color vision provides a host of research issues, both for psychology and computer vision. We briefly discuss two aspects of color vision: color spaces and color perception. Several models of the human visual system not only include color but have proven useful in applications [Granrath 1981].

Color Spaces

Color spaces are a way of organizing the colors perceived by human beings. It happens that weighted combinations of stimuli at three principal wavelengths are sufficient to define almost all the colors we perceive. These wavelengths form a natural basis or coordinate system from which the color measurement process can be described. Color perception is not related in a simple way to color measurement, however.

Color is a perceptual phenomenon related to human response to different wavelengths in the visible *electromagnetic spectrum* [400 (blue) to 700 nanometers (red); a nanometer (nm) is 10^{-9} meter]. The sensation of color arises from the sensitivities of three types of neurochemical sensors in the retina to the visible spectrum. The relative response of these sensors is shown in Fig. 2.6. Note that each sensor responds to a range of wavelengths. The illumination source has its own spectral composition $f(\lambda)$ which is modified by the reflecting surface. Let $r(\lambda)$ be this reflectance function. Then the measurement R produced by the "red" sensor is given by

$$R = \int f(\lambda) r(\lambda) h_R(\lambda) \, d\lambda \qquad (2.32)$$

So the sensor output is actually the integral of three different wavelength-dependent components: the source f, the surface reflectance r, and the sensor h_R.

Surprisingly, only weighted combinations of three delta-function approximations to the different $f(\lambda) h(\lambda)$, that is, $\delta(\lambda_R)$, $\delta(\lambda_G)$, and $\delta(\lambda_B)$, are necessary to

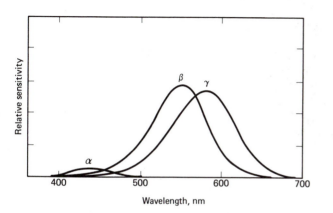

Fig. 2.6 Spectral response of human color sensors.

produce the sensation of nearly all the colors. This result is displayed on a *chromaticity diagram*. Such a diagram is obtained by first normalizing the three sensor measurements:

$$r = \frac{R}{R + G + B}$$
$$g = \frac{G}{R + G + B}$$
$$b = \frac{B}{R + G + B}$$

(2.33)

and then plotting perceived color as a function of any two (usually red and green). Chromaticity explicitly ignores intensity or brightness; it is a section through the three-dimensional color space (Fig. 2.7). The choice of $(\lambda_R, \lambda_G, \lambda_B) = (410, 530, 650)\, nm$ maximizes the realizable colors, but some colors still cannot be realized since they would require negative values for some of r, g, and b.

Another more intuitive way of visualizing the possible colors from the *RGB* space is to view these measurements as Euclidean coordinates. Here any color can be visualized as a point in the unit cube. Other coordinate systems are useful for different applications; computer graphics has proved a strong stimulus for investigation of different color space bases.

Color Perception

Color perception is complex, but the essential step is a transformation of three input intensity measurements into another basis. The coordinates of the new

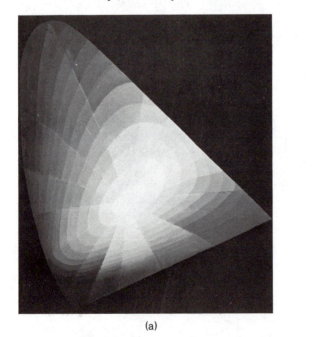

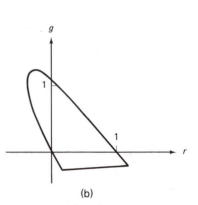

(a)   (b)

Fig. 2.7 (a) An artist's conception of the chromaticity diagram—*see color insert*; (b) a more useful depiction. Spectral colors range along the curved boundary; the straight boundary is the line of purples.

basis are more directly related to human color judgments.

Although the *RGB* basis is good for the acquisition or display of color information, it is not a particularly good basis to explain the perception of colors. Human vision systems can make good judgments about the relative surface reflectance $r(\lambda)$ despite different illuminating wavelengths; this reflectance seems to be what we mean by surface color.

Another important feature of the color basis is revealed by an ability to perceive in "black and white," effectively deriving intensity information from the color measurements. From an evolutionary point of view, we might expect that color perception in animals would be compatible with preexisting noncolor perceptual mechanisms.

These two needs—the need to make good color judgments and the need to retain and use intensity information—imply that we use a transformed, non-*RGB* basis for color space. Of the different bases in use for color vision, all are variations on this theme: Intensity forms one dimension and color is a two-dimensional subspace. The differences arise in how the color subspace is described. We categorize such bases into two groups.

1. *Intensity/Saturation/Hue (IHS).* In this basis, we compute intensity as

$$\text{intensity:} = R + G + B \tag{2.34}$$

The saturation measures the lack of whiteness in the color. Colors such as "fire engine" red and "grass" green are saturated; pastels (e.g., pinks and pale blues) are desaturated. Saturation can be computed from *RGB* coordinates by the formula [Tenenbaum and Weyl 1975]

$$\text{saturation:} = 1 - \frac{3 \min (R,\ G,\ B)}{\text{intensity}} \tag{2.35}$$

Hue is roughly proportional to the average wavelength of the color. It can be defined using *RGB* by the following program fragment:

$$\text{hue:} = \cos^{-1} \left\{ \frac{\{\frac{1}{2}[(R - G) + (R - B)]\}}{\sqrt{(R - G)^2 + (R - B)(G - B)}} \right\} \tag{2.36}$$

$$\text{If } B > G \text{ then hue:} = 2pi - \text{hue}$$

The IHS basis transforms the *RGB* basis in the following way. Thinking of the color cube, the diagonal from the origin to (1, 1, 1) becomes the intensity axis. Saturation is the distance of a point from that axis and hue is the angle with regard to the point about that axis from some reference (Fig. 2.8).

This basis is essentially that used by artists [Munsell 1939], who term saturation *chroma*. Also, this basis has been used in graphics [Smith 1978; Joblove and Greenberg 1978].

One problem with the IHS basis, particularly as defined by (2.34) through (2.36), is that it contains essential singularities where it is impossible to define the color in a consistent manner [Kender 1976]. For example, hue has an essential singularity for all values of $(R,\ G,\ B)$, where $R = G = B$. This means that special care must be taken in algorithms that use hue.

2. *Opponent processes.* The opponent process basis uses Cartesian rather than

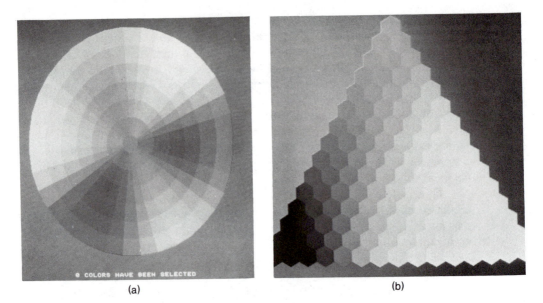

(a)   (b)

Fig. 2.8 An IHS Color Space. (a) Cross section at one intensity; (b) cross section at one hue—*see color inserts.*

cylindrical coordinates for the color subspace, and was first proposed by Hering [Teevan and Birney 1961]. The simplest form of basis is a linear transformation from *R, G, B* coordinates. The new coordinates are termed "*R − G*", "*Bl − Y*", and "*W − Bk*":

$$\begin{bmatrix} R - G \\ Bl - Y \\ W - Bk \end{bmatrix} = \begin{bmatrix} 1 & -2 & 1 \\ -1 & -1 & 2 \\ 1 & 1 & 1 \end{bmatrix} \begin{bmatrix} R \\ G \\ B \end{bmatrix}$$

The advocates of this representation, such as [Hurvich and Jameson 1957], theorize that this basis has neurological correlates and is in fact the way human beings represent ("name") colors. For example, in this basis it makes sense to talk about a "reddish blue" but not a "reddish green." Practical opponent process models usually have more complex weights in the transform matrix to account for psychophysical data. Some startling experiments [Land 1977] show our ability to make correct color judgments even when the illumination consists of only two principal wavelengths. The opponent process, at the level at which we have developed it, does not demonstrate how such judgments are made, but does show how stimulus at only two wavelengths will project into the color subspace. Readers interested in the details of the theory should consult the references.

Commercial television transmission needs an intensity, or "*W − Bk*" component for black-and-white television sets while still spanning the color space. The National Television Systems Committee (NTSC) uses a "YIQ" basis extracted from *RGB* via

$$\begin{bmatrix} I \\ Q \\ Y \end{bmatrix} = \begin{bmatrix} 0.60 & -0.28 & -0.32 \\ 0.21 & -0.52 & 0.31 \\ 0.30 & 0.59 & 0.11 \end{bmatrix} \begin{bmatrix} R \\ G \\ B \end{bmatrix}$$

This basis is a weighted form of

$$(I, \ Q, \ Y) = (\text{``} R - \text{cyan,''} \ \text{``magenta} - \text{green,''} \ \text{``} W - Bk \text{''})$$

2.2.6 Digital Images

The *digital images* with which computer vision deals are represented by m-vector discrete-valued image functions $f(\mathbf{x})$, usually of one, two, three, or four dimensions.

Usually $m = 1$, and both the domain and range of $f(\mathbf{x})$ are discrete. The domain of f is finite, usually a rectangle, and the range of f is positive and bounded: $0 \leqslant f(\mathbf{x}) \leqslant M$ for some integer M. For all practical purposes, the image is a continuous function which is represented by measurements or *samples* at regularly spaced intervals. At the time the image is sampled, the intensity is usually *quantized* into a number of different *gray levels*. For a discrete image, $f(\mathbf{x})$ is an integer gray level, and $\mathbf{x} = (x, y)$ is a pair of *integer* coordinates representing a sample point in a two-dimensional image plane. Sampling involves two important choices: (1) the *sampling interval*, which determines in a basic way whether all the information in the image is represented, and (2) the *tesselation* or spatial pattern of sample points, which affects important notions of connectivity and distance. In our presentation, we first show qualitatively the effects of sampling and gray-level quantization. Second, we discuss the simplest kinds of tesselations of the plane. Finally, and most important, we describe the sampling theorem, which specifies how close the image samples must be to represent the image unambiguously.

The choice of integers to represent the gray levels and coordinates is dictated by limitations in sensing. Also, of course, there are hardware limitations in representing images arising from their sheer size. Table 2.3 shows the storage required for an image in 8-bit bytes as a function of m, the number of bits per sample, and N, the linear dimension of a square image.

For reasons of economy (and others discussed in Chapter 3) we often use images of considerably less spatial resolution than that required to preserve fidelity to the human viewer. Figure 2.9 provides a qualitative idea of image degradation with decreasing spatial resolution.

As shown in Table 2.3, another way to save space besides using less spatial resolution is to use fewer bits per gray level sample. Figure 2.10 shows an image represented with different numbers of bits per sample. One striking effect is the "contouring" introduced with small numbers of gray levels. This is, in general, a problem for computer vision algorithms, which cannot easily discount the false contours. The choice of spatial and gray-level resolution for any particular computer vision task is an important one which depends on many factors. It is typical in

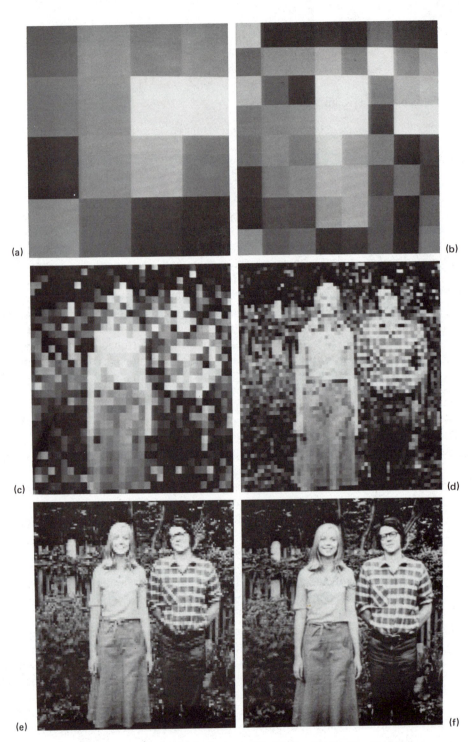

(a)

(b)

(c)

(d)

(e)

(f)

Fig. 2.9 Using different numbers of samples. (a) $N = 16$; (b) $N = 32$; (c) $N = 64$; (d) $N = 128$; (e) $N = 256$; (f) $N = 512$.

Table 2.3

**NUMBER OF 8-BIT BYTES OF STORAGE FOR
VARIOUS VALUES OF N AND M**

N m	32	64	128	256	512
1	128	512	2,048	8,192	32,768
2	256	1,024	4,096	16,384	65,536
3	512	2,048	8,192	32,768	131,072
4	512	2,048	8,192	32,768	131,072
5	1,024	4,096	16,384	65,536	262,144
6	1,024	4,096	16,384	65,536	262,144
7	1,024	4,096	16,384	65,536	262,144
8	1,024	4,096	16,384	65,536	262,144

computer vision to have to balance the desire for increased resolution (both gray scale and spatial) against its cost. Better data can often make algorithms easier to write, but a small amount of data can make processing more efficient. Of course, the image domain, choice of algorithms, and image characteristics all heavily influence the choice of resolutions.

Tesselations and Distance Metrics

Although the spatial samples for $f(\mathbf{x})$ can be represented as points, it is more satisfying to the intuition and a closer approximation to the acquisition process to think of these samples as finite-sized cells of constant gray-level partitioning the image. These cells are termed *pixels*, an acronym for *picture elements*. The pattern into which the plane is divided is called its *tesselation*. The most common regular tesselations of the plane are shown in Fig. 2.11.

Although rectangular tesselations are almost universally used in computer vision, they have a structural problem known as the "connectivity paradox." Given a pixel in a rectangular tesselation, how should we define the pixels to which it is connected? Two common ways are *four-connectivity* and *eight-connectivity*, shown in Fig. 2.12.

However, each of these schemes has complications. Consider Fig. 2.12c, consisting of a black object with a hole on a white background. If we use four-connectedness, the figure consists of four disconnected pieces, yet the hole is separated from the "outside" background. Alternatively, if we use eight-connectedness, the figure is one connected piece, yet the hole is now connected to the outside. This paradox poses complications for many geometric algorithms. Triangular and hexagonal tesselations do not suffer from connectivity difficulties (if we use three-connectedness for triangles); however, *distance* can be more difficult to compute on these arrays than for rectangular arrays.

The distance between two pixels in an image is an important measure that is fundamental to many algorithms. In general, a distance d is a *metric*. That is,

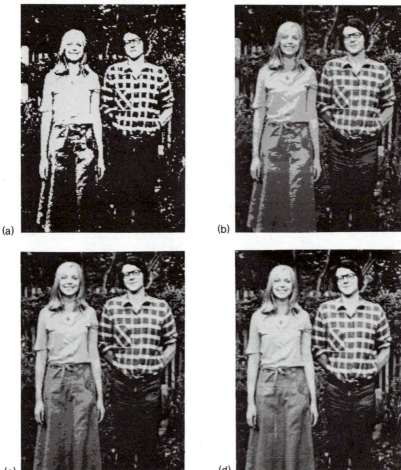

Fig. 2.10 Using different numbers of bits per sample. (a) $m = 1$; (b) $m = 2$; (c) $m = 4$; (d) $m = 8$.

(1) $d(\mathbf{x}, \mathbf{y}) = 0$ iff $\mathbf{x} = \mathbf{y}$

(2) $d(\mathbf{x}, \mathbf{y}) = d(\mathbf{y}, \mathbf{x})$

(3) $d(\mathbf{x}, \mathbf{y}) + d(\mathbf{y}, \mathbf{z}) \geqslant d(\mathbf{x}, \mathbf{z})$

For square arrays with unit spacing between pixels, we can use any of the following common distance metrics (Fig. 2.13) for two pixels $\boldsymbol{x} = (x_1, y_1)$ and $\boldsymbol{y} = (x_2, y_2)$.

Euclidean:

$$d_e(\mathbf{x}, \mathbf{y}) = \sqrt{(x_1 - x_2)^2 + (y_1 - y_2)^2} \tag{2.37}$$

City block:

$$d_{cb}(\mathbf{x}, \mathbf{y}) = |x_1 - x_2| + |y_1 - y_2| \tag{2.38}$$

Ch. 2   Image Formation

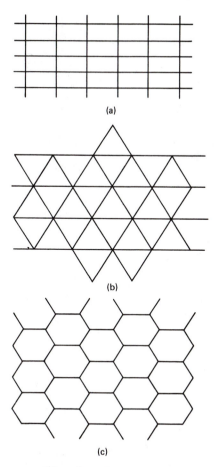

(a)

(b)

(c)

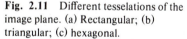

Fig. 2.11 Different tesselations of the image plane. (a) Rectangular; (b) triangular; (c) hexagonal.

Chessboard:

$$d_{ch}(\mathbf{x}, \mathbf{y}) = \max\left\{|x_1 - x_2|, |y_1 - y_2|\right\} \qquad (2.39)$$

Other definitions are possible, and all such measures extend to multiple dimensions. The tesselation of higher-dimensional space into pixels usually is confined to (n-dimensional) cubical pixels.

The Sampling Theorem

Consider the one-dimensional "image" shown in Fig. 2.14. To digitize this image one must sample the image function. These samples will usually be separated at regular intervals as shown. How far apart should these samples be to allow reconstruction (to a given accuracy) of the underlying continuous image from its samples? This question is answered by the Shannon sampling theorem. An excellent rigorous presentation of the sampling theorem may be found in [Rosenfeld and Kak 1976]. Here we shall present a shorter graphical interpretation using the results of Table 2.2. For simplicity we consider the image to be periodic in order to avoid small edge effects introduced by the finite image domain. A more rigorous

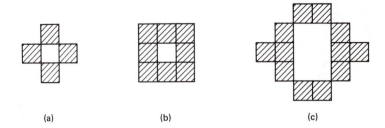

(a)   (b)   (c)

Fig. 2.12 Connectivity paradox for rectangular tesselations. (a) A central pixel and its 4-connected neighbors; (b) a pixel and its 8-connected neighbors; (c) a figure with ambiguous connectivity.

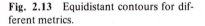

```
  3 3 3          3          3 3 3 3 3 3 3
 3 2 2 2 3       3 2 3       3 2 2 2 2 2 3
3 2 1 1 1 2 3   3 2 1 2 3    3 2 1 1 1 2 3
3 2 1 0 1 2 3  3 2 1 0 1 2 3  3 2 1 0 1 2 3
3 2 1 1 1 2 3   3 2 1 2 3    3 2 1 1 1 2 3
 3 2 2 2 3       3 2 3       3 2 2 2 2 2 3
  3 3 3          3          3 3 3 3 3 3 3
```

(a)   (b)   (c)

Fig. 2.13 Equidistant contours for different metrics.

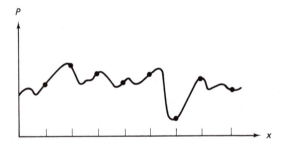

Fig. 2.14 One-dimensional image and its samples.

treatment, which considers these effects, is given in [Andrews and Hunt 1977].

Suppose that the image is sampled with a "comb" function of spacing x_0 (see Table 2.2). Then the sampled image can be modeled by

$$f_s(x) = f(x)\sum_n \delta(x - nx_0) \qquad (2.40)$$

where the image function modulates the comb function. Equivalently, this can be written as

$$f_s(x) = \sum_n f(nx_0)\,\delta(x - nx_0) \qquad (2.41)$$

The right-hand side of Eq. (2.40) is the product of two functions, so that property

(6) in Table 2.1 is appropriate. The Fourier transform of $f_s(x)$ is equal to the convolution of the transforms of each of the two functions. Using this result yields

$$F_s(u) = F(u) * \frac{1}{x_0} \sum_n \delta(u - \frac{n}{x_0}) \tag{2.42}$$

But from Eq. (2.3),

$$F(u) * \delta(u - \frac{n}{x_0}) = F(u - \frac{n}{x_0}) \tag{2.43}$$

so that

$$F_s(u) = \frac{1}{x_0} \sum_n F(u - \frac{n}{x_0}) \tag{2.44}$$

Therefore, sampling the image function $f(x)$ at intervals of x_0 is equivalent in the frequency domain to replicating the transform of f at intervals of $\frac{1}{x_0}$. This limits the recovery of $f(x)$ from its sampled representation, $f_s(x)$. There are two basic situations to consider. If the transform of $f(x)$ is *bandlimited* such that $F(u) = 0$ for $|u| > 1/(2x_0)$, then there is no overlap between successive replications of $F(u)$ in the frequency domain. This is shown for the case of Fig. 2.15a, where we have arbitrarily used a triangular-shaped image transform to illustrate the effects of sampling. Incidentally, note that for this transform $F(u) = F(-u)$ and that it has no imaginary part; from Table 2.2, the one-dimensional image must also be real and even. Now if $F(u)$ is not bandlimited, i.e., there are $u > \frac{1}{2x_0}$ for which $F(u) \neq 0$, then components of different replications of $F(u)$ will interact to produce the composite function $F_s(u)$, as shown in Fig. 2.15b. In the first case $f(x)$ can be recovered from $F_s(u)$ by multiplying $F_s(u)$ by a suitable $G(u)$:

$$G(u) = \begin{cases} 1 & |u| < \frac{1}{2x_0} \\ 0 & \text{otherwise} \end{cases} \tag{2.45}$$

Then

$$f(x) = \mathscr{F}^{-1}[F_s(u) G(u)] \tag{2.46}$$

However, in the second case, $F_s(u) G(u)$ is very different from the original $F(u)$. This is shown in Fig. 2.15c. Sampling a $F(u)$ that is not bandlimited allows information at high spatial frequencies to interfere with that at low frequencies, a phenomenon known as *aliasing*.

Thus the sampling theorem has this very important result: As long as the image contains no spatial frequencies greater than one-half the sampling frequency, the underlying continuous image is unambiguously represented by its samples. However, lest one be tempted to insist on images that have been so sampled, note that it may be useful to sample at lower frequencies than would be required for total reconstruction. Such sampling is usually preceded by some form of blurring of

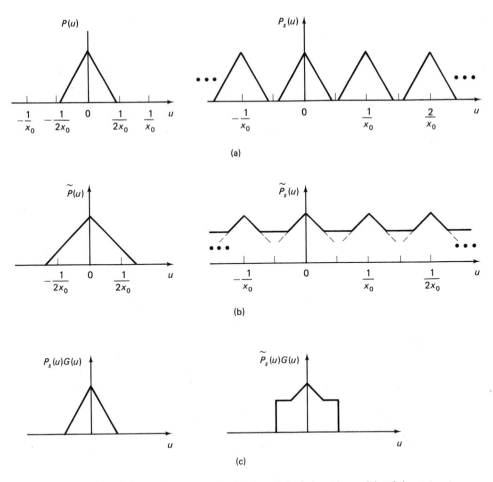

Fig. 2.15 (a) $F(u)$ bandlimited so that $F(u) = 0$ for $|u| > 1/2 x_0$. (b) $F(u)$ not bandlimited as in (a). (c) reconstructed transform.

the image, or can be incorporated with such blurring (by integrating the image intensity over a finite area for each sample). Image blurring can bury irrelevant details, reduce certain forms of noise, and also reduce the effects of aliasing.

2.3 IMAGING DEVICES FOR COMPUTER VISION

There is a vast array of methods for obtaining a digital image in a computer. In this section we have in mind only "traditional" images produced by various forms of radiation impinging on a sensor after having been affected by physical objects.

Many sensors are best modeled as an *analog* device whose response must be *digitized* for computer representation. The types of imaging devices possible are limited only by the technical ingenuity of their developers; attempting a definitive

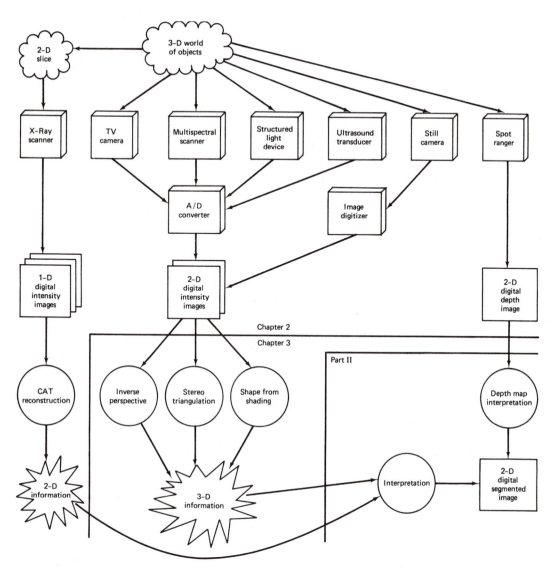

Fig. 2.16 Imaging devices (boxes), information structures (rectangles), and processes (circles).

taxonomy is probably unwise. Figure 2.16 is a flowchart of devices, information structures, and processes addressed in this and succeeding sections.

When the image already exists in some form, or physical considerations limit choice of imaging technology, the choice of digitizing technology may still be open. Most images are carried on a permanent medium, such as film, or at least are available in (essentially) analog form to a digitizing device. Generally, the relevant technical characteristics of imaging or digitizing devices should be foremost in mind when a technique is being selected. Such considerations as the signal-to-noise ratio of the device, its resolution, the speed at which it works, and its expense are important issues.

2.3.1 Photographic Imaging

The camera is the most familiar producer of optical images on a permanent medium. We shall not address here the multitudes of still- and movie-camera options; rather, we briefly treat the characteristics of the photographic film and of the digitizing devices that convert the image to machine-readable form. More on these topics is well presented in the References.

Photographic (black-and-white) film consists of an emulsion of silver halide crystals on a film base. (Several other layers are identifiable, but are not essential to an understanding of the relevant properties of film.) Upon exposure to light, the silver halide crystals form *development centers,* which are small grains of metallic silver. The photographic development process extends the formation of metallic silver to the entire silver halide crystal, which thus becomes a binary ("light" or "no light") detector. Subsequent processing removes undeveloped silver halide. The resulting film *negative* is dark where many crystals were developed and light where few were. The resolution of the film is determined by the *grain* size, which depends on the original halide crystals and on development techniques. Generally, the *faster* the film (the less light needed to expose it), the coarser the grain. Film exists that is sensitive to infrared radiation; x-ray film typically has two emulsion layers, giving it more gray-level range than that of normal film.

A repetition of the negative-forming process is used to obtain a photographic *print*. The negative is projected onto photographic paper, which responds roughly in the same way as the negative. Most photographic print paper cannot capture in one print the range of densities that can be present in a negative. Positive films do exist that do not require printing; the most common example is color slide film.

The response of film to light is not completely linear. The photographic *density* obtained by a negative is defined as the logarithm (base 10) of the ratio of incident light to transmitted light.

$$D = \log_{10}\left(\frac{I}{I_t}\right)$$

The *exposure* of a negative dictates (approximately) its response. Exposure is defined as the energy per unit area that exposed the film (in its sensitive spectral range). Thus exposure is the product of the *intensity* and the time of exposure. This mathematical model of the behavior of the photographic exposure process is correct for a wide operating range of the film, but *reciprocity failure* effects in the film keep one from being able always to trade light level for exposure time. At very low light levels, longer exposure times are needed than are predicted by the product rule.

The response of film to light is usually plotted in an "H&D curve" (named for Hurter and Driffield), which plots density versus exposure. The H&D curve of film displays many of its important characteristics. Figure 2.17 exhibits a typical H&D curve for a black and white film.

The *toe* of the curve is the lower region of low slope. It expresses reciprocity failure and the fact that the film has a certain bias, or *fog* response, which dominates its behavior at the lowest exposure levels. As one would expect, there is an upper limit to the density of the film, attained when a maximum number of silver

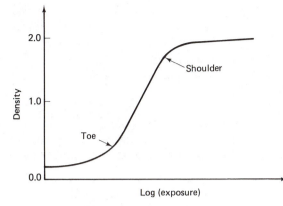

Fig. 2.17 Typical H & D curve.

halide crystals are rendered developable. Increasing exposure beyond this maximum level has little effect, accounting for the *shoulder* in the H&D curve, or its flattened upper end.

In between the toe and shoulder, there is typically a linear operating region of the curve. High-contrast films are those with high slope (traditionally called *gamma*); they respond dramatically to small changes in exposure. A high-contrast film may have a gamma between about 1.5 and 10. Films with gammas of approximately 10 are used in graphics arts to copy line drawings. General-purpose films have gammas of about 0.5 to 1.0.

The resolution of a general film is about 40 lines/mm, which means that a 1400×1400 image may be digitized from a 35mm slide. At any greater sampling frequency, the individual film grains will occupy more than a pixel, and the resolution will thus be grain-limited.

Image Digitizers (Scanners)

Accuracy and speed are the main considerations in converting an image on film into digital form. Accuracy has two aspects: spatial resolution, loosely the level of image spatial detail to which the digitizer can respond, and gray-level resolution, defined generally as the range of densities or reflectances to which the digitizer responds and how finely it divides the range. Speed is also important because usually many data are involved; images of 1 million samples are commonplace.

Digitizers broadly take two forms: mechanical and "flying spot." In a mechanical digitizer, the film and a sensing assembly are mechanically transported past one another while readings are made. In a flying-spot digitizer, the film and sensor are static. What moves is the "flying spot," which is a point of light on the face of a cathode-ray tube, or a laser beam directed by mirrors. In all digitizers a very narrow beam of light is directed through the film or onto the print at a known coordinate point. The light transmittance or reflectance is measured, transformed from analog to digital form, and made available to the computer through interfacing electronics. The location on the medium where density is being measured may also be transmitted with each reading, but it is usually determined by relative offset from positions transmitted less frequently. For example, a "new scan line" impulse is transmitted for TV output; the position along the current scan line yields an x position, and the number of scan lines yields a y position.

The mechanical scanners are mostly of two types, *flat-bed* and *drum*. In a flat-bed digitizer, the film is laid flat on a surface over which the light source and the sensor (usually a very accurate photoelectric cell) are transported in a raster fashion. In a drum digitizer, the film is fastened to a circular drum which revolves as the sensor and light source are transported down the drum parallel to its axis of rotation.

Color mechanical digitizers also exist; they work by using colored filters, effectively extracting in three scans three "color overlays" which when superimposed would yield the original color image. Extracting some "composite" color signal with one reading presents technical problems and would be difficult to do as accurately.

Satellite Imagery

LANDSAT and ERTS (Earth Resources Technology Satellites) have similar scanners which produce images of 2340 x 3380 7-bit pixels in four spectral bands, covering an area of 100×100 nautical miles. The scanner is mechanical, scanning six horizontal scan lines at a time; the rotation of the earth accounts for the advancement of the scan in the vertical direction.

A set of four images is shown in Fig. 2.18. The four spectral bands are numbered 4, 5, 6, and 7. Band 4 [0.5 to 0.6 μm (green)] accentuates sediment-laden water and shallow water, band 5 [0.6 to 0.7 μm (red)] emphasizes cultural features such as roads and cities, band 6 [0.7 to 0.8 μm (near infrared)] emphasizes vegetation and accentuates the contrast between land and water, band 7 [0.8 to 1.1 μm (near infrared)] is like band 6 except that it is better at penetrating atmospheric haze.

The LANDSAT images are available at nominal cost from the U.S. government (The EROS Data Center, Sioux Falls, South Dakota 57198). They are furnished on tape, and cover the entire surface of the earth (often the buyer has a choice of the amount of cloud cover). These images form a huge data base of multispectral imagery, useful for land-use and geological studies; they furnish something of an image analysis challenge, since one satellite can produce some 6 billion bits of image data per day.

Television Imaging

Television cameras are appealing devices for computer vision applications for several reasons. For one thing, the image is immediate; the camera can show events as they happen. For another, the image is already in electrical, if not digital form. "Television camera" is basically a nontechnical term, because many different technologies produce video signals conforming to the standards set by the FCC and NTSC. Cameras exist with a wide variety of technical specifications.

Usually, TV cameras have associated electronics which scan an entire "picture" at a time. This operation is closely related to broadcast and receiver standards, and is more oriented to human viewing than to computer vision. An entire image (of some 525 scan lines in the United States) is called a *frame*, and consists of two *fields*, each made up of alternate scan lines from the frame. These fields are generated and transmitted sequentially by the camera electronics. The transmitted image is thus *interlaced*, with all odd-numbered scan lines being "painted" on the

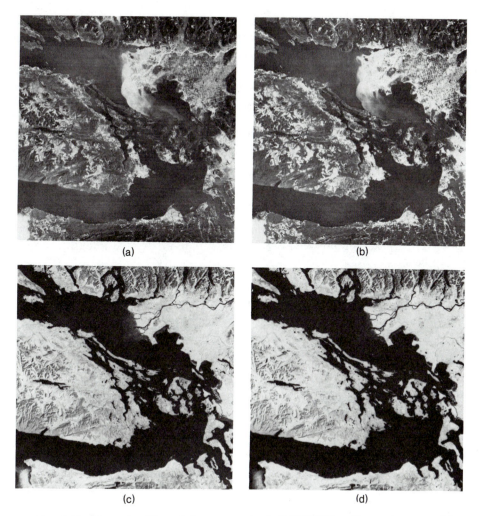

Fig. 2.18 The straits of Juan de Fuca as seen by the LANDSAT multispectral scanner. (a) Band 4; (b) band 5; (c) band 6; (d) band 7.

screen alternating with all even-numbered scan lines. In the United States, each field takes $\frac{1}{60}$ sec to scan, so a whole frame is scanned every $\frac{1}{30}$ sec. The interlacing is largely to prevent flickering of the image, which would become noticeable if the frame were painted from top to bottom only once in $\frac{1}{30}$ sec. These automatic scanning electronics may be replaced or overridden in many cameras, allowing "random access" to the image. In some technologies, such as the image dissector, the longer the signal is collected from any location, the better the signal-to-noise performance.

There are a number of different systems used to generate television images. We discuss five main methods below.

Image orthicon tube. This is one of the two main methods in use today (in addition to the vidicon). It offers very stable performance at all incident light levels and is widely used in commercial television. It is a storage-type tube, since it depends on the neutralization of positive charges by a scanning electron beam.

The image orthicon (Fig. 2.19) is divided into an imaging and readout section. In the imaging section, light from the scene is focused onto a semitransparent photocathode. This photocathode operates the same way as the cathode in a phototube. It emits electrons which are magnetically focused by a coil and are accelerated toward a positively charged target. The target is a thin glass disk with a fine-wire-mesh screen facing the photocathode. When electrons strike it, secondary emission from the glass takes place. As electrons are emitted from the photocathode side of the disk, positive charges build up on the scanning side. These charges correspond to the pattern of light intensity in the scene being viewed.

In the readout section, the back of the target is scanned by a low velocity electron beam from an electron gun at the rear of the tube. Electrons in this beam are absorbed by the target in varying amounts, depending on the charge on the target. The image is represented by the amplitude-modulated intensity of the returned beam.

Vidicon tube. The vidicon is smaller, lighter, and more rugged than the image orthicon, making it ideal for portable use. Here the target (the inner surface of the face plate) is coated with a transparent conducting film which forms a video signal electrode (Fig. 2.20). A thin photosensitive layer is deposited on the film, consisting of a large number of tiny resistive globules whose resistance decreases on illumination. This layer is scanned in raster fashion by a low velocity electron beam from the electron gun at the rear of the tube. The beam deposits electrons on the layer, thus reducing its surface potential. The two surfaces of the target essentially form a capacitor, and the scanning action of the beam produces a capacitive current at the video signal electrode which represents the video signal.

The plumbicon is essentially a vidicon with a lead oxide photosensitive layer. It offers the following advantages over the vidicon: higher sensitivity, lower dark current, and negligible persistence or lag.

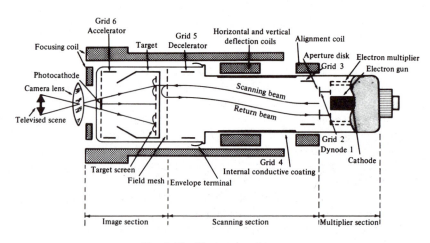

Fig. 2.19   The image orthicon.

Ch. 2   Image Formation

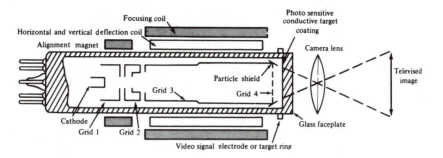

Fig. 2.20 The vidicon.

Iconoscope tube. The iconoscope is now largely of historical interest. In it, an electron beam scans a target consisting of a thin mica sheet or mosaic coated with a photosensitive layer. In contrast to the vidicon and orthicon, the electron beam and the light both strike the same side of the target surface. The back of the mosaic is covered with a conductive film connected to an output load. The arrangement is equivalent to a matrix of small capacitors which discharge through a common lead.

Image dissector tube. The image dissector tube operates on instantaneous scanning rather than by neutralizing positive charges. Light from the scene is focused on a cathode coated with a photosensitive layer (Fig. 2.21). The cathode emits electrons in proportion to the amount of light striking it. These electrons are accelerated toward a target by the anode. The target is an electron multiplier covered by a small aperture which allows only a small part of the "electron image" emitted by the cathode to reach the target. The electron image is focused by a focusing coil that produces an axial magnetic field. The deflection coils then scan the electron image past the target aperture, where the electron multiplier produces a varying voltage representing the video signal. The image is thus "dissected" as it is scanned past the target, in an electronic version of a flat-bed digitizing process.

Charge transfer devices. A more recent development in image formation is that of solid-state image sensors, known as charge transfer devices (CTDs). There are two main classes of CTDs: charge-coupled devices (CCDs) and charge-injection devices (CIDs).

CCDs resemble MOSFETs (metal-oxide semiconductor field-effect transistor) in that they contain a "source" region and a "drain" region coupled by a depletion-region channel (Fig. 2.22). For imaging purposes, they can be considered as a monolithic array of closely spaced MOS capacitors forming a shift register (Fig. 2.23). Charges in the depletion region are transferred to the output by applying a series of clocking pulses to a row of electrodes between the source and the drain.

Photons incident on the semiconductor generate a series of charges on the CCD array. They are transferred to an output register either directly one line at a time (line transfer) or via a temporary storage area (frame transfer). The storage

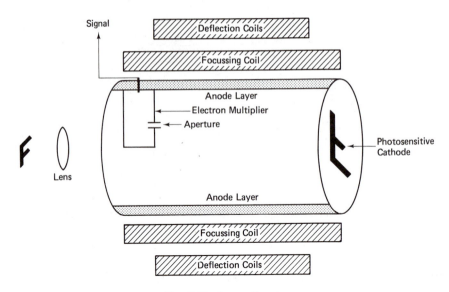

Fig. 2.21 Image dissector.

area is needed in frame transfer because the CCD array is scanned more rapidly than the output can be directly accommodated.

Charge injection devices (CIDs) resemble CCDs except that during sensing the charge is confined to the image site where it was generated (Fig. 2.24). The charges are read using an *X-Y* addressing technique similar to that used in computer memories. Basically, the stored charge is "injected" into the substrate and the resulting displacement current is detected to create the video signal.

CTD technology offers a number of advantages over conventional-tube-type cameras: light weight, small size, low power consumption, resistance to burn-in, low blooming, low dark current, high sensitivity, wide spectral and dynamic range, and lack of persistence. CIDs have the further advantages over CCDs of tolerance to processing defects, simple mechanization, avoidance of charge transfer losses, and minimized blooming. CTD cameras are now available commercially.

Analog-to-Digital Conversion

With current technology, the representation of an image as an analog electrical waveform is usually an unavoidable precursor to further processing. Thus the operation of deriving a digital representation of an analog voltage is basic to computer vision input devices.

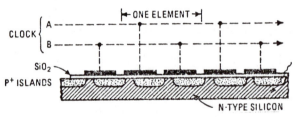

Fig. 2.22 Charge coupled device.

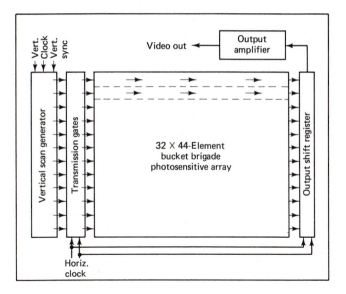

Fig. 2.23 A CCD array (line transfer).

The function of an analog-to-digital (A/D) converter is to take as input a voltage such as a video signal and to produce as output a representation of the voltage in digital memory, suitable for reading by an interface to a digital computer. The quality of an A/D converter is measured by its temporal resolution (the speed at which it can perform conversions) and the accuracy of its digital output. Analog-

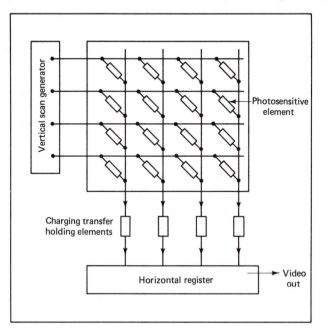

Fig. 2.24 A CID array.

to-digital converters are being produced as integrated circuit chips, but high-quality models are still expensive. The output precision is usually in the 8- to 12-bit range.

It is quite possible to digitize an entire frame of a TV camera (i.e., approximately 525 scan lines by 300 or so samples along a scan line) in a single frame time (1/30 sec in the United States). Several commercial systems can provide such fast digitization into a "frame buffer" memory, along with raster graphics display capabilities from the same frame buffer, and "video rate processing" of the digital data. The latter term refers to any of various low-level operations (such as averaging, convolution with small templates, image subtraction) which may be performed as fast as the images are acquired.

One inexpensive alternative to digitizing entire TV frames at once is to use an interface that acquires the TV signal for a particular point when the scan passes the requested location. With efficient programming, this point-by-point digitization can acquire an entire frame in a few seconds.

2.3.2 Sensing Range

The third dimension may be derived from binocular images by triangulation, as we saw earlier, or inferred from single monocular visual input by a variety of "depth cues," such as size and occlusion. Specialized technology exists to acquire "depth images" directly and reliably. Here we outline two such techniques: "light striping," which is based on triangulation, and "spot ranging," which is based on different principles.

Light Striping

Light striping is a particularly simple case of the use of *structured light* [Will and Pennington 1971]. The basic idea is to use geometric information in the illumination to help extract geometric information from the scene. The spatial frequencies and angles of bars of light falling on a scene may be clustered to find faces; randomly structured light may allow blank, featureless surfaces to be matched in stereo views; and so forth.

Many researchers [Popplestone et al. 1975; Agin 1972; Sugihara 1977] have used striping to derive three dimensions. In light striping, a single plane of light is projected onto a scene, which causes a stripe of light to appear on the scene (Fig. 2.25). Only the part of the scene illuminated by the plane is sensed by the vision system. This restricts the "image" to be an essentially one-dimensional entity, and simplifies matching corresponding points. The plane itself has a known position (equation in world coordinates), determinable by any number of methods involving either the measurement of the projecting device or the measurement of the final resulting plane of light. Every image point determines a single "line of sight" in three-space upon which the world point that produces the image point must lie. This line is determined by the focal point of the imaging system and the image point upon which the world point projects. In a light-striping system, any point that is sensed in the image is also guaranteed to lie on the light plane in three-space. But the light plane and the line of sight intersect in just one point (as long as

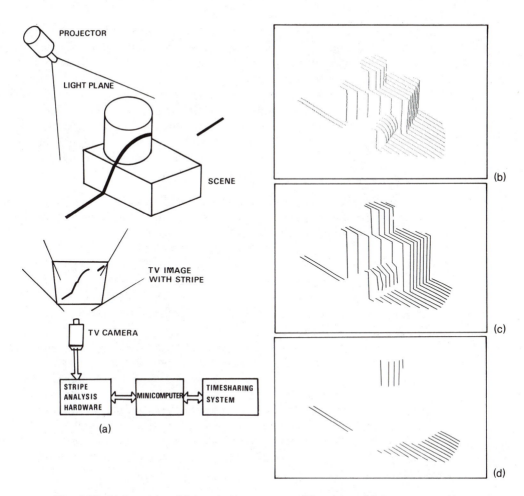

Fig. 2.25 Light striping. (a) A typical arrangement; (b) raw data; (c) data segmented into strips; (d) strips segmented into two surfaces.

the camera's focal point is not in the light plane). Thus by computation of the intersection of the line of sight with the plane of light, we derive the three-dimensional point that corresponds to any image point visible as part of a stripe.

The plane of light may result from a laser or from the projection of a slit. Only the light stripe should be visible to the imaging device; unless a laser is used, this implies a darkened room. If a camera is fitted with the proper filter, a laser-based system can be operated in normal light. Another advantage of the laser is that it can be focused into a narrower plane than can a slit image.

The only points whose three-dimensional coordinates can be computed are those that can be "seen" by both the light-stripe source and the camera at once. Since there must be a nonzero baseline if triangulation is to derive three-dimensional information, the camera cannot be too close to the projector, and thus concavities in the scene are potential trouble spots, since both the striper and the camera may not be able to "see" into them. Surfaces in the scene that are nearly parallel with the light plane will have a relatively small number of stripes projected onto them by any uniform stripe placement strategy. This problem is ameliorated by striping with two sets of parallel planes at right angles to each other [Agin 1972]. A major advantage of light striping over spot ranging is that (barring shadows) its continuity and discontinuity indicate similar conditions on the surface. It is easy to "segment" stripe images (Part II): Stripes falling on the same surface may easily be gathered together. This set of related stripes may be used in a number of ways to derive further information on the characteristics of the surface (Fig. 2.25b).

Spot Ranging

Civil engineers have used laser-based "spot range finders" for some time. In laboratory-size environments, they are a relatively new development. There are two basic techniques. First, one can emit a very sharp pulse and time its return ("lidar," the light equivalent of radar). This requires a sophisticated laser and electronics, since light moves 1 ft every billionth of a second, approximately. The second technique is to modulate the laser light in amplitude and upon its return compare the phase of the returning light with that of the modulator. The phase differences are related to the distance traveled [Nitzan et al. 1977]. A representative image is shown in Fig. 2.26.

Both these techniques produce results that are accurate to within about 1% of the range. Both of them allow the laser to be placed close to a camera, and thus "intensity maps" (images) and range maps may be produced from single viewpoints. The laser beam can easily poke into holes, and the return beam may be sensed close to the emitted one, so concavities do not present a serious problem. Since the laser beam is attenuated by absorption, it can yield intensity information as well. If the laser produces light of several wavelengths, it is possible to use filters and obtain multispectral reflectance information as well as depth information from the same device [Garvey 1976; Nitzan et al. 1977].

The usual mode of use of a spot ranging device is to produce a range map that corresponds to an intensity map. This has its advantages in that the correspondence may be close. The structural properties of light stripes are lost: It can be hard to "segment" the image into surfaces (to tell which "range pixels" are associated with the same surface). Range maps are amenable to the same sorts of segmentation techniques that are used for intensity images: Hough techniques, region growing, or differentiation-based methods of edge finding (Part II).

Ultrasonic Ranging

Just as light can be pulsed to determine range, so can sound and ultrasound (frequencies much higher than the audible range). Ultrasound has been used extensively in medicine to produce images of human organs (e.g., [Waag and Gramiak 1976]). The time between the transmitted and received signal determines range; the sound signal travels much slower than light, making the problem of timing the returning signal rather easier than it is in pulsed laser devices. However, the signal is severely attenuated as it travels through biological tissue, so that the detection apparatus must be very sensitive.

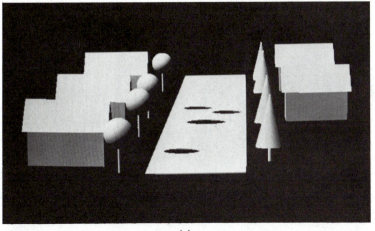

(a)

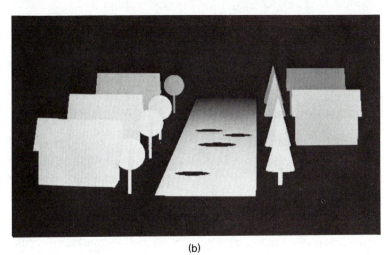

(b)

Fig. 2.26 Intensity and range images. (a) A (synthesized) intensity image of a street scene with potholes. The roofs all have the same intensity, which is different from the walls; (b) a corresponding range image. The wall and roof of each house have similar ranges, but the ranges differ from house to house.

One basic difference between sound and visible light ranging is that a light beam is usually reflected off just one surface, but that a sound beam is generally partially transmitted and partially reflected by "surfaces." The returning sound pulse has structure determined by the discontinuities in impedence to sound found in the medium through which it has passed. Roughly, a light beam returns information about a spot, whereas a sound beam can return information about the medium in the entire column of material. Thus, although sound itself travels relatively slowly, the data rate implicit in the returning structured sound pulse is quite high. Figure 2.27 shows an image made using the range data from ultrasound. The

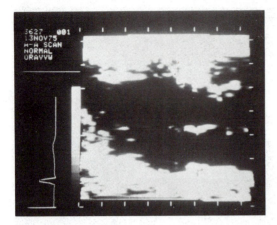

Fig. 2.27 Image made from ultrasound ranging.

sound pulses emanate from the top of the image and proceed toward the bottom, being partially reflected and transmitted along the way. In the figure, it is as if we were looking perpendicular to the beams, which are being displayed as brighter where strong reflectance is taking place. A single "scan line" of sound thus produces an image of an entire planar slice of medium.

2.3.3 Reconstruction Imaging

Two-dimensional reconstruction has been the focus of much research attention because of its important medical applications. High-quality images such as that shown in Fig. 1.2b can be formed by multiple images of x-ray projection data. This section contains the principles behind the most important reconstruction algorithms. These techniques are discussed in more detail with an expanded list of references in [Gordon and Herman 1974]. For a view of the many applications of two-dimensional reconstruction other than transmission scanning, the reader is referred to [Gordon et al. 1975].

Figure 2.28 shows the basic geometry to collect one-dimensional projections of two-dimensional data. (Most systems construct the image in a plane and repeat this technique for other planes; there are few true three-dimensional reconstruction systems that use planes of projection data simultaneously to construct volumes.)

In many applications sensors can measure the one-dimensional *projection* of two-dimensional image data. The projection $g(x')$ of an ideal image $f(x, y)$ in the direction θ is given by $\int f(x', y') \, dy'$ where $\mathbf{x}' = R_\theta \mathbf{x}$. If enough different projections are obtained, a good approximation to the image can be obtained with two-dimensional reconstruction techniques.

From Fig. 2.28, with the source at the first position along line AA', we can obtain the first projection datum from the detector at the first position along BB'. The line AB is termed a ray and the measurement at B a ray sum. Moving the source

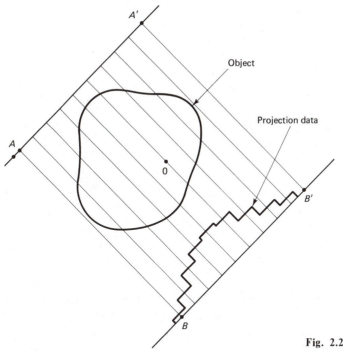

Fig. 2.28 Projection geometry.

and detector along lines AA' and BB' in synchrony allows us to obtain the entire data for projection 1. Now the lines AA' and BB' are rotated by a small angle $d\theta$ about 0 and the process is repeated. In the original x-ray systems $d\theta$ was 1° of angle, and 180 projections were taken. Each projection comprised 160 transmission measurements. The reconstruction problem is simply this: Given the projection data $g_k(x')$, $k = 0, \ldots, N - 1$, construct the original image $f(\mathbf{x})$.

Systems in use today use a fan beam rather than the parallel rays shown. However, the mathematics is simpler for parallel rays and illustrates the fundamental ideas. We describe three related techniques: summation, Fourier interpolation, and convolution.

The Summation Method

The summation method is simple: Distribute every ray sum $g_k(x')$ over the image cells along the ray. Where there are N cells along a ray, each such cell is incremented by $\frac{1}{N}g(x')$. This step is termed *back projection*. Repeating this process for every ray results in an approximate version of the original [DeRosier 1971]. This technique is equivalent (within a scale factor) to blurring the image, or convolving it with a certain point-spread function. In the continuous case of infinitely many projections, this function is simply the radically symmetric $h(r) = 1/r$.

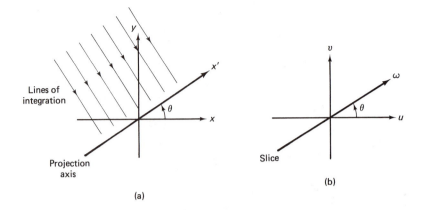

Fig. 2.29 Basis of Fourier techniques. (a) Projection axis x′; (b) corresponding axis in Fourier Space.

Fourier Algorithms

If a projection is Fourier-transformed, it defines a line through the origin in frequency space (Fig. 2.29). To show this formally, consider the expression for the two-dimensional transform

$$F(\mathbf{u}) = \int\int f(x, y) \exp\left[-j2\pi(ux + vy)\right] dx\, dy \tag{2.47}$$

Now consider $y = 0$ (projection onto the x axis): $x' = x$ and

$$g_0(x') = \int f(x, y)\, dy \tag{2.48}$$

The Fourier transform of this equation is

$$\mathcal{F}[g_0(x')] = \int\int [f(x, y)\, dy] \exp - j2\pi ux\ dx \tag{2.49}$$

$$= \int\int f(x, y) \exp - j2\pi ux\ dy\, dx$$

which, by comparison with (2.47), is

$$\mathcal{F}[g_0(x')] = F(u, 0) \tag{2.50}$$

Generalizing to any θ, the transform of an arbitrary $g(x')$ defines a line in the Fourier space representation of the cross section. Where $S_k(\omega)$ is the cross section of the Fourier transform along this line,

$$S_k(\omega) = F(u\cos\theta,\ u\sin\theta) \tag{2.51}$$

$$= \int g_k(x') \exp\left[j2\pi u(x')\right] dx'$$

Thus one way of reconstructing the original image is to use the Fourier transform of the projections to define points in the transform of $f(x)$, interpolate the undefined points of the transform from the known points, and finally take the inverse transform to obtain the reconstructed image.

Ch. 2 Image Formation

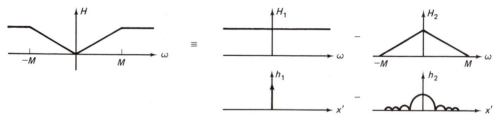

Fig. 2.30 Convolution method.

This technique can be applied with transforms other than the Fourier transform, and such methods are discussed in [DeRosier 1971; Crowther and Klug 1971].

The Convolution Method

The convolution method is the natural extension of the summation method. Since the summation method produces an image degraded from its convolution with some function h, one can remove the degradation by a "deconvolution." The straightforward way to accomplish this is to Fourier-transform the degraded image, multiply the result by an estimate of the transformed h^{-1}, and inverse-Fourier-transform the result. However, since all the operations are linear, a faster approach is to deconvolve the projections before performing the back projection. To show this formally, we use the inverse transform

$$f(\mathbf{x}) = \iint F(u, v) \exp[j2\pi(ux + vy)]du\,dv \qquad (2.52)$$

Changing to cylindrical coordinates (ω, θ) yields

$$f(\mathbf{x}) = \iint F_\theta(\omega) \exp[j2\pi\omega(x\cos\theta + y\sin\theta)]|\omega|d\omega\,d\theta \qquad (2.53)$$

Since $x' = x\cos\theta + y\sin\theta$, rewrite Eq. (2.53) as

$$f(\mathbf{x}) = \int \mathcal{F}^{-1}\{F_\theta(\omega)H(\omega)\}d\theta \qquad (2.54)$$

Since the image is bandlimited at some interval $(-\omega_m, \omega_m)$ one can define $H(\omega)$ arbitrarily outside of this interval. Therefore, $H(\omega)$ can be defined as a constant minus a triangular peak as shown in Fig. 2.30. Finally, the operation inside the integral in Eq. (2.54) is a convolution. Using the transforms shown in Fig. 2.30,

$$f(\mathbf{x}) = \int [f_\theta(x') - f_\theta(x')\omega_m\text{sinc}^2(\omega_m x')]\,d\theta \qquad (2.55)$$

Owing to its speed and the fact that the deconvolutions can be performed while the data are being acquired, the convolution method is the method employed in the majority of systems.

EXERCISES

2.1 In a binocular animal vision system, assume a focal length f of an eye of 50 mm and a separation distance d of 5 cm. Make a plot of Δx vs. $-z$ using Eq. (2.9). If the resolution of each eye is on the order of 50 line pairs/mm, what is the useful range of the binocular system?

2.2 In an opponent-process color vision system, assume that the following relations hold:

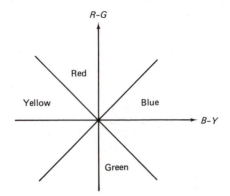

For example, if the $(R-G, B-Y, W-Bk)$ components of the opponent-process system are $(0.5, 3, 4)$, the perceived color will be blue.

Work out the perceived colors for the following (R,G,B) measurements:

(a) $(0.2, 0.3, 0.4)$ (b) $(0.2, 0.3, 0)$ (c) $(7, 4, 1)$

2.3 Develop an indexing scheme for a hexagonal array and define a Euclidean distance measure between points in the array.

2.4 Assume that a one-dimensional image has the following form:

$$f(x) = \cos(2\pi u_o x)$$

and is sampled with $u_s = u_o$. Using the graphical method of Section 2.2.6, find an expression for $f(x)$ as given by Eq. (2.49). Is this expression equal to the original image? Explain.

2.5 A certain image has the following Fourier transform:

$$F(\mathbf{u}) = \begin{cases} \text{nonzero} & \text{inside a hexagonal domain} \\ 0 & \text{otherwise} \end{cases}$$

(a) What are the smallest values for u and v so that $F(\mathbf{u})$ can be reconstructed from $F_x(\mathbf{u})$?

(b) Suppose now that rectangular sampling is *not* used but that now the u and v directions subtend an angle of $\pi/3$. Does this change your answer as to the smallest u and v? Explain.

2.6 Extend the binocular imaging model of Fig. 2.3 to include convergence: Let the two imaging systems pivot in the $y = 0$ plane about the viewpoint. Let the system have a baseline of $2d$ and be converged at some angle θ such that a point (x, y, z) appears at the origin of each image plane.

(a) Solve for z in terms of r and θ.

(b) Solve for z in this situation for points with nonzero disparity.

2.7 Compute the convolution of two Rect functions, where

$$\text{Rect}(x) = \begin{cases} 1 & 0 < x < 1 \\ 0 & \text{otherwise} \end{cases}$$

Show the steps in your calculations.

2.8

$$\text{Rect}(x) = \begin{cases} b & \text{for } |x| < a \\ 0 & \text{otherwise} \end{cases}$$

(a) What is $\text{Rect}(x) * \delta(x-a)$?

(b) What is the Fourier transform of $f(x)$ where $f(x) = \text{Rect}(x+c) + \text{Rect}(x-c)$ and $c > a$?

2.9 A digitizer has a sampling interval of $\Delta x = \Delta y = \Delta$. Which of the following images can be represented unambiguously by their samples? (Assume that effects of a finite image domain can be neglected.)

(a) $(sin(\pi x/\Delta))/(\pi x/\Delta)$

(b) $cos(\pi x/2\Delta)\cos(3\pi x/4\Delta)$

(c) $\text{Rect}(x)$ (see Problem 2.8)

(d) e^{-ax^2}

REFERENCES

AGIN, G. J. "Representation and description of curved objects" (Ph.D. dissertation). AIM-173, Stanford AI Lab, October 1972.

ANDREWS, H. C. and B. R. HUNT. *Digital Image Restoration.* Englewood Cliffs, NJ: Prentice-Hall, Inc., 1977.

CROWTHER, R. A. and A. KLUG. "ART and science, or, conditions for 3-d reconstruction from electron microscope images." *J. Theoretical Biology 32*, 1971.

DEROSIER, D. J. "The reconstruction of three-dimensional images from electron micrographs." *Contemporary Physics 12*, 1971.

DUDA, R. O. and P. E. HART. *Pattern Recognition and Scene Analysis.* New York: Wiley, 1973.

GARVEY, T. D. "Perceptual strategies for purposive vision." Technical Note 117, AI Center, SRI International, September 1976.

GONZALEZ, R. C. and P. WINTZ. *Digital Image Processing.* Reading, MA: Addison-Wesley, 1977.

GORDON, R. and G. T. HERMAN. "Three-dimensional reconstruction from projections: a review of algorithms." *International Review of Cytology 38*, 1974, 111-151.

GORDON, R., G. T. HERMAN, and S. A. JOHNSON. "Image reconstruction from projections." *Scientific American*, October 1975.

HERING, E. "Principles of a new theory of color sense." *In Color Vision*, R.C. Teevan and R.C. Birney (Eds.). Princeton, NJ: D. Van Nostrand, 1961.

HORN, B. K. P. "Understanding image intensities." *Artificial Intelligence 8*, 2, April 1977, 201-231.

HORN, B. K. P. and R. W. SJOBERG. "Calculating the reflectance map." *Proc.*, DARPA IU Workshop, November 1978, 115-126.

HURVICH, L. M. and D. JAMESON. "An opponent-process theory of color vision." *Psychological Review 64*, 1957, 384-390.

JAIN, A. K. "Advances in mathematical models for image processing." *Proc. IEEE 69*, 5, May 1981, 502-528.

JOBLOVE, G. H. and D. GREENBERG. "Color spaces for computer graphics." *Computer Graphics 12*, 3, August 1978, 20-25.

KENDER, J. R. "Saturation, hue, and normalized color: calculation, digitization effects, and use." Technical Report, Dept. of Computer Science, Carnegie-Mellon Univ., November 1976.

LAND, E. H. "The retinex theory of color vision." *Scientific American,* December 1977, 108–128.

MUNSELL, A. H. *A Color Notation*, 8th ed. Baltimore, MD: Munsell Color Co., 1939.

NICODEMUS, F. E., J. C. RICHMOND, J. J. HSIA, I. W. GINSBERG, and T. LIMPERIS. "Geometrical considerations and nomenclature for reflectance." NBS Monograph 160, National Bureau of Standards, U.S. Department of Commerce, Washington, DC, October 1977.

NITZAN, D., A. BRAIN, and R. DUDA. "The measurement and use of registered reflectance and range data in scene analysis." *Proc. IEEE 65,* 2, February 1977.

POPPLESTONE, R. J., C. M. BROWN, A. P. AMBLER, and G. F. CRAWFORD. "Forming models of plane-and-cylinder faceted bodies from light stripes." *Proc.*, 4th IJCAI, September 1975, 664-668.

PRATT, W. K. *Digital Image Processing.* New York: Wiley-Interscience, 1978.

ROSENFIELD A. and A. C. KAK. *Digital Picture Processing.* New York: Academic Press, 1976.

SMITH, A. R. "Color gamut transform pairs." *Computer Graphics 12,* 3, August 1978, 12–19.

SUGIHARA, K. "Dictionary-guided scene analysis based on depth information." In *Progress Report on 3-D Object Recognition.* Bionics Research Section, ETL, Tokyo, March 1977.

TENENBAUM, J. M. and S. WEYL. "A region-analysis subsystem for interactive scene analysis." *Proc.*, 4th IJCAI, September 1975, 682–687.

WAAG, R. B. and R. GRAMIAK. "Methods for ultrasonic imaging of the heart." *Ultrasound in Medicine and Biology 2*, 1976, 163–170.

WILL, P. M. and K. S. PENNINGTON. "Grid coding: a preprocessing technique for robot and machine vision." *Artificial Intelligence 2*, 3/4, Winter 1971, 319–329.

Early Processing 3

3.1 RECOVERING INTRINSIC STRUCTURE

The imaging process confounds much useful physical information into the gray-level array. In this respect, the imaging process is a collection of degenerate transformations. However, this information is not irrevocably lost, because there is much spatial redundancy: Neighboring pixels in the image have the same or nearly the same physical parameters. A collection of techniques, which we call *early processing*, exploits this redundancy in order to undo the degeneracies in the imaging process. These techniques have the character of transformations for changing the image into "parameter images" *or intrinsic images* [Barrow and Tenenbaum 1978; 1981] which reflect the spatial properties of the scene. Common intrinsic parameters are surface discontinuities, range, surface orientation, and velocity.

In this chapter we neglect high-level internal model information even though it is important and can affect early processing. Consider the case of the perceived central edge in Fig. 3.1a. As shown by Fig. 3.1b, which shows portions of the same image, the central edge of Fig. 3.1a is not present in the data. Nevertheless, the human perceiver "sees" the edge, and one reasonable explanation is that it is a product of an internal block model. Model-directed activity is taken up in later chapters. These examples show how high level models (e.g., circles) can affect low-level processors (e.g., edge finders). However, for the purposes of study it is often helpful to neglect these effects. These simplifications make it easier to derive the fundamental constraints between the physical parameters and gray levels. Once these are understood, they can be modified using the more abstract structures of later chapters.

Most early computer vision processing can be done with parallel computations whose inputs tend to be spatially localized. When computing intrinsic images

(a) (b)

Fig. 3.1 (a) A perceived edge. (b) Portions of image in (a) showing the lack of image data.

the parallel computations are iterated until the intrinsic parameter measurements converge to a set of values. A computation that falls in the parallel-iterative category is known in computer vision as *relaxation* [Rosenfeld et al. 1976]. Relaxation is a very general computational technique that is useful in computer vision. Specific examples of relaxation computations appear throughout the book; general observations on relaxation appear in Chapter 12.

This chapter covers six categories of early processing techniques:

1. *Filtering* is a generic name for techniques of changing image gray levels to enhance the appearance of objects. Most often this means transformations that make the intensity discontinuities between regions more prominent. These transformations are often dependent on gross object characteristics. For example, if the objects of interest are expected to be relatively large, the image can be blurred to erase small intensity discontinuities while retaining those of the object's boundary. Conversely, if the objects are relatively small, a transformation that selectively removes large discontinuities may be appropriate. Filtering can also compensate for spatially varying illumination.

2. *Edge operators* detect and measure very local discontinuities in intensity or its gradient. The result of an edge operator is usually the magnitude and orientation of the discontinuity.

3. *Range transforms* use known geometry about stereo images to infer the distance of points from the viewer. These transforms make use of the inverse perspective transform to interpret how points in three-dimensional space project onto stereo pairs. A correspondence between points in two stereo images of known geometry determines the range of those points. Relative range may also be derived from local correspondences without knowing the imaging geometry precisely.

4. *Surface orientation* can be calculated if the source illumination and reflectance properties of the surface are known. This calculation is sometimes called

"shape from shading." Surface orientation is particularly simple to calculate when the source illumination can be controlled.

5. *Optical flow*, or velocity fields of image points, can be calculated from local temporal and spatial variations in sequences of gray-level images.

6. A *pyramid* is a general structure for representing copies of the image at multiple resolutions. A pyramid is a "utility structure" which can dramatically improve the speed and effectiveness of many early processing and later segmentation algorithms.

3.2 FILTERING THE IMAGE

Filtering is a very general notion of transforming the image intensities in some way so as to enhance or deemphasize certain features. We consider only transforms that leave the image in its original format: a spatial array of gray levels. Spurred on by the needs of planetary probes and aerial reconnaissance, filtering initially received more attention than any other area of image processing and there are excellent detailed reference works (e.g., [Andrews and Hunt 1977; Pratt 1978; Gonzalez and Wintz 1977]). We cannot afford to examine these techniques in great detail here; instead, our intent is to describe a set of techniques that conveys the principal ideas.

Almost without exception, the best time to filter an image is at the image formation stage, before it has been sampled. A good example of this is the way chemical stains improve the effectiveness of microscopic tissue analysis by changing the image so that diagnostic features are obvious. In contrast, filtering after sampling often emphasizes random variations in the image, termed *noise*, that are undesirable effects introduced in the sampling stage. However, for cases where the image formation process cannot be changed, digital filtering techniques do exist. For example, one may want to suppress low spatial frequencies in an image and sharpen its edges. An image filtered in this way is shown in Fig. 3.2.

Note that in Fig. 3.2 the work of recognizing real-world objects still has to be done. Yet the edges in the image, which constitute object boundaries, have been made more prominent by the filtering operation. Good filtering functions are not easy to define. For example, one hazard with Fourier techniques is that sharp edges in the filter will produce unwanted "ringing" in the spatial domain, as evidenced by Fig. 2.5. Unfortunately, it would be too much of a digression to discuss techniques of filter design. Instead, the interested reader should refer to the references cited earlier.

3.2.1 Template Matching

Template matching is a simple filtering method of detecting a particular feature in an image. Provided that the appearance of this feature in the image is known accu-

(a)   (b)

Fig. 3.2   Effects of high frequency filtering. (a) Original image. (b) Filtered image.

rately, one can try to detect it with an operator called a *template*. This template is, in effect, a subimage that looks just like the image of the object. A similarity measure is computed which reflects how well the image data match the template for each possible template location. The point of maximal match can be selected as the location of the feature. Figure 3.3 shows an industrial image and a relevant template.

Correlation

One standard similarity measure between a function $f(\mathbf{x})$ and a template $t(\mathbf{x})$ is the Euclidean distance $d(\mathbf{y})$ squared, given by

$$d^2(\mathbf{y}) = \sum_{\mathbf{x}} [f(\mathbf{x}) - t(\mathbf{x} - \mathbf{y})]^2 \qquad (3.1)$$

By $\sum_{\mathbf{x}}$ we mean $\sum_{x=-M}^{M} \sum_{y=-N}^{N}$, for some $M,\ N$ which define the size of the template extent. If the image at point $\mathbf{y}$ is an exact match, then $d(\mathbf{y}) = 0$; otherwise, $d(\mathbf{y}) > 0$. Expanding the expression for d^2, we can see that

$$d^2(\mathbf{y}) = \sum_{\mathbf{x}} [f^2(\mathbf{x}) - 2f(\mathbf{x})t(\mathbf{x} - \mathbf{y}) + t^2(\mathbf{x} - \mathbf{y})] \qquad (3.2)$$

Notice that $\sum_{\mathbf{x}} t^2(\mathbf{x} - \mathbf{y})$ is a constant term and can be neglected. When $\sum_{\mathbf{x}} f^2(\mathbf{x})$ is approximately constant it too can be discounted, leaving what is called the *cross correlation* between f and t.

$$R_{ft}(\mathbf{y}) = \sum_{\mathbf{x}} f(\mathbf{x})t(\mathbf{x} - \mathbf{y}) \qquad (3.3)$$

This is maximized when the portion of the image "under" t is identical to t.

Template

Industrial Image

Fig. 3.3 An industrial image and template for a hexagonal nut.

One may visualize the template-matching calculations by imagining the template being *shifted* across the image to different offsets; then the superimposed values at this offset are *multiplied* together, and the products are *added*. The resulting sum of products forms an entry in the "correlation array" whose coordinates are the offsets attained by the source template.

If the template is allowed to take *all* offsets with respect to the image such that some overlap takes place, the correlation array is larger than either the template or the image. An $n \times n$ image with an $m \times m$ template yields an $(n + m - 1 \times n + m - 1)$ correlation array. If the template is not allowed to shift off the image, the correlation array is $(n - m + 1 \times n - m + 1)$; for $m < n$. Another form of correlation results from computing the offsets modulo the size of the image; in other words, the template "wraps around" the image. Being shifted off to the right, its right portion reappears on the left of the image. This sort of correlation is called *periodic* correlation, and those with no such wraparound properties are called *aperiodic*. We shall be concerned exclusively with aperiodic correlation. One can always modify the input to a periodic correlation algorithm by padding the outside with zeros so that the output is the aperiodic correlation.

Figure 3.4 provides an example of (aperiodic) "shift, add, multiply" template matching. This figure illustrates some difficulties with the simple correlation measure of similarity. Many of the advantages and disadvantages of this measure stem from the fact that it is linear. The advantages of this simplicity have mainly to do with the existence of algorithms for performing the calculation efficiently (in a transform domain) for the entire set of offsets. The disadvantages have to do with

Template	Image	Correlation
1 1 1	1 1 0 0 0	7 4 2 x x
1 1 1	1 1 1 0 0	5 3 2 x x
1 1 1	1 0 1 0 0	2 1 9 x x
	0 0 0 0 0	x x x x x
	0 0 0 0 8	x x x x x
		x = undefined

Fig. 3.4 (a) A simple template. (b) An image with noise. (c) The aperiodic correlation array of the template and image. Ideally peaks in the correlation indicate positions of good match. Here the correlation is only calculated for offsets that leave the template entirely within the image. The correct peak is the upper left one at 0, 0 offset. The "false alarm" at offset 2, 2 is caused by the bright "noise point" in the lower right of the image.

the fact that the metric is sensitive to properties of the image that may vary with the offset, such as its average brightness. Slight changes in the shape of the object, its size, orientation, or intensity values can also disturb the match.

Nonetheless, the idea of template matching is important, particularly if Eq. (3.3) is viewed as a *filtering* operation instead of an algorithm that does all the work of object detection. With this viewpoint one chooses one or more templates (filters) that transform the image so that certain features of an object are more readily apparent. These templates generally highlight subparts of the objects. One such class of templates is edge templates (discussed in detail in Section 3.3).

We showed in Section 2.2.4 that convolution and multiplication are Fourier transform pairs. Now note that the correlation operation in (3.3) is essentially the same as a convolution with a function $t'(\mathbf{x}) \equiv t(-\mathbf{x})$. Thus in a mathematical sense cross correlation and convolution are equivalent. Consequently, if the size of the template is sufficiently large, it is cheaper to perform the template matching operation in the spatial frequency domain, by the same transform techniques as for filtering.

Normalized Correlation

A crucial assumption in the development of Eq. (3.3) was that the image energy covered by the matching template at any offset was constant; this leads to a linear correlation matching technique. This assumption is approximately correct if the average image intensity varies slowly compared to the template size, but a bright spot in the image can heavily influence the correlation by affecting the sum of products violently in a small area (Fig. 3.4). Even if the image is well behaved, the range of values of the metric can vary with the size of the matching template. Are there ways of normalizing the correlation metric to make it insensitive to these variations?

There is a well-known treatment of the normalized correlation operation. It has been used for a variety of tasks involving registration and stereopsis of images [Quam and Hannah 1974]. Let us say that two input images are being matched to find the best offset that aligns them.

Let $f_1(\mathbf{x})$ and $f_2(\mathbf{x})$ be the images to be matched. q_2 is the patch of f_2 (possibly all of it) that is to be matched with a similar-sized patch of f_1. q_1 is the patch of f_1 that is covered by q_2 when q_2 is offset by $\mathbf{y}$.

Let $E()$ be the expectation operator. Then

$$\sigma(q_1) = [E(q_1^2) - (E(q_1))^2]^{1/2} \tag{3.4}$$

$$\sigma(q_2) = [E(q_2^2) - (E(q_2))^2]^{1/2} \tag{3.5}$$

give the standard deviations of points in patches q_1 and q_2. (For notational convenience, we have dropped the spatial arguments of q_1 and q_2.) Finally, the normalized correlation is

$$N(\mathbf{y}) = \frac{E(q_1 q_2) - E(q_1)E(q_2)}{\sigma(q_1)\sigma(q_2)} \tag{3.6}$$

and $E(q_1 q_2)$ is the expected value of the product of intensities of points that are superimposed by the translation by $\mathbf{y}$.

The normalized correlation metric is less dependent on the local properties of the reference and input images than is the unnormalized correlation, but it is sensitive to the signal-to-noise content of the images. High uncorrelated noise in the two images, or the image and the reference, decreases the value of the correlation. As a result, one should exercise some care in interpreting the metric. If the noise properties of the image are known, one indication of reliability is given by the "(signal + noise)-to-noise" ratio. For the normalized correlation to be useful, the standard deviation of the patches of images to be matched (i.e., of the areas of image including noise) should be significantly greater than that of the noise. Then a correlation value may be considered significant if it is approximately equal to the theoretically expected one. Consider uncorrelated noise of identical standard deviation, in a patch of true value $f(x, y)$. Let the noise component of the image be $n(x, y)$. Then the theoretical maximum correlation is

$$1 - \frac{\sigma^2(n)}{\sigma^2(f + n)} \tag{3.7}$$

In matching an idealized, noise-free reference pattern, the best expected value of the cross correlation is

$$\frac{\sigma(f)}{\sigma(f + n)} \tag{3.8}$$

If the noise and signal characteristics of the data are known, the patch size may be optimized by using that information and the simple statistical arguments above. However, such considerations leave out the effects of systematic, nonstatistical error (such as imaging distortions, rotations, and scale differences between images). These systematic errors grow with patch size, and may swamp the statistical advantages of large patches. In the worst case, they may vitiate the advantages of the correlation process altogether.

Since correlation is expensive, it is advantageous to ensure that there is enough information in the patches chosen for correlation before the operation is done. One way to do this is to apply a cheap "interest operator" before the relatively expensive correlation. The idea here is to make sure that the image varies enough to give a usable correlation image. If the image is of uniform intensity, even its correlation with itself (autocorrelation) is flat everywhere, and no information about where the image is registered with itself is derivable. The "interest operator" is a way of finding areas of image with high variance. In fact, a common and useful interest measure is exactly the (directional) variance over small areas of image. One directional variance algorithm works as follows.

The Moravec interest operator [Moravec 1977] produces candidate match points by measuring the distinctness of a local piece of the image from its surround. To explain the operator, we first define a variance measure at a pixel $(\mathbf{x})$ as

$$\mathrm{var}(x, y) = \left\{ \sum_{k,\, l \text{ in } s} [f(x, y) - f(x + k, y + l)]^2 \right\}^{\frac{1}{2}} \tag{3.9}$$

$$s = \left\{ (0, a), (0, -a), (a, 0), (-a, 0) \right\}$$

where a is a parameter. Now the interest operator value is initially the minimum of itself and surrounding points:

$$\text{IntOpVal } (\mathbf{x}) := \min_{y \leq 1} [\text{var} (\mathbf{x} + \mathbf{y})] \qquad (3.10)$$

Next a check is made to see if the operator is a local maximum by checking neighbors again. Only local maxima are kept.

$$\text{IntOpVal}(\mathbf{x}) := 0 \text{ unless}$$

$$\text{IntOpVal}(\mathbf{x}) \geq \text{IntOpVal}(\mathbf{x} + \mathbf{y}) \qquad (3.11)$$

$$\text{for } \mathbf{y} \leq 1$$

Finally, candidate points are chosen from the IntOpVal array by thresholding.

$$\mathbf{x} \text{ is a candidate point iff IntOpVal } (\mathbf{x}) > T \qquad (3.12)$$

The threshold is chosen empirically to produce some fraction of the total image points.

3.2.2 Histogram Transformations

A gray-level histogram of an image is a function that gives the frequency of occurrence of each gray level in the image. Where the gray levels are quantized from 0 to n, the value of the histogram at a particular gray level p, denoted $h(p)$, is the number or fraction of pixels in the image with that gray level. Figure 3.5 shows an image with its histogram.

A histogram is useful in many different ways. In this section we consider the histogram as a tool to guide gray-level transformation algorithms that are akin to filtering. A very useful image transform is called *histogram equalization*. Histogram equalization defines a mapping of gray levels p into gray levels q such that the distribution of gray levels q is uniform. This mapping stretches contrast (expands the

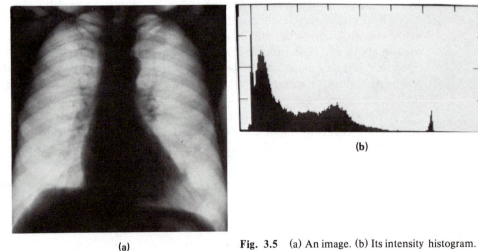

(a)

(b)

Fig. 3.5 (a) An image. (b) Its intensity histogram.

range of gray levels) for gray levels near histogram maxima and compresses contrast in areas with gray levels near histogram minima. Since contrast is expanded for most of the image pixels, the transformation usually improves the detectability of many image features.

The histogram equalization mapping may be defined in terms of the *cumulative* histogram for the image. To see this, consider Fig. 3.6a. To map a small interval of gray levels dp onto an interval dq in the general case, it must be true that

$$g(q)\, dq = h(p)\, dp \qquad (3.13)$$

where $g(q)$ is the new histogram. If, in the histogram equalization case, $g(q)$ is to be uniform, then

$$g(q) = \frac{N^2}{M} \qquad (3.14)$$

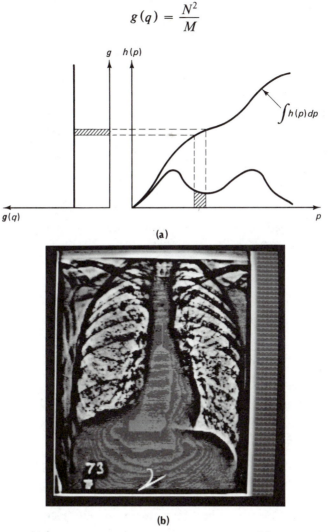

(a)

(b)

Fig. 3.6 (a) Basis for a histogram equalization technique. (b) Results of histogram equalization.

where N^2 is the number of pixels in the image and M is the number of gray levels. Thus combining Eqs. (3.13) and (3.14) and integrating, we have

$$q(p) = \frac{M}{N^2} \left(\int_0^p h(s)\,ds \right) \qquad (3.15)$$

But Eq. (3.15) is simply the equation for the normalized cumulative histogram. Figure 3.6b shows the histogram-equalized image.

3.2.3 Background Subtraction

Background subtraction can be another important filtering step in early processing. Many images can have slowly varying background gray levels which are incidental to the task at hand. Examples of such variations are:

- Solution gradients in cell slides
- Lighting variations on surfaces in office scenes
- Lung images in a chest radiograph

Note that the last example is only a "background" in the context of looking for some smaller variations such as tumors or pneumoconiosis.

Background subtraction attempts to remove these variations by first approximating them (perhaps analytically) with a background image f_b and then subtracting this approximation from the original image. That is, the new image f_n is

$$f_n(\mathbf{x}) = f(\mathbf{x}) - f_b(\mathbf{x}) \qquad (3.16)$$

Various functional forms have been tried for analytic representations of slowly varying backgrounds. In the simplest cases, $f_b(x)$ may be a constant,

$$f_b(\mathbf{x}) = c \qquad (3.17)$$

or linear,

$$f_b(\mathbf{x}) = \mathbf{m} \cdot \mathbf{x} + c \qquad (3.18)$$

A more sophisticated background model is to use a low-pass filtered variant of the original image:

$$f_b(\mathbf{x}) = \mathcal{F}^{-1}[H(\mathbf{u})\,F(\mathbf{u})] \qquad (3.19)$$

where $H(\mathbf{u})$ is a low-pass filtering function. The problem with this technique is that it is global; one cannot count on the "best" effect in any local area since the filter treats all parts of the image identically. For the same reason, it is difficult to design a Fourier filter that works for a number of very different images.

A workable alternative is to approximate $f_b(\mathbf{x})$, using *splines*, which are piecewise polynomial approximation functions. The mathematics of splines is treated in Chapter 8 since they find more general application as representations of shape. The filtering application is important but specialized. The attractive feature of a spline approximation for filtering is that it is *variation diminishing* and *spatially variant*. The spline approximation is guaranteed to be "smoother" than the original function and will approximate the background differently in different parts of the image. The latter feature distinguishes the method from Fourier-domain techniques which are spatially invariant. Figure 3.7 shows the results of spline filtering.

3.2.4 Filtering and Reflectance Models

Leaving the effects of imaging geometry implicit (Section 2.2.2), the definitions in Section 2.2.3 imply that the image irradiance (gray level) at the image point x' is proportional to the product of the scene irradiance E and the *reflectance* r at its corresponding world point $\mathbf{x}$.

$$f(\mathbf{x}') = E(\mathbf{x})r(\mathbf{x}) \qquad (3.20)$$

The irradiance at x is the sum of contributions from all illumination sources, and the reflectance is that portion of the irradiance which is reflected toward the observer (camera). Usually E changes slowly over a scene, whereas r changes quickly over edges, due to varying face angles, paint, and so forth. In many cases one would like to detect these changes in r while ignoring changes in E. One way of doing this is to filter the image $f(\mathbf{x}')$ to eliminate the slowly varying component. However, as f is the *product* of illumination and reflectance, it is difficult to define an operation that selectively diminishes E while retaining r. Furthermore, such an operation must retain the positivity of f. One solution is to take the logarithm of Eq. (3.20). Then

$$\log f = \log E + \log r \qquad (3.21)$$

Equation (3.21) shows two desirable properties of the logarithmic transformation: (1) the logarithmic image is unrestricted in sign, and (2) the image is a superposition of the irradiance component and reflectance component. Since reflectance is an in-

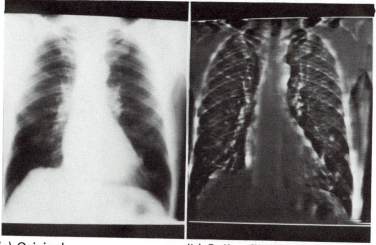

(a) Original.   (b) Spline filtered image.

Fig. 3.7 The results of spline filtering to remove background variation.

trinsic characteristic of objects, the obvious goal of image analysis is to recognize the reflectance component under various conditions of illumination. Since the separation of two components is preserved under linear transformations and the irradiance component is usually of low spatial frequency compared to the reflectance component, filtering techniques can suppress the irradiance component of the signal relative to the reflectance component.

If the changes in r occur over very short distances in the images, r may be isolated by a three-step process [Horn 1974]. First, to enhance reflectance changes, the image function is differentiated (Section 3.3.1). The second step removes the low irradiance gradients by thresholding. Finally, the resultant image is integrated to obtain an image of perceived "lightness" or reflectance. Figure 3.8 shows these steps for the one-dimensional case.

A basic film parameter is density, which is proportional to the logarithm of transmitted intensity; the logarithmically transformed image is effectively a *density image*. In addition to facilitating the extraction of lightness, another advantage of the density image is that it is well matched to our visual experience. The ideas for many image analysis programs stem from our visual inspection of the image. However, the human visual system responds logarithmically to light intensity and also enhances high spatial frequencies [Stockham 1972]. Algorithms derived from

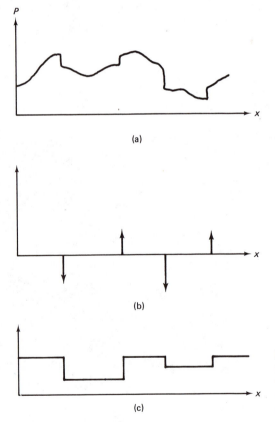

Fig. 3.8 Steps in processing an image to detect reflectance. (a) Original image. (b) Differentiation followed by thresholding. (c) Integration of function in (b).

introspective reasoning about the perceived image (which has been transformed by our visual system) will not necessarily be successful when applied to an unmodified intensity image. Thus one argument for using a density transformation followed by high spatial frequency emphasis filtering is that the computer is then "seeing" more like the human image analyzer.

3.3 FINDING LOCAL EDGES

Boundaries of objects tend to show up as intensity discontinuities in an image. Experiments with the human visual system show that boundaries in images are extremely important; often an object can be recognized from only a crude outline [Attneave 1954]. This fact provides the principal motivation for representing objects by their boundaries. Also, the boundary representation is easy to integrate into a large variety of object recognition algorithms.

One might expect that algorithms could be designed that find the boundaries of objects directly from the gray-level values in the image. But when the boundaries have complicated shapes, this is difficult. Much greater success has been obtained by first transforming the image into an intermediate image of *local* gray-level discontinuities, or edges, and then composing these into a more elaborate boundary. This strategy reflects the principle: When the gap between representations becomes too large, introduce intermediate representations. In this case, boundaries that are highly model-dependent may be decomposed into a series of local edges that are highly model-independent.

A local edge is a small area in the image where the local gray levels are changing rapidly in a simple (e.g., monotonic) way. An *edge operator* is a mathematical operator (or its computational equivalent) with a small spatial extent designed to detect the presence of a local edge in the image function.

It is difficult to specify a priori which local edges correspond to relevant boundaries in the image. Depending on the particular task domain, different local changes will be regarded as likely edges. Plots of gray level versus distance along the direction perpendicular to the edge for some hypothetical edges (Fig. 3.9a-d) demonstrate some different kinds of "edge profiles" that are commonly encountered. Of course, in most practical cases, the edge is noisy (Fig. 3.9d) and may appear as a composite of profile types. The fact that different kinds of edge operators perform best in different task domains has prompted the development of a variety of operators. However, the unifying feature of most useful edge operators is that they compute a *direction* which is aligned with the direction of maximal gray-level change, and a *magnitude* describing the severity of this change. Since edges are a high-spatial-frequency phenomenon, edge finders are also usually sensitive to high-frequency noise, such as "snow" on a TV screen or film grain.

Operators fall into three main classes: (1) operators that approximate the mathematical gradient operator, (2) template matching operators that use multiple templates at different orientations, and (3) operators that fit local intensities with parametric edge models. Representative examples from the first two of these categories appear in this section. The computer vision literature abounds with edge

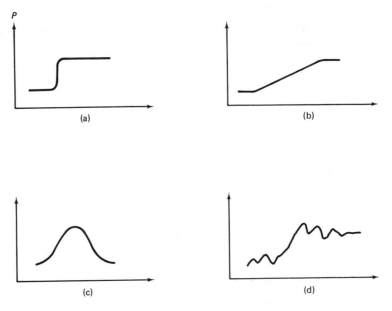

Fig. 3.9 Edge profiles.

operators, and we make no attempt to summarize them all here. For a guide to this literature, see [Rosenfeld and Kak 1976].

Parametric models generally capture more detailed edge structure than the two-parameter direction and magnitude vector; as a result, they can be more computationally complicated. For this reason and others discussed in Section 3.3.4, we shall omit a detailed discussion of these kinds of edge operators. One of the best known parametric models is Hueckel's [Hueckel 1971, 1973], but several others have been developed since [Mero and Vassy 1975; Nevatia 1977; Abdou 1978; Tretiak 1979].

3.3.1 Types of Edge Operators

Gradient and Laplacian

The most common and historically earliest edge operator is the gradient [Roberts 1965]. For an image function $f(\mathbf{x})$, the gradient magnitude $s(\mathbf{x})$ and direction $\phi(\mathbf{x})$ can be computed as

$$s(\mathbf{x}) = (\Delta_1^2 + \Delta_2^2)^{\frac{1}{2}} \tag{3.22}$$

$$\phi(\mathbf{x}) = \tan^{-1}(\Delta_2/\Delta_1) \tag{3.23}$$

where

$$\Delta_1 = f(x + n, y) - f(x, y) \tag{3.24}$$

$$\Delta_2 = f(x, y + n) - f(x, y)$$

n is a small integer, usually unity, and atan (x, y) and $\tan^{-1}(x/y)$ must be the proper quadrant. The parameter *n* is called the "span" of the gradient. Roughly, *n* should be small enough so that the gradient is a good approximation to the local changes in the image function, yet large enough to overcome the effects of small variations in *f*.

Equation (3.24) is only one *difference operator*, or way of measuring gray-level intensities along orthogonal directions using Δ_1 and Δ_2. Figure 3.10 shows the gradient difference operators compared to other operators [Roberts 1965; Prewitt 1970]. The reason for the modified operators of Prewitt and Sobel is that the local averaging tends to reduce the effects of noise. These operators do, in fact, perform better than the Roberts operator for a step edge model.

One way to study an edge operator's performance is to use an ideal edge such as the step edge shown in Fig. 3.11. This edge has two gray levels: zero and h units. If the edge goes through the finite area associated with a pixel, the pixel is given a value between zero and h, depending on the proportion of its area covered. Comparative edge operator performance has been carried out [Abdou 1978]. In the case of the Sobel operator (Fig. 3.10c) the measured orientation ϕ' is given by

Δ_1 $\qquad\qquad\qquad$ Δ_2

0	1
−1	0

1	0
0	−1

(a)

−1	0	1
−1	0	1
−1	0	1

1	1	1
0	0	0
−1	−1	−1

(b)

−1	0	1
−2	0	2
−1	0	1

1	2	1
0	0	0
−1	−2	−1

(c)

Fig. 3.10 Gradient operators.

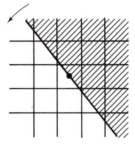

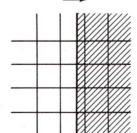

Fig. 3.11 Edge models for orientation and displacement sensitivity analyses.

$$
\phi' = \begin{cases} \phi & \text{if } 0 \leqslant \phi \leqslant \tan^{-1}\left(\dfrac{1}{3}\right) \\[2ex] \tan^{-1}\left(\dfrac{7\tan^2\phi + 6\tan\phi - 1}{-9\tan^2\phi + 22\tan\phi - 1}\right) & \text{if } \tan^{-1}\left(\dfrac{1}{3}\right) \leqslant \phi \leqslant \pi/4 \quad (3.25) \end{cases}
$$

Arguments from symmetry show that only the $0 \leqslant \phi < \pi/4$ cases need be examined. Similar studies could be made using ramp edge models.

A rather specialized kind of gradient is that taken "between pixels." This scheme is shown in Fig. 3.12. Here a pixel may be thought of as having four *crack edges* surrounding it, whose directions are fixed by the pixel to be multiples of $\pi/2$. The magnitude of the edge is determined by $|f(\mathbf{x}) - f(\mathbf{y})|$, where $\mathbf{x}$ and $\mathbf{y}$ are the coordinates of the pixels that have the edge in common. One advantage of this formulation is that it provides an effective way of separating regions and their boundaries. The disadvantage is that the edge orientation is crude.

The *Laplacian* is an edge detection operator that is an approximation to the mathematical Laplacian $\partial^2 f/\partial x^2 + \partial^2 f/\partial y^2$ in the same way that the gradient is an approximation to the first partial derivatives. One version of the discrete Laplacian is given by

"Crack" edge   **Fig. 3.12** "Crack" edge representation.

$$L(x, y) = f(x, y) - \tfrac{1}{4}[f(x, y+1) + f(x, y-1) \qquad (3.26)$$
$$+ f(x+1, y) + f(x-1, y)]$$

The Laplacian has two disadvantages as an edge measure: (1) useful directional information is not available, and (2) the Laplacian, being an approximation to the second derivative, doubly enhances any noise in the image. Because of these disadvantages, the Laplacian has fallen into disuse, although some authors have used it as an adjunct to the gradient [Wechsler and Sklansky 1977; Akatsuka 1974] in the following manner: There is an edge at x with magnitude $g(x)$ and direction $\phi(x)$ if $g(x) > T_1$ and $L(x) > T_2$.

Edge Templates

The *Kirsch operator* [Kirsch 1971] is related to the edge gradient and is given by

$$S(x) = \max\left[1, \max_k \sum_{k-1}^{k+1} |f(\mathbf{x}_k) - f(x)|\right] \qquad (3.27)$$

where $f(\mathbf{x}_k)$ are the eight neighboring pixels to x and where subscripts are computed modulo 8. A 3-bit direction can also be extracted from the value of k that yields the maximum in (3.27). In practice, "pure" template matching has replaced the use of (3.27). Four separate templates are matched with the image and the operator reports the magnitude and direction associated with the maximum match. As one might expect, the operator is sensitive to the magnitude of $f(x)$, so that in practice variants using large templates are generally used. Figure 3.13 shows Kirsch-motivated templates with different spans.

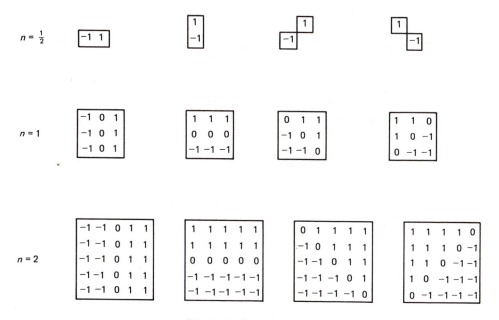

Fig. 3.13 Kirsch templates.

This brief discussion of edge templates should not be construed as a comment on their appropriateness or popularity. In fact, they are widely used, and the template-matching concept is the essence of the other approaches. There is also evidence that the mammalian visual system responds to edges through special low-level template-matching edge detectors [Hubel and Wiesel 1979].

3.3.2 Edge Thresholding Strategies

For most images there will be but few places where the gradient magnitude is equal to zero. Furthermore, in the absence of any special context, small magnitudes are most likely to be due to random fluctuations, or noise in the image function f. Thus in practical cases one may use the expedient of only reporting an edge element at $\mathbf{x}$ if $g(\mathbf{x})$ is greater than some threshold, in order to reduce these noise effects.

This strategy is computationally efficient but may not be the best. An alternative thresholding strategy [Frei and Chen 1977] views difference operators as part of a set of orthogonal basis functions analogous to the Fourier basis of Section 2.2.4. Figure 3.14 shows the nine Frei–Chen basis functions. Using this basis, the image near a point $\mathbf{x}_0$ can be represented as

$$f(\mathbf{x}) = \sum_{k=0}^{8} (f, h_k) h_k (\mathbf{x} - \mathbf{x}_0) / (h_k, h_k) \tag{3.28}$$

where the (f, h_k) is the correlation operation given by

$$(f, h_k) = \sum_D f(\mathbf{x}_0) h_k (\mathbf{x} - \mathbf{x}_0) \tag{3.29}$$

and D is the nonzero domain of the basis functions. This operation is also regarded as the *projection* of the image into the basis function h_k. When the image can be reconstructed from the basis functions and their coefficients, the basis functions span the space. In the case of a smaller set of functions, the basis functions span a subspace.

The value of a projection into any basis function is highest when the image function is identical to the basis function. Thus one way of measuring the "edgeness" of a local area in an image is to measure the relative projection of the image

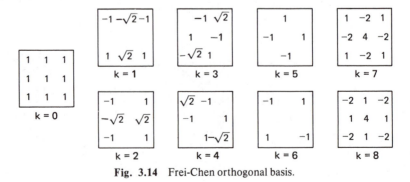

Fig. 3.14 Frei-Chen orthogonal basis.

into the edge basis functions. The relative projection into the particular "edge subspace" is given by

$$\cos \theta = \left(\frac{E}{S}\right)^{1/2}$$ (3.30)

where

$$E = \sum_{k=1}^{2} (f, h_k)^2$$

and

$$S = \sum_{k=0}^{8} (f, h_k)^2$$

Thus if $\theta < T$, report an edge; otherwise, not. Figure 3.15 shows the potential advantage of this technique compared to the technique of thresholding the gradient magnitude, using two hypothetical projections B_1 and B_2. Even though B_2 has a small magnitude, its relative projection into edge subspace is large and thus would be counted as an edge with the Frei–Chen criterion. This is not true for B_1.

Under many circumstances it is appropriate to use model information about the image edges. This information can affect the way the edges are interpreted after they have been computed or it may affect the computation process itself. As an example of the first case, one may still use a gradient operator, but vary the threshold for reporting an edge. Many versions of the second, more extreme strategies of using special spatially variant detection methods have been tried [Pingle and Tenenbaum 1971; Griffith 1973; Shirai 1975]. The basic idea is illustrated in Fig. 3.16. Knowledge of the orientation of an edge allows a special orientation-sensitive operator to be brought to bear on it.

3.3.3 Three-Dimensional Edge Operators

In many imaging applications, particularly medicine, the images are three-dimensional. Consider the examples of the reconstructed planes described in Sections 1.1 and 2.3.4. The medical scanner that acquires these data follows several parallel image planes, effectively producing a three-dimensional volume of data.

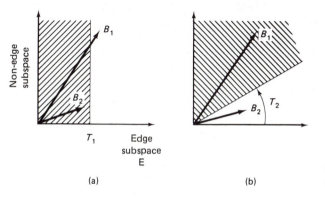

(a)

(b)

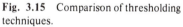

Fig. 3.15 Comparison of thresholding techniques.

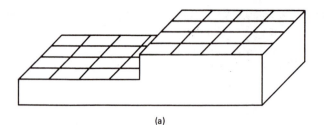

(a)

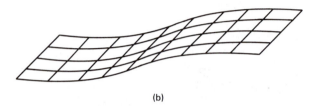

(b)

Fig. 3.16 Model-directed edge detection.

In three-dimensional data, boundaries of objects are surfaces. Edge elements in two dimensions become surface elements in three dimensions. The two-dimensional image gradient, when generalized to three dimensions, is the local surface normal. Just as in the two-dimensional case, many different basis operators can be used [Liu 1977; Zucker and Hummel 1979]. That of Zucker and Hummel uses an optimal basis assuming an underlying continuous model. We shall just describe the operator here; the proof of its correctness given the continuous image model may be found in the reference. The basis functions for the three-dimensional operator are given by

$$g_1(x, y, z) = \frac{x}{r} \tag{3.31}$$

$$g_2(x, y, z) = \frac{y}{r}$$

$$g_3(x, y, z) = \frac{z}{r}$$

where $r = (x^2 + y^2 + z^2)^{1/2}$. The discrete form of these operators is shown in Fig. 3.17 for a $3 \times 3 \times 3$ pixel domain D. Only g_1 is shown since the others are obvious by symmetry. To apply the operator at a point x_0, y_0, z_0 compute projections a, b, and c, where

$$a = (g_1, f) = \sum_D g_1(\mathbf{x}) f(\mathbf{x} - \mathbf{x}_0)$$

$$b = (g_2, f) \tag{3.32}$$

$$c = (g_3, f)$$

The result of these computations is the surface normal $\mathbf{n} = (a, b, c)$ at (x_0, y_0, z_0). Surface thresholding is analogous to edge thresholding: Report a surface element

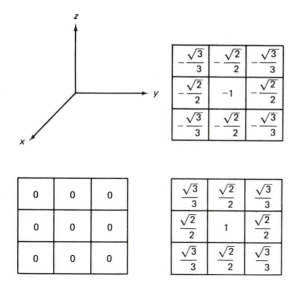

Fig. 3.17 The $3 \times 3 \times 3$ edge basis function $g_1(x, y, z)$.

only if $s(x, y, z) = |n|$ exceeds some threshold. Figure 3.18 shows the results of applying the operator to a synthetic three-dimensional image of a torus. The display shows small detected surface patches.

3.3.4 How Good are Edge Operators?

The plethora of edge operators is very difficult to compare and evaluate. For example, some operators may find most edges but also respond to noise; others may be

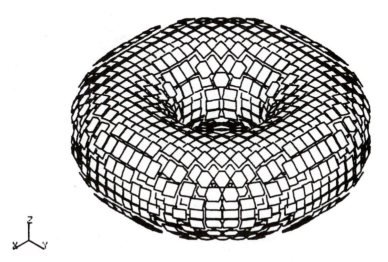

Fig. 3.18 Results of applying the Zucker-Hummel 3-D operator to synthetic image data in the shape of a torus.

noise-insensitive but miss some crucial edges. The following figure of merit [Pratt 1978] may be used to compare edge operators:

$$F = \frac{1}{\max\,(N_A,\,N_I)} \sum_{i=1}^{N_A} \frac{1}{1 + (ad_i^2)}$$ (3.33)

where N_A and N_I represent the number of actual and ideal edge points, respectively, a is a scaling constant, and d is the signed separation distance of an actual edge point normal to a line of ideal edge points. The term ad_i^2 penalizes detected edges which are offset from their true position; the penalty can be adjusted via a. Using this measure, all operators have surprisingly similar behaviors. Unsurprisingly, the performance of each deteriorates in the presence of noise [Abdou 1978]. (Pratt defines a signal-to-noise ratio as the square of the step edge amplitude divided by the standard deviation of Gaussian white noise.) Figure 3.19 shows some typical curves for different operators. To make this figure, the threshold for reporting an edge was chosen independently for each operator so as to maximize Eq. (3.33).

These comparisons are important as they provide a gross measure of differences in performance of operators even though each operator embodies a specific edge model and may be best in special circumstances. But perhaps the more important point is that since all real-world images have significant amounts of noise, all edge operators will generally produce imperfect results. This means that in considering the overall computer vision problem, that of building descriptions of objects, the efforts are usually best spent in developing methods that can use or improve the measurements from unreliable edges rather than in a search for the ideal edge detector.

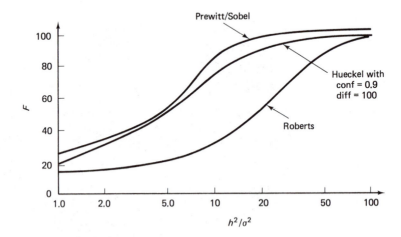

Fig. 3.19   Edge operator performance using Pratt's measure (Eq. 3.33).

3.3.5 Edge Relaxation

One way to improve edge operator measurements is to adjust them based on measurements of neighboring edges. This is a natural thing to want to do: If a weak horizontal edge is positioned between two strong horizontal edges, it should gain credibility. The edges can be adjusted based on local information using parallel-iterative techniques. This sort of process is related to more global analysis and is complementary to sequential approaches such as edge tracking (Chapter 4).

Early cooperative edge detection techniques used pairwise measurements between pixels [Zucker et al. 1977]. A later version [Prager 1980] allows for more complicated adjustment formulas. In describing the edge relaxation scheme, we essentially follow Prager's development and use the crack edges described at the end of the discussion on gradients (Sec. 3.31). The development can be extended to the other kinds of edges and the reader is invited to do just this in the Exercises.

The overall strategy is to recognize local edge patterns which cause the confidence in an edge to be modified. Prager recognizes three groups of patterns: patterns where the confidence of an edge can be increased, decreased, or left the same. The overall structure of the algorithm is as follows:

Algorithm 3.1   Edge Relaxation

0.   Compute the initial confidence of each edge $C^0(e)$ as the normalized gradient magnitude normalized by the maximum gradient magnitude in the image.

1. $k = 1$;

2.   Compute each edge type based on the confidence of edge neighbors;

3.   Modify the confidence of each edge $C^k(e)$ based on its edge type and its previous confidence $C^{k-1}(e)$;

4.   Test the $C^k(e)$'s to see if they have all converged to either 0 or 1. If so, stop; else, increment k and go to 2.

The two important parts of the algorithm are step 2, computing the edge type, and step 3, modifying the edge confidence.

The edge-type classification relies on the notation for edges (Fig. 3.20). The edge type is a concatenation of the left and right vertex types. Vertex types are computed from the strength of edges emanating from a vertex. Vertical edges are handled in the same way, exploiting the obvious symmetries with the horizontal case. Besides the central edge e, the left vertex is the end point for three other possible edges. Classifying these possible edges into "edge" and "no-edge" provides the underpinnings for the vertex types in Fig. 3.21.

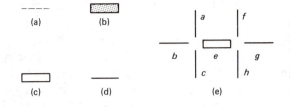

Fig. 3.20 Edge notation. (a) Edge position with no edge. (b) Edge position with edge. (c) Edge to be updated. (d) Edge of unknown strength. (e) Configuration of edges around a central edge e.

To compute vertex type number (the number of edges impinging on the edge of interest), choose the maximum confidence vertex, i.e., the vertex is type j where j maximizes conf (j)

and

$$\text{conf}(0) = (m - a)(m - b)(m - c)$$
$$\text{conf}(1) = a(m - b)(m - c)$$
$$\text{conf}(2) = ab(m - c)$$
$$\text{conf}(3) = abc$$

where

$m = \max(a, b, c, q)$

q is a constant (0.1 is about right)

and a, b, and c are the normalized gradient magnitudes for the three edges. Without loss of generality, $a \geqslant b \geqslant c$. The parameter m adjusts the vertex classification so that it is relative to the local maximum. Thus $(a, b, c) = (0.25, 0.01, 0.01)$ is a type 1 vertex. The parameter q forces weak vertices to type zero [e.g., $(0.01, 0.001, 0.001)$ is type zero].

Once the vertex type has been computed, the edge type is simple. It is merely the concatenation of the two vertex types. That is, the edge type is (ij), where i and j are the vertex types. (From symmetry, only consider $i \geqslant j$.)

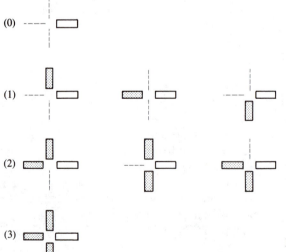

Fig. 3.21 Classification of vertex type of left-hand endpoint of edge e, Fig. 3.20.

Decisions in the second step of modifying edge confidence based on edge type appear in Table 3.1. The updating formula is:

increment: $\quad C^{k+1}(e) = \min\,(1,\,C^k(e) + \delta)$

decrement: $\quad C^{k+1}(e) = \max\,(0,\,C^k(e) - \delta)$

leave as is: $\quad C^{k+1}(e) = C^k(e)$

where δ is a constant (values from 0.1 to 0.3 are appropriate). The result of using the relaxation scheme is shown in Fig. 3.22. The figures on the left-hand side show

Fig. 3.22 Edge relaxation results. (a) Raw edge data. Edge strengths have been thresholded at 0.25 for display purposes only. (b) Results after five iterations of relaxation applied to (a). (c) Different version of (a). Edge strengths have been thresholded at 0.25 for display purposes only. (d) Results after five iterations of relaxation applied to (c).

the edges with normalized magnitudes greater than 0.25. Weak edges cause many gaps in the boundaries. The figures on the right side show the results of five iterations of edge relaxation. Here the confidence of the weak edges has been increased owing to the proximity of other edges, using the rules in Table 3.1.

Table 3.1

Decrement	Increment	Leave as is
0-0	1-1	0-1
0-2	1-2	2-2
0-3	1-3	2-3
		3-3

3.4 RANGE INFORMATION FROM GEOMETRY

Neither the perspective or orthogonal projection operations, which take the three-dimensional world to a two-dimensional image, is invertible in the usual sense. Since projection maps an infinite line onto a point in the image, information is lost. For a fixed viewpoint and direction, infinitely many continuous and discontinuous three-dimensional configurations of points could project on our retina in an image of, say, our grandmother. Simple cases are grandmothers of various sizes cleverly placed at varying distances so as to project onto the same area. An astronomer might imagine millions of points distributed perhaps through light-years of space which happen to line up into a "grandmother constellation." All that can be mathematically guaranteed by imaging geometry is that the image point corresponds to one of the infinite number of points on that three-dimensional line of sight. The "inverse perspective" transformation (Appendix 1) simply determines the equation of the infinite line of sight from the parameters of the imaging process modeled as a point projection.

However, a line and a plane not including it intersect in just one point. Lines of sight are easy to compute, and so it is possible to tell where any image point projects on to any known plane (the supporting ground or table plane is a favorite). Similarly, if two images from different viewpoints can be placed in correspondence, the intersection of the lines of sight from two matching image points determines a point in three-space. These simple observations are the basis of light-striping ranging (Section 2.3.3) and are important in stereo imaging.

3.4.1. Stereo Vision and Triangulation

One of the first ideas that occurs to one who wants to do three-dimensional sensing is the biologically motivated one of stereo vision. Two cameras, or one camera from two positions, can give relative depth or absolute three-dimensional location, depending on the elaboration of the processing and measurement. There has been considerable effort in this direction [Moravec 1977; Quam and Hannah 1974; Binford 1971; Turner 1974; Shapira 1974]. The technique is conceptually simple:

1. Take two images separated by a baseline.
2. Identify points between the two images.
3. Use the inverse perspective transform (Appendix 1) or simple triangulation (Section 2.2.2) to derive the two lines on which the world point lies.
4. Intersect the lines.

The resulting point is in three-dimensional world coordinates.

The hardest part of this method is step 2, that of identifying corresponding points in the two images. One way of doing this is to use correlation, or template matching, as described in Section 3.2.1. The idea is to take a patch of one image and match it against the other image, finding the place of best match in the second image, and assigning a related "disparity" (the amount the patch has been displaced) to the patch.

Correlation is a relatively expensive operation, its naive implementation requiring $0(n^2 m^2)$ multiplications and additions for an $m \times m$ patch and $n \times n$ image. This requirement can be drastically improved by capitalizing on the idea of variable resolution; the improved technique is described in Section 3.7.2.

Efficient correlation is of technological concern, but even if it were free and instantaneous, it would still be inadequate. The basic problems with correlation in stereo imaging have to do with the fact that things can look significantly different from different points of view. It is possible for the two stereo views to be sufficiently different that corresponding areas may not be matched correctly. Worse, in scenes with much obscuration, very important features of the scene may be present in only one view. This problem is alleviated by decreasing the baseline, but of course then the accuracy of depth determinations suffers; at a baseline length of zero there is no problem, but no stereo either. One solution is to identify world features, not image appearance, in the two views, and match those (the nose of a person, the corner of a cube). However, if three-dimensional information is sought as a help in perception, it is unreasonable to have to do perception first in order to do stereo.

3.4.2 A Relaxation Algorithm for Stereo

Human *stereopsis*, or fusing the inputs from the eyes into a stereo image, does not necessarily involve being aware of features to match in either view. Most human beings can fuse quite efficiently stereo pairs which individually consist of randomly placed dots, and thus can perceive three-dimensional shapes without recognizing monocular clues in either image. For example, consider the stereo pair of Fig. 3.23. In either frame by itself, nothing but a randomly speckled rectangle can be perceived. All the stereo information is present in the relative displacement of dots in the two rectangles. To make the right-hand member of the stereo pair, a patch of

Fig. 3.23 A random-dot stereogram.

the randomly placed dots of the left-hand image is displaced sideways. The dots which are thus covered are lost, and the space left by displacing the patch is filled in with random dots.

Interestingly enough, a very simple algorithm [Marr and Poggio 1976] can be formulated that computes disparity from random dot stereograms. First consider the simpler problem of matching one-dimensional images of four points as depicted in Fig. 3.24. Although only one depth plane allows all four points to be placed in correspondence, lesser numbers of points can be matched in other planes.

The crux of the algorithm is the rules, which help determine, on a local basis, the appropriateness of a match. Two rules arise from the observation that most images are of opaque objects with smooth surfaces and depth discontinuities only at object boundaries:

1. Each point in an image may have only one depth value.
2. A point is almost sure to have a depth value near the values of its neighbors.

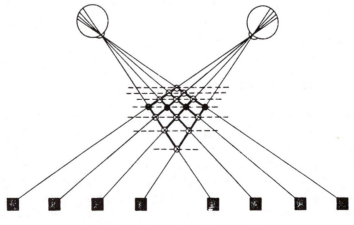

Fig. 3.24 The stereo matching problem.

Figure 3.24 can be viewed as a binary network where each possible match is represented by a binary state. Matches have value 1 and nonmatches value 0. Figure 3.25 shows an expanded version of Fig. 3.24. The connections of alternative matches for a point inhibit each other and connections between matches of equal depth reinforce each other. To extend this idea to two dimensions, use parallel arrays for different values of y where equal depth matches have reinforcing connections. Thus the extended array is modeled as the matrix $C(x, y, d)$ where the point x, y, d corresponds to a particular match between a point (x_1, y_1) in the right image and a point (x_2, y_2) in the left image. The stereopsis algorithm produces a series of matrices C_n which converges to the correct solution for most cases. The initial matrix $C_0(x, y, d)$ has values of one where x, y, d correspond to a match in the original data and has values of zero or otherwise.

Algorithm 3.2 [Marr and Poggio 1976]

Until C satisfies some convergence criterion, do

$$C_{n+1}(x, y, d) = \left\{ \sum_{x',y',d' \in S} C_n(x', y', d') - \sum_{x',y',d' \in \theta} C_n(x', y', d') + C_0(x, y, d) \right\} (3.34)$$

where the term in braces is handled as follows:

$$\{ t \} = \begin{cases} 1 & \text{if } t > T \\ 0 & \text{otherwise} \end{cases}$$

S = set of points x', y', d' such that $|x - x'| \leqslant 1$ and $d = d'$

θ = set of points x', y', d' such that $|x - x'| \leqslant 1$ and $|d - d'| = 1$

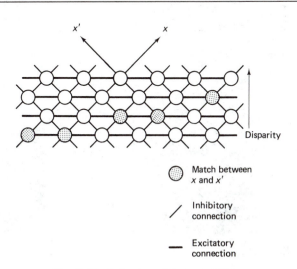

Fig. 3.25 Extension of stereo matching.

One convergence criterion is that the number of points modified on an iteration must be less than some threshold T. Fig. 3.26 shows the results of this computation; the disparity is encoded as a gray level and displayed as an image for different values of n.

A more general version of this algorithm matches image features such as edges rather than points (in the random-dot stereogram, the only features are

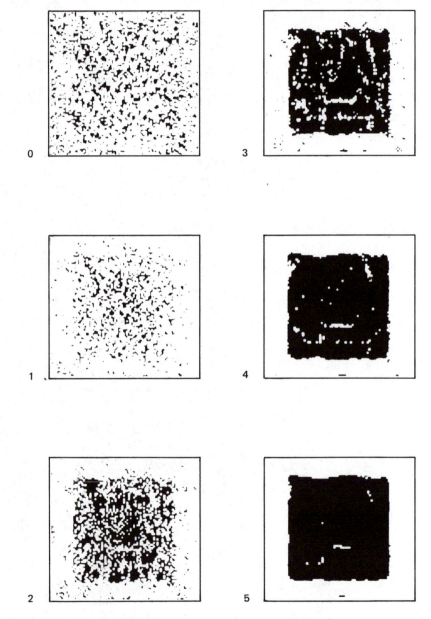

Fig. 3.26 The results of relaxation computations for stereo.

points), but the principles are the same. The extraction of features more complicated than edges or points is itself a thorny problem and the subject of Part II. It
should be mentioned that Marr and Poggio have refined their stereopsis algorithm
to agree better with psychological data [Marr and Poggio 1977].

3.5 SURFACE ORIENTATION FROM REFLECTANCE MODELS

The ordinary visual world is mostly composed of opaque three-dimensional objects. The intensity (gray level) of a pixel in a digital image is produced by the light
reflected by a small area of surface near the corresponding point on the object.

It is easiest to get consistent shape (orientation) information from an image if
the lighting and surface reflectance do not change from one scene location to
another. Analytically, it is possible to treat such lighting as uniform illumination, a
point source at infinity, or an infinite linear source. Practically, the human shape-
from-shading transform is relatively robust. Of course, the perception of shape
may be manipulated by changing the surface shading in calculated ways. In part,
cosmetics work by changing the reflectivity properties of the skin and misdirecting
our human shape-from-shading algorithms.

The recovery transformation to obtain information about surface orientation
is possible if some information about the light source and the object's reflectivity is
known. General algorithms to obtain and quantify this information are complicated but practical simplifications can be made [Horn 1975; Woodham 1978; Ikeuchi 1980]. The main complicating factor is that even with mathematically tractable
object surface properties, a single image intensity does not uniquely define the surface orientation. We shall study two ways of overcoming this difficulty. The first algorithm uses intensity images as input and determines the surface orientation by
using multiple light source positions to remove ambiguity in surface orientation.
The second algorithm uses a single source but exploits constraints between neighboring surface elements. Such an algorithm assigns initial ranges of orientations to
surface elements (actually to their corresponding image pixels) on the basis of intensity. The neighboring orientations are "relaxed" against each other until each
converges to a unique orientation (Section 3.5.4).

3.5.1 Reflectivity Functions

For all these derivations, consider a distant point source of light impinging on a
small patch of surface; several angles from this situation are important (Fig. 3.27).

A surface's reflectance is the fraction of a given incident energy flux (irradiance) it reflects in any given direction. Formally, the *reflectivity function* is defined
as $r = \dfrac{dL}{dE}$, where L is exitant radiance and E is incident flux. In general, for an-
isotropic reflecting surfaces, the reflectivity function (hence L) is a function of all
three angles i, e, and g. The quantity of interest to us is image irradiance, which is
proportional to scene radiance, given by $L = \int r \, dE$. In general, the evaluation of
this integral can be quite complicated, and the reader is referred to [Horn and

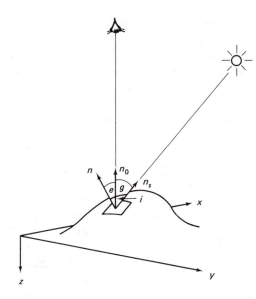

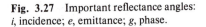

Fig. 3.27 Important reflectance angles: i, incidence; e, emittance; g, phase.

Sjoberg 1978] for a more detailed study. For our purposes we consider surfaces with simple reflectivity functions.

Lambertian surfaces, those with an ideal matte finish, have a very simple reflectivity function which is proportional only to the cosine of the incident angle. These surfaces have the property that under uniform or collimated illumination they look equally bright from any direction. This is because the amount of light reflected from a unit area goes down as the cosine of the viewing angle, but the amount of area seen in any solid angle goes up as the reciprocal of the cosine of the viewing angle. Thus the perceived intensity of a surface element is constant with respect to viewer position. Other surfaces with simple reflectivity functions are "dusty" and "specular" surfaces. An example of a dusty surface is the lunar surface, which reflects in all directions equally. Specular (purely mirror-like) surfaces such as polished metal reflect only at the angle of reflection = angle of incidence, and in a direction such that the incidence, normal, and emittance vectors are coplanar.

Most smooth things have a specular component to their reflection, but in general some light is reflected at all angles in decreasing amounts from the specular angle. One way to achieve this effect is to use the cosine of the angle between the predicted specular angle and the viewing angle, which is given by C where

$$C = 2\cos(i)\cos(e) - \cos(g)$$

This quantity is unity in the pure specular direction and falls off to zero at $\frac{\pi}{2}$ radians away from it. Convincing specular contributions of greater or less sharpness are produced by taking powers of C. A simple radiance formula that allows the simulation of both matte and specular effects is

$$L(i, e, g) = s(C)^n + (1 - s)\cos(i) \tag{3.35}$$

Here s varies between 0 and 1 and determines the fraction of specularly reflected light; n determines the sharpness of specularity peaks. As n increases, the specular peak gets sharper and sharper. Computer graphics research is constantly extending the frontiers of realistic and detailed reflectance, refractance, and illumination calculations [Blinn 1978; Phong 1975; Whitted 1980].

3.5.2 Surface Gradient

The reflectance functions described above are defined in terms of angles measured with respect to a local coordinate frame. For our development, it is more useful to relate the reflectivity function to surface gradients measured with respect to a viewer-oriented coordinate frame.

The concept of *gradient space*, which is defined in a viewer-oriented frame [Horn 1975], is extremely useful in understanding the recovery transformation algorithm for the surface normal. This gradient refers to the orientation of a physical surface, *not* to local intensities. It must not be confused with the *intensity* gradients discussed in Section 3.3 and elsewhere in this book.

Gradient space is a two-dimensional space of slants of scene surfaces. It measures a basic "intrinsic" (three-dimensional) property of surfaces. Consider the point-projection imaging geometry of Fig. 2.2, with the viewpoint at infinity (far from the scene relative to the scene dimensions). The image projection is then orthographic, not perspective.

The surface gradient is defined for a surface expressed as $-z = f(x, y)$. The gradient is a vector (p, q), where

$$p = \frac{\partial(-z)}{\partial x} \tag{3.36}$$

$$q = \frac{\partial(-z)}{\partial y}$$

Any plane in the image (such as the face plane of a polyhedral face) may be expressed in terms of its gradient. The general plane equation is

$$Ax + By + Cz + D = 0 \tag{3.37}$$

Thus

$$-z = \frac{A}{C}x + \frac{B}{C}y + \frac{D}{C} \tag{3.38}$$

and from (3.36) the gradient may be related to the plane equation:

$$-z = px + qy + K \tag{3.39}$$

Gradient space is thus the two-dimensional space of (p, q) vectors. The p and q axes are often considered to be superimposed on the x and y image plane coordinate axes. Then the (p, q) vector is "in the direction" of the surface slant of imaged surfaces. Any plane perpendicular to the viewing direction has a (p, q) vector of $(0,0)$. Vectors on the q (or y) axis correspond to planes tilted about the x axis in an "upward" or "downward" ("*y*ward") direction (like the tilt of a dressing table mirror). The direction arctan (q/p) is the direction of fastest change of surface depth $(-z)$ as x and y change. $(p^2 + q^2)^{1/2}$ is the rate of this change. For instance, a vertical plane "edge on" to the viewer has a (p, q) of $(\infty, 0)$.

The *reflectance map* $R(p, q)$ represents this variation of perceived brightness with surface orientation. $R(p, q)$ gives scene radiance (Section 2.2.3) as a function of surface gradient (in our usual viewer-centered coordinate system). (Figure 3.27 showed the situation and defined some important angles.) $R(p, q)$ is usually shown as contours of constant scene radiance (Fig. 3.28). The following are a few useful cases.

In the case of a Lambertian surface with the source in the direction of the viewer $(i = e)$, the gradient space image looks like Fig. 3.28. Remember that Lambertian surfaces have constant intensity for constant illumination angle; these constant angles occur on the concentric circles of Fig. 3.28, since the direction of tilt does not affect the magnitude of the angle. The brightest surfaces are those illuminated from a normal direction—they are facing the viewer and so their gradients are $(0, 0)$.

Working this out from first principles, the incident angle and emittance angle are the same in this case, since the light is near the viewer. Both are the angle between the surface normal and the view vector. Looking at the $x-y$ plane means a vector to the light source of $(0, 0, -1)$, and at a gradient point (p, q), the surface normal is $(p, q, -1)$. Also,

$$R = r_o \cos i \qquad (3.40)$$

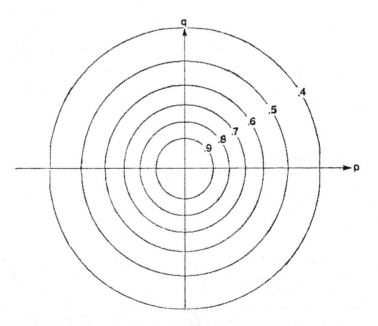

Fig. 3.28 Contours of constant radiance in gradient space for Lambertian surfaces; single light source near the viewpoint.

where r_o is a proportionality constant, and we conventionally use R to denote radiance in a viewer-centered frame. Let $\mathbf{n}_s$ and $\mathbf{n}$ be unit vectors in the source and surface normal directions. Since $\cos i = \mathbf{n}_s \cdot \mathbf{n}$

$$R = \frac{r_0}{(1 + p^2 + q^2)^{1/2}} \tag{3.41}$$

Thus $\cos (i)$ determines the image brightness, and so a plot of it is the gradient space image (Figs. 3.29 and 3.30).

For a more general light position, the mathematics is the same; if the light source is in the $(p_s, q_s, -1)$ direction, take the dot product of this direction and the surface normal.

$$R = r_0 \mathbf{n} \cdot \mathbf{n}_s \tag{3.42}$$

Or, in other words,

$$R = \frac{r_0 (p_s p + q_s q + 1)}{[(1 + p^2 + q^2)(1 + p_s^2 + q_s^2)]^{1/2}}$$

The phase angle g is constant throughout gradient space with orthographic projection (viewer distant from scene) and light source distant from scene.

Setting R constant to obtain contour lines gives a second-order equation, producing conic sections. In fact, the contours are produced by a set of cones of varying angles, whose axis is in the direction of the light source, intersecting a plane at unit distance from the origin. The resulting contours appear in Fig. 3.29. Here the straight line is the terminator, and represents all those planes that are

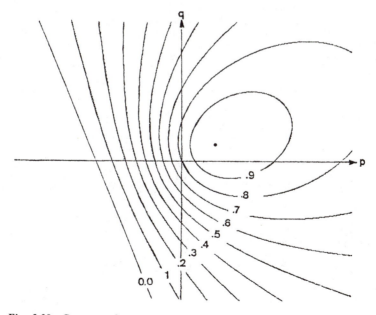

Fig. 3.29 Contours of constant radiance in gradient space. Lambertian surfaces; light not near viewpoint.

edge-on to the light source; gradients on the back side of the terminator represent self-shadowed surfaces (facing away from the light). One intensity determines a contour and so gives a cone whose tangent planes all have that emittance. For a surface with specularity, contours of constant L (i, e, g) could appear as in Fig. 3.30.

The point of specularity is between the matte component maximum brightness gradient and the origin. The brightest matte surface normal points at the light source and the origin points at the viewer. Pure specular reflection can occur if the vector tilts halfway toward the viewer maintaining the direction of tilt. Thus its gradient is on a line between the origin and the light-source direction gradient point.

3.5.3 Photometric Stereo

The reflectance equation (3.42) constrains the possible surface orientation to a locus on the reflectance map. Multiple light-source positions can determine the orientation uniquely [Woodham 1978]. Each separate light position gives a separate value for the intensity (proportional to radiance) at each point $f(\mathbf{x})$. If the surface reflectance r_0 is unknown, three equations are needed to determine the reflectance together with the unit normal $\mathbf{n}$. If each source position vector is denoted by $\mathbf{n}_k$, $k = 1, \ldots, 3$, the following equations result:

$$I_k(x, y) = r_0(\mathbf{n}_k \cdot \mathbf{n}), \qquad k = 1, \ldots, 3 \qquad (3.43)$$

where I is normalized intensity. In matrix form

$$\mathbf{I} = r_0 N \mathbf{n} \qquad (3.44)$$

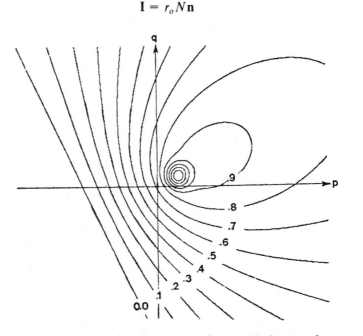

Fig. 3.30 Contours of constant radiance for a specular/matte surface.

where

$$\mathbf{I} = [I_1(x, y), I_2(x, y), I_3(x, y)]^T,$$

and

$$N = \begin{bmatrix} n_{11} & n_{12} & n_{13} \\ n_{21} & n_{22} & n_{23} \\ n_{31} & n_{32} & n_{33} \end{bmatrix} \tag{3.45}$$

and $I = fc$ where c is the appropriate normalization constant. If c is not known, it can be regarded as being part of r_o without affecting the normal direction calculation. As long as the three source positions $\mathbf{n}_1, \mathbf{n}_2, \mathbf{n}_3$ are not coplanar, the matrix N will have an inverse. Then solve for r_o and $\mathbf{n}$ by using (3.44), first using the fact that $\mathbf{n}$ is a unit vector to derive

$$r_0 = |N^{-1}\mathbf{I}| \tag{3.46}$$

and then solving for $\mathbf{n}$ to obtain

$$\mathbf{n} = \frac{1}{r_o} N^{-1}\mathbf{I} \tag{3.47}$$

Examples of a particular solution are shown in Fig. 3.31. Of course, a prerequisite for using this method is that the surface point not be in shadows for any of the sources.

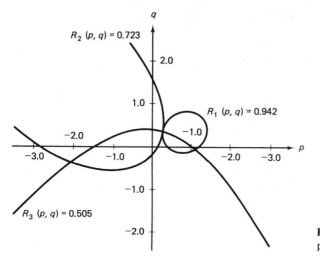

Fig. 3.31 A particular solution for photometric stereo.

3.5.4 Shape from Shading by Relaxation

Combining local information allows improved estimates for edges (Section 3.3.5) and for disparity (Section 3.4.2). In a similar manner local information can help in computing surface orientation [Ikeuchi 1980]. Basically, the reflectance equation provides one constraint on the surface orientation and another is provided by the heuristic requirement that the surface be smooth.

Suppose there is an estimate of the surface normal at a point $(p(x, y), q(x, y))$. If the normal is not accurate, the reflectivity equation $I(x, y) = R(p, q)$ will not hold. Thus it seems reasonable to seek p and q that minimize $(I - R)^2$. The other requirement is that $p(x, y)$ and $q(x, y)$ be smooth, and this can be measured by the partial derivatives squared, i.e., $p_x^2, p_y^2, q_x^2, q_y^2$. For a smooth curve both of these terms should be small. The goal is to minimize the error at a point,

$$E(x, y) = (I(x, y) - R(p, q))^2 + \lambda (p_x^2 + p_y^2 + q_x^2 + q_y^2) \qquad (3.48)$$

where the Lagrange multiplier λ [Russell 1976] incorporates the smoothness constraint. Differentiating $E(x, y)$ with respect to p and q and approximating derivatives numerically gives the following equations for $p(x, y)$ and $q(x, y)$:

$$p(x, y) = p_{av}(x, y) + T(x, y, p, q) \frac{\partial R}{\partial p} \qquad (3.49)$$

$$q(x, y) = q_{av}(x, y) + T(x, y, p, q) \frac{\partial R}{\partial q} \qquad (3.50)$$

where

$$T(x, y, p, q) = (1/\lambda)[I(x, y) - R(p, q)]$$

using

$$p_{av}(x, y) = \frac{1}{4}[p(x + 1, y) + p(x - 1, y) + p(x, y + 1) + p(x, y - 1)] \qquad (3.51)$$

and a similar expression for q_{av}. Now Eqs. (3.49) and (3.50) lend themselves to solution by the Gauss-Seidel method: calculate the left-hand sides with an estimate for p and q and use them to derive a new estimate for the right-hand sides. More formally,

---

Algorithm 3.3: Shape from Shading [Ikeuchi 1980].

Step 0. $k = 0$. Pick an initial $p^0(x, y)$ and $q^0(x, y)$ near boundaries.

Step 1. $k = k + 1$; compute

$$p^k = p_{av}^{k-1} + T \frac{\partial R}{\partial p}$$

$$q^k = q_{av}^{k-1} + T \frac{\partial R}{\partial q}$$

Step 2. If the sum of all the E's is sufficiently small, stop. Else, go to step 1.

A loose end in this algorithm is that boundary conditions must be specified. These are values of p and q determined a priori that remain constant throughout each iteration. The simplest place to specify a surface gradient is at an occluding contour (see Fig. 3.32) where the gradient is nearly 90° to the line of sight. Unfortunately, p and q are infinite at these points. Ikeuchi's elegant solution to this is to use a different coordinate system for gradient space, that of a Gaussian sphere (Appendix 1). In this system, the surface normal is described relative to where it intersects the sphere if the tail of the normal is at the sphere's origin. This is the point at which a plane perpendicular to the normal would touch the sphere if translated toward it (Fig. 3.32b).

In this system the radiance may be described in terms of the spherical coordinates θ, ϕ. For a Lambertian surface

$$R(\theta, \phi) = \cos \theta \ \cos \theta_s + \sin \theta \sin \theta_s \ \cos(\phi - \phi_s) \qquad (3.52)$$

At an occluding contour $\phi = \pi/2$ and θ is given by $\tan^{-1}(\partial y / \partial x)$, where the derivatives are calculated at the occluding contour (Fig. 3.32c).

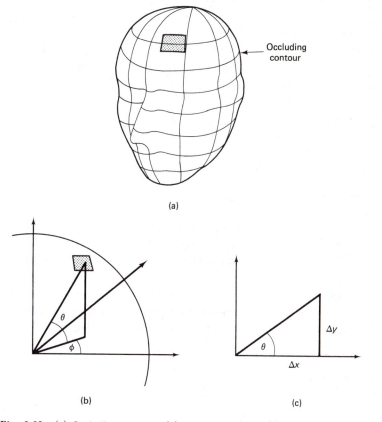

(a)

(b)

(c)

Fig. 3.32 (a) Occluding contour. (b) Gaussian sphere. (c) Calculating θ from occluding contour.

To use the (θ, ϕ) formulation instead of the (p, q) formulation is an easy matter. Simply substitute θ for p and ϕ for q in all instances of the formula in Algorithm 3.3.

3.6 OPTICAL FLOW

Much of the work on computer analysis of visual motion assumes a stationary observer and a stationary background. In contrast, biological systems typically move relatively continuously through the world, and the image projected on their retinas varies essentially continuously while they move. Human beings perceive smooth continuous motion as such.

Although biological visual systems are discrete, this quantization is so fine that it is capable of producing essentially continuous outputs. These outputs can mirror the continuous flow of the imaged world across the retina. Such continuous information is called *optical flow*. Postulating optical flow as an input to a perceptual system leads to interesting methods of motion perception.

The optical flow, or instantaneous velocity field, assigns to every point on the visual field a two-dimensional "retinal velocity" at which it is moving across the visual field. This section describes how approximations to instantaneous flow may be computed from the usual input situation in a sequence of discrete images. Methods of using optical flow to compute the observer's motion, a relative depth map, surface normals of his or her surroundings, and other useful information are given in Chapter 7.

3.6.1 The Fundamental Flow Constraint

One of the important features of optical flow is that it can be calculated simply, using local information. One way of doing this is to model the motion image by a continuous variation of image intensity as a function of position and time, then expand the intensity function $f(x, y, t)$ in a Taylor series.

$$f(x + dx, y + dy, t + dt) = \qquad (3.53)$$

$$f(x, y, t) + \frac{\partial f}{\partial x} dx + \frac{\partial f}{\partial y} dy + \frac{\partial f}{\partial t} dt + \text{higher-order terms}$$

As usual, the higher-order terms are henceforth ignored. The crucial observation to be exploited is the following: If indeed the image at some time $t + dt$ is the result of the original image at time t being moved translationally by dx and dy, then in fact

$$f(x + dx, y + dy, t + dt) = f(x, y, t) \qquad (3.54)$$

Consequently, from Eqs. (3.53) and (3.54),

$$-\frac{\partial f}{\partial t} = \frac{\partial f}{\partial x} \frac{dx}{dt} + \frac{\partial f}{\partial y} \frac{dy}{dt} \qquad (3.55)$$

Now $\frac{\partial f}{\partial t}$, $\frac{\partial f}{\partial x}$, and $\frac{\partial f}{\partial y}$ are all measurable quantities, and $\frac{dx}{dt}$ and $\frac{dy}{dt}$ are estimates of what we are looking for—the velocity in the x and y directions. Writing

$$\frac{dx}{dt} = u, \qquad \frac{dy}{dt} = v$$

gives

$$-\frac{\partial f}{\partial t} = \frac{\partial f}{\partial x}u + \frac{\partial f}{\partial y}v \qquad (3.56)$$

or equivalently,

$$-\frac{\partial f}{\partial t} = \nabla f \cdot \mathbf{u} \qquad (3.57)$$

where ∇f is the spatial gradient of the image and $\mathbf{u} = (u, v)$ the velocity.

The implications of (3.57) are interesting. Consider a fixed camera with a scene moving past it. The equations say that the *time* rate of change in intensity of a point in the image is (to first order) explained as the *spatial* rate of change in the intensity of the scene multiplied by the *velocity* that points of the scene move past the camera.

This equation also indicates that the velocity (u, v) must lie on a line perpendicular to the vector (f_x, f_y) where f_x and f_y are the partial derivatives with respect to x and y, respectively (Fig. 3.33). In fact, if the partial derivatives are very accurate the magnitude component of the velocity in the direction (f_x, f_y) is (from 3.57):

$$\frac{-f_t}{[(f_x^2 + f_y^2)]^{1/2}}$$

3.6.2 Calculating Optical Flow by Relaxation

Equation (3.57) constrains the velocity but does not determine it uniquely. The development of Section 3.5.4 motivates the search for a solution that satisfies Eq.

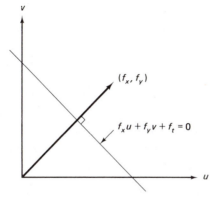

Fig. 3.33 Relation between (u, v) and (f_x, f_y).

(3.57) as closely as possible and also is locally smooth [Horn and Schunck 1980]. In this case as well, smoothness is measured using partial derivatives squared as error terms, i.e., $u_x{}^2$, $u_y{}^2$, $v_x{}^2$, $v_y{}^2$.

By standard squared-error techniques, minimize the flow error

$$E^2(x, y) = (f_x u + f_y v + f_t)^2 + \lambda (u_x{}^2 + u_y{}^2 + v_x{}^2 + v_y{}^2) \qquad (3.58)$$

Differentiating this equation with respect to u and v provides equations for the change in error with respect to u and v, which must be zero for a minimum. Writing $\nabla^2 u$ as $u - u_{av}$ and $\nabla^2 v$ as $v - v_{av}$, these equations are

$$(\lambda^2 + f_x^2)u + f_x f_y v = \lambda^2 u_{av} - f_x f_t \qquad (3.59)$$

$$f_x f_y u + (\lambda^2 + f_y^2)v = \lambda^2 v_{av} - f_y f_t \qquad (3.60)$$

These equations may be solved for u and v, yielding

$$u = u_{av} - f_x \frac{P}{D} \qquad (3.61)$$

$$v = v_{av} - f_y \frac{P}{D} \qquad (3.62)$$

where

$$P = f_x u_{av} + f_y v_{av} + f_t$$

$$D = \lambda^2 + f_x^2 + f_y^2$$

To turn this into an iterative equation for solving $u(x, y)$ and $v(x, y)$, again use the Gauss-Seidel method.

Algorithm 3.4: Optical Flow [Horn and Schunck 1980].

$k = 0$.
Initialize all u^k and v^k to zero.
Until some error measure is satisfied, do

$$u^k = u_{av}^{k-1} - f_x \frac{P}{D}$$

$$v^k = v_{av}^{k-1} - f_y \frac{P}{D}$$

As Horn and Schunck demonstrate, this method derives the flow for two time frames, but it can be improved by using several time frames and using the final solution after one iteration at one time for the initial solution at the following time frame. That is:

Algorithm 3.5: Multiframe Optical Flow.

$t = 0$.
Initialize all $u(x, y, 0)$, $v(x, y, 0)$
for $t = 1$ *until* maxframes *do*

$$u(x, y, t) = u_{av}(x, y, t-1) - f_x \frac{P}{D}$$

$$v(x, y, t) = v_{av}(x, y, t-1) - f_y \frac{P}{D}$$

The results of using synthetic data from a rotating checkered sphere are shown in Fig. 3.34.

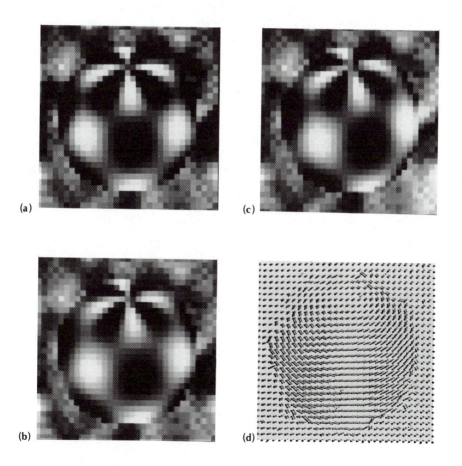

Fig. 3.34 Optical flow results. (a), (b) and (c) are three frames from the rotating sphere, (d) is the derived three-dimensional flow after 32 such time frames.

3.7 RESOLUTION PYRAMIDS

What is the best spatial resolution for an image? The sampling theorem states that the maximum spatial frequency in the image data must be less than half the sampling frequency in order that the sampled image represent the original unambiguously. However, the sampling theorem is not a good predictor of how easily objects can be recognized by computer programs. Often objects can be more easily recognized in images that have a very low sampling rate. There are two reasons for this. First, the computations are fewer because of the reduction in dimensionality. Second, confusing detail present in the high-resolution versions of the images may not appear at the reduced resolution. But even though some objects are more easily found at low resolutions, usually an object description needs detail only revealed at the higher resolutions. This leads naturally to the notion of a *pyramidal* image data structure in which the search for objects is begun at a low resolution, and refined at ever-increasing resolutions until one reaches the highest resolution of interest. Figure 3.35 shows the correspondence between pixels for the pyramidal structure.

In the next three sections, pyramids are applied to gray-level images and edge images. Pyramids, however, are a very general tool and can be used to represent any image at varying levels of detail.

3.7.1 Gray-level Consolidation

In some applications, redigitizing the image with a different sampling rate is a way to reduce the number of samples. However, most digitizer parameters are difficult to change, so that often computational means of reduction are needed. A straightforward method is to partition the digitized image into nonoverlapping

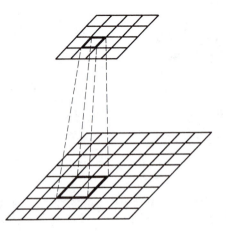

Fig. 3.35 Pyramidal image structure.

neighborhoods of equal size and shape and to replace each of those neighborhoods by the average pixel densities in that neighborhood. This operation is *consolidation*. For an $n \times n$ neighborhood, consolidation is equivalent to averaging the original image over the neighborhood followed by sampling at intervals n units apart.

Consolidation tends to offset the aliasing that would be introduced by sampling the sensed data at a reduced rate. This is due to the effects of the averaging step in the consolidation process. For the one-dimensional case where

$$f'(x) = \frac{1}{2}[f(x) + f(x + \Delta)] \tag{3.63}$$

the corresponding Fourier transform [Steiglitz 1974] is

$$H(u) = \frac{1}{2}\left(1 + e^{-j2\pi u \Delta}\right)F(u) \tag{3.64}$$

which has magnitude $|H(u)| = \cos[\pi(u/u_o)]$ and phase $-\pi(u/u_o)$. The sampling frequency $u_o = 1/\Delta$ where Δ is the spacing between samples. Thus the averaging step has the effect of attenuating the higher frequencies of $F(u)$ as shown in Fig. 3.36. Since the higher frequencies are involved in aliasing, attenuating these frequencies reduces the aliasing effects.

3.7.2 Pyramidal Structures in Correlation

With correlation matching, the use of multiple resolution techniques can sometimes provide significant functional and computational advantages [Moravec 1977]. Binary search correlation uses pyramids of the input image and reference

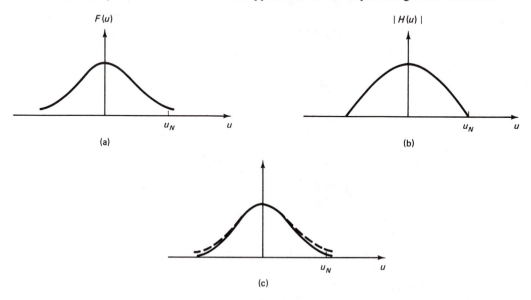

Fig. 3.36 Consolidation effects viewed in the spatial frequency domain. (a) Original transform. (b) Transform of averaging operator. (c) Transform of averaged image.

patterns. The algorithm partakes of the computational efficiency of binary (as opposed to linear) search [Knuth 1973]. Further, the low-resolution correlation operations at high levels in the pyramid ensure that the earlier correlations are on gross image features rather than details.

In binary search correlation a feature to be located is at some unknown location in the input image. The reference version of the feature originates in another image, the reference image. The feature in the reference image is contained in a window of $n \times n$ pixels. The task of the correlator is to find an $n \times n$ window in the input image that best matches the reference image window containing the feature. The details of the correlation processes are given in the following algorithm.

Algorithm 3.6: Binary Search Correlation Control Algorithm

Definitions

OrigReference: an $N \times N$ image containing a feature centered at (FeatureX, FeatureY).

OrigInput: an $M \times M$ array in which an instance of the Feature is to be located. For simplicity, assume that it is at the same resolution as OrigReference.

n: a window size; an $n \times n$ window in OrigReference is large enough to contain the Feature.

Window: an $n \times n$ array containing a varying-resolution subimage of OrigReference centered on the Feature.

Input: a $2n \times 2n$ array containing a varying-resolution subimage of OrigInput, centered on the best match for the Feature.

Reference: a temporary array.

Algorithm

1. Input := Consolidate OrigInput by a factor of $2n/M$ to size $2n \times 2n$.

2. Reference := Consolidate OrigReference by the same factor $2n/M$ to size $2nN/M \times 2nN/M$. This consolidation takes the Feature to a new (FeatureX, FeatureY).

3. Window := $n \times n$ window from Reference centered on the new (FeatureX, FeatureY).

4. Calculate the match metric of the window at the $(n + 1)^2$ locations in Input at which it is wholly contained. Say that the best match occurs at (BestMatchX, BestMatchY) in Input.

5. Input := $n \times n$ window from Input centered at (BestMatch$X$, BestMatch$Y$), enlarged by a factor of 2.

6. Reference := Reference enlarged by a factor of 2. This takes Feature to a new (Feature$X$, Feature$Y$).

7. *Go to 3.*

Through time, the algorithm uses a reference image for matching that is always centered on the feature to be matched, but that homes in on the feature by being increased in resolution and thus reduced in linear image coverage by a factor of 2 each time. In the input image, a similar homing-in is going on, but the search area is usually twice the linear dimension of the reference window. Further, the center of the search area varies in the input image as the improved resolution refines the point of best match.

Binary search correlation is for matching features with context. The template at low resolution possibly corresponds to much of the area around the feature, while the feature may be so small in the initial consolidated images as to be invisible. The coarse-to-fine strategy is perfect for such conditions, since it allows gross features to be matched first and to guide the later high-resolution search for best match. Such matching with context is less useful for locating several instances of a shape dotted at random around an image.

3.7.3 Pyramidal Structures in Edge Detection

As an example of the use of pyramidal structures in processing, consider the use of such structures in edge detection. This application, after [Tanimoto and Pavlidis 1975], uses two pyramids, one to store the image and another to store the image edges. The idea of the algorithm is that a neighborhood in the low-resolution image where the gray-level values are the same is taken to imply that in fact there is no gray-level change (edge) in the neighborhood. Of course, the low-resolution levels in the pyramid tend to blur the image and thus attenuate the gray-level changes that denote edges. Thus the starting level in the pyramid must be picked judiciously to ensure that the important edges are detected.

Algorithm 3.7: Hierarchical Edge Detection

recursive procedure refine (k, x, y)
  begin
  if $k \leqslant$ MaxLevel *then*
    for $dx = 0$ *until* 1 *do*
    for $dy = 0$ *until* 1 *do*
      if EdgeOp $(k, x + dx, y + dy) >$ Threshold(k)
      then refine $(k + 1, x + dx, y + dy)$
  end;

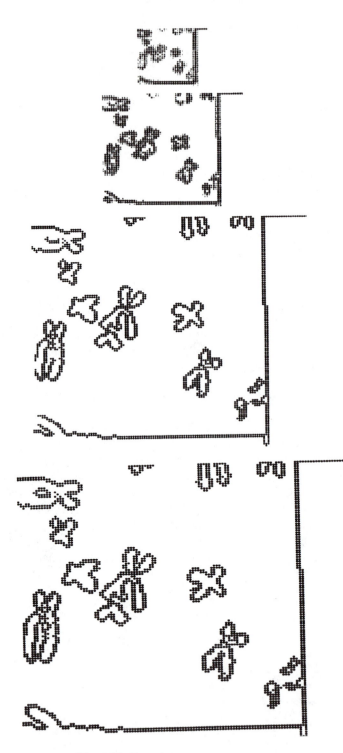

Fig. 3.37 Pyramidal edge detection.

```
procedure FindEdges:
    begin
    comment apply operator to every pixel in the
        starting level s, refining where necessary;
        for x := 0 until 2^S − 1 do
        for y := 0 until 2^S − 1 do
            if EdgeOp (s, x, y) > Threshold(s)
            then refine (s + 1, x, y);
    end;
```

Figure 3.37 shows Tanimoto's results for a chromosome image.

Similar kinds of edge detection strategies based on pyramids have been pursued by [Levine 1978; Hanson and Riseman 1978]. The latter effort is a little different in that processing within the pyramid is bidirectional; information from edges detected at a high-resolution level is projected to low-resolution levels of the pyramid.

EXERCISES

3.1 Derive an analytical expression for the response of the Sobel operator to a vertical step edge as a function of the distance of the edge to the center of the operator.

3.2 Use the formulas of Eqs. (3.31) to derive the digital template function for g_1 in a 5^3 pixel domain.

3.3 Specify a version of Algorithm 3.1 that uses the gradient edge operator instead of the "crack" edge operator.

3.4 In photometric stereo, three or more light source positions are used to determine a surface orientation. The dual of this problem uses surface orientations to determine light source position. What is the usefulness of the latter formulation? In particular, how does it relate to Algorithm 3.3?

3.5 Using any one of Algorithms 3.1 through 3.4 as an example, show how it could be modified to use pyramidal data structures.

3.6 Write a reflectance function to capture the "grazing incidence" phenomenon—surfaces become more mirror-like at small angles of incidence (and reflectance).

3.7 Equations 3.49 and 3.50 were derived by minimizing the local error. Show how these equations are modified when total error [i.e., $\sum_{x,y} E(x, y)$] is minimized.

REFERENCES

ABDOU, I. E. "Quantitative methods of edge detection." USCIPI Report 830, Image Processing Institute, Univ. Southern California, July 1978.

AKATSUKA, T., T. ISOBE, and O. TAKATANI. "Feature extraction of stomach radiograph." *Proc.*, 2nd IJCPR, August 1974, 324-328.

ANDREWS, H. C. and B. R. HUNT. *Digital Image Restoration.* Englewood Cliffs, NJ: Prentice-Hall, 1977.

ATTNEAVE, F. "Some informational aspects of visual perception." *Psychological Review 61,* 1954.

BARROW, H. G. and J. M. TENENBAUM. "Computational Vision." *Proc. IEEE 69,* 5, May 1981, 572-595

BARROW, H. G. and J. M. TENENBAUM. "Recovering intrinsic scene characteristics from images." Technical Note 157, AI Center, SRI International, April 1978.

BINFORD, T. O. "Visual perception by computer." *Proc.,* IEEE Conf. on Systems and Control, Miami, December 1971.

BLINN, J. E. "Computer display of curved surfaces." Ph.D. dissertation, Computer Science Dept., Univ. Utah, 1978.

FREI, W. and C. C. CHEN. "Fast boundary detection: a generalization and a new algorithm." *IEEE Trans. Computers 26,* 2, October 1977, 988-998.

GONZALEZ, R. C. and P. WINTZ. *Digital Image Processing.* Reading, MA: Addison-Wesley, 1977.

GRIFFITH, A. K. "Edge detection in simple scenes using a priori information." *IEEE Trans. Computers 22,* 4, April 1973.

HANSON, A. R. and E. M. RISEMAN (Eds.). *Computer Vision Systems (CVS).* New York: Academic Press, 1978.

HORN, B. K. P. "Determining lightness from an image." *CGIP 3,* 4, December 1974, 277-299.

HORN, B. K. P. "Shape from shading." In *PCV,* 1975.

HORN, B. K. P. and B. G. SCHUNCK. "Determining optical flow." AI Memo 572, AI Lab, MIT, April 1980.

HORN, B. K. P. and R. W. SJOBERG. "Calculating the reflectance map." *Proc.,* DARPA IU Workshop, November 1978, 115-126.

HUBEL, D. H. and T. N. WIESEL. "Brain mechanisms of vision." *Scientific American,* September 1979, 150-162.

HUECKEL, M. "An operator which locates edges in digitized pictures." *J. ACM 18,* 1, January 1971, 113-125.

HUECKEL, M. "A local visual operator which recognizes edges and lines." *J. ACM 20,* 4, October 1973, 634-647.

IKEUCHI, K. "Numerical shape from shading and occluding contours in a single view." AI Memo 566, AI Lab, MIT, revised February 1980.

KIRSCH, R. A. "Computer determination of the constituent structure of biological images." *Computers and Biomedical Research 4,* 3, June 1971, 315-328.

KNUTH, D. E. *The Art of Computer Programming.* Reading, MA: Addison-Wesley, 1973.

LEVINE, M. D. "A knowledge-based computer vision system." In *CVS,* 1978.

LIU, H. K. "Two- and three-dimensional boundary detection." *CGIP 6,* 2, 1977, 123-134.

MARR, D. and T. POGGIO. "Cooperative computation of stereo disparity." *Science 194,* 1976, 283-287.

MARR, D. and T. POGGIO. "A theory of human stereo vision." AI Memo 451, AI Lab, MIT, November 1977.

MERO, L. and Z. VASSY. "A simplified and fast version of the Hueckel operator for finding optimal edges in pictures." *Proc.,* 4th IJCAI, September 1975, 650-655.

MORAVEC, H. P. "Towards automatic visual obstacle avoidance." *Proc.,* 5th IJCAI, August 1977, 584.

NEVATIA, R. "Evaluation of a simplified Hueckel edge-line detector." Note, *CGIP 6,* 6, December 1977, 582-588.

PHONG, B-T. "Illumination for computer generated pictures." *Commun. ACM 18,* 6, June 1975, 311-317.

PINGLE, K. K. and J. M. TENENBAUM. "An accommodating edge follower." *Proc.,* 2nd IJCAI, September 1971, 1-7.

PRAGER, J. M. "Extracting and labeling boundary segments in natural scenes." *IEEE Trans. PAMI 2*, 1, January 1980, 16-27.

PRATT, W. K. *Digital Image Processing*. New York: Wiley-Interscience, 1978.

PREWITT, J. M. S. "Object enhancement and extraction." In *Picture Processing and Psychopictorics*, B. S. Lipkin and A. Rosenfeld (Eds.). New York: Academic Press, 1970.

QUAM, L. and M. J. HANNAH. "Stanford automated photogrammetry research." AIM-254, Stanford AI Lab, November 1974.

ROBERTS, L. G. "Machine perception of three-dimensional solids." In *Optical and Electro-optical Information Processing*, J. P. Tippett et al. (Eds.). Cambridge, MA: MIT Press, 1965.

ROSENFELD, A. and A. C. KAK. *Digital Picture Processing*. New York: Academic Press, 1976.

ROSENFELD, A., R. A. HUMMEL, and S. W. ZUCKER. "Scene labelling by relaxation operations." *IEEE Trans. SMC 6*, 1976, 430.

RUSSELL, D. L. (Ed.). *Calculus of Variations and Control Theory*. New York: Academic Press, 1976.

SHAPIRA, R. "A technique for the reconstruction of a straight-edge, wire-frame object from two or more central projections." *CGIP 3*, 4, December 1974, 318-326.

SHIRAI, V. "Analyzing intensity arrays using knowledge about scenes." In *PCV*, 1975.

STEIGLITZ, K. *An Introduction to Discrete Systems*. New York: Wiley, 1974.

STOCKHAM, T. J., Jr. "Image processing in the context of a visual model." *Proc. IEEE 60*, 7, July 1972, 828-842.

TANIMOTO, S. and T. PAVLIDIS. "A hierarchical data structure for picture processing." *CGIP 4*, 2, June 1975, 104-119.

TRETIAK, O. J. "A parameteric model for edge detection." *Proc.*, 3rd COMPSAC, November 1979, 884-887.

TURNER, K. J. "Computer perception of curved objects using a television camera." Ph.D. dissertation, Univ. Edinburgh, 1974.

WECHSLER, H. and J. SKLANSKY. "Finding the rib cage in chest radiographs." *Pattern Recognition 9*, 1977, 21-30.

WHITTED, T. "An improved illumination model for shaded display." *Comm. ACM 23*, 6, June 1980, 343-349.

WOODHAM, R. J. "Photometric stereo: A reflectance map technique for determining surface orientation from image intensity." *Proc.*, 22nd International Symp., Society of Photo-optical Instrumentation Engineers, San Diego, CA, August 1978, 136-143.

ZUCKER, S. W. and R. A. HUMMEL. "An optimal three-dimensional edge operator." Report 79-10, McGill Univ., April 1979.

ZUCKER, S. W., R. A. HUMMEL, and A. ROSENFELD. "An application of relaxation labeling to line and curve enhancement." *IEEE Trans. Computers 26*, 1977.

SEGMENTED IMAGES

II

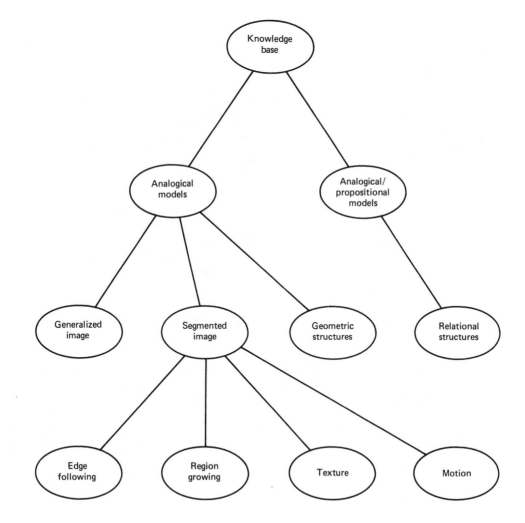

The idea of segmentation has its roots in work by the Gestalt psychologists (e.g., Kohler), who studied the preferences exhibited by human beings in grouping or organizing sets of shapes arranged in the visual field. Gestalt principles dictate certain grouping preferences based on features such as proximity, similarity, and continuity. Other results had to do with figure/ground discrimination and optical illusions. The latter have provided a fertile ground for vision theories to post-Gestaltists such as Gibson and Gregory, who emphasize that these grouping mechanisms organize the scene into *meaningful units* that are a significant step toward image understanding.

In computer vision, grouping parts of a generalized image into units that are homogeneous with respect to one or more characteristics (or features) results in a *segmented image*. The segmented image extends the generalized image in a crucial respect: it contains the beginnings of domain-dependent interpretation. At this descriptive level the internal domain-dependent models of objects begin to influence the grouping of generalized image structures into units meaningful in the domain. For instance, the model may supply crucial parameters to segmentation procedures.

In the segmentation process there are two important aspects to consider: one is the data structure used to keep track of homogeneous groups of features; the other is the transformation involved in computing the features.

Two basic sorts of segments are natural: boundaries and regions. These can be combined into a single descriptive structure, a set of nodes (one per region), connected by arcs representing the "adjacency" relation. The "dual" of this structure has arcs corresponding to boundaries connecting nodes representing points where several regions meet. Chapters 4 and 5 describe segmentation with respect to boundaries and regions respectively, emphasizing gray levels and gray-level differences as indicators of segments. Of course, from the standpoint of the algorithms involved, it is irrelevant whether the features are intensity gray levels or intrinsic image values perhaps representing motion, color, or range.

Texture and motion images are addressed in Chapters 6 and 7. Each has several computationally difficult aspects, and neither has received the attention given static, nontextured images. However, each is very important in the segmentation enterprise.

Boundary
Detection

4

4.1 ON ASSOCIATING EDGE ELEMENTS

Boundaries of objects are perhaps the most important part of the hierarchy of structures that links raw image data with their interpretation [Marr 1975]. Chapter 3 described how various operators applied to raw image data can yield primitive edge elements. However, an image of only disconnected edge elements is relatively featureless; additional processing must be done to group edge elements into structures better suited to the process of interpretation. The goal of the techniques in this chapter is to perform a level of *segmentation*, that is, to make a coherent one-dimensional (*edge*) feature from many individual local edge elements. The feature could correspond to an object boundary or to any meaningful boundary between scene entities. The problems that edge-based segmentation algorithms have to contend with are shown by Fig. 4.1, which is an image of the local edge elements yielded by one common edge operator applied to a chest radiograph. As can be seen, the edge elements often exist where no meaningful scene boundary does, and conversely often are absent where a boundary is. For example, consider the boundaries of ribs as revealed by the edge elements. Missing edge elements and extra edge elements both tend to frustrate the segmentation process.

The methods in this chapter are ordered according to the amount of knowledge incorporated into the grouping operation that maps edge elements into boundaries. "Knowledge" means implicit or explicit constraints on the likelihood of a given grouping. Such constraints may arise from general physical arguments or (more often) from stronger restrictions placed on the image arising from domain-dependent considerations. If there is much knowledge, this implies that the global form of the boundary and its relation to other image structures is very constrained. Little prior knowledge means that the segmentation must proceed more on the basis of local clues and evidence and general (domain-dependent) assumptions with fewer expectations and constraints on the final resulting boundary.

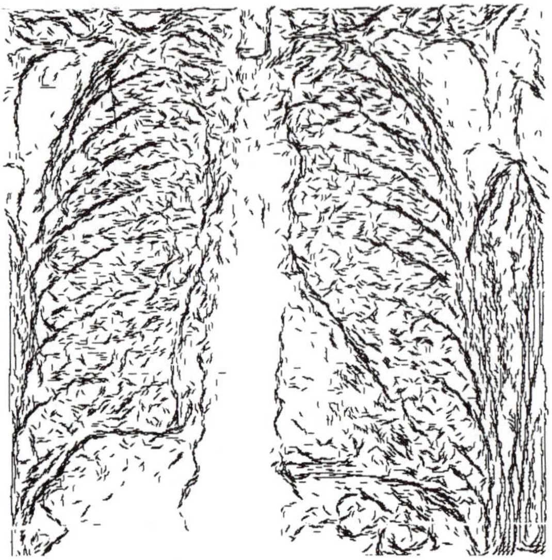

Fig. 4.1 Edge elements in a chest radiograph.

These constraints take many forms. Knowledge of where to expect a boundary allows very restricted searches to verify the edge. In many such cases, the domain knowledge determines the type of curve (its parameterization or functional form) as well as the relevant "noise processes." In images of polyhedra, only straight-edged boundaries are meaningful, and they will come together at various sorts of vertices arising from corners, shadows of corners, and occlusions. Human rib boundaries appear approximately like conic sections in chest radiographs, and radiographs have complex edge structures that can compete with rib edges. All this specific knowledge can and should guide our choice of grouping method.

If less is known about the specific image content, one may have to fall back on general world knowledge or heuristics that are true for most domains. For instance, in the absence of evidence to the contrary, the shorter line between two points might be selected over a longer line. This sort of general principle is easily built into evaluation functions for boundaries, and used in segmentation algorithms that proceed by methodically searching for such groupings. If there are no a priori restrictions on boundary shapes, a general contour-extraction method is called for, such as edge following or linking of edge elements.

The methods we shall examine are the following:

1. *Searching near an approximate location.* These are methods for refining a boundary given an initial estimate.

2. *The Hough transform.* This elegant and versatile technique appears in various guises throughout computer vision. In this chapter it is used to detect boundaries whose shape can be described in an analytical or tabular form.

3. *Graph searching.* This method represents the image of edge elements as a graph. Thus a boundary is a path through a graph. Like the Hough transform, these techniques are quite generally applicable.

4. *Dynamic programming.* This method is also very general. It uses a mathematical formulation of the globally best boundary and can find boundaries in noisy images.

5. *Contour following.* This hill-climbing technique works best with good image data.

4.2 SEARCHING NEAR AN APPROXIMATE LOCATION

If the approximate or a priori likely location of a boundary has been determined somehow, it may be used to guide the effort to refine that boundary [Kelly 1971]. The approximate location may have been found by one of the techniques below applied to a lower resolution image, or it may have been determined using high-level knowledge.

4.2.1 Adjusting A Priori Boundaries

This idea was described by [Bolles 1977] (see Fig. 4.2). Local searches are carried out at regular intervals along directions perpendicular to the approximate (a priori) boundary. An edge operator is applied to each of the discrete points along each of these perpendicular directions. For each such direction, the edge with the highest magnitude is selected from among those whose orientations are nearly parallel to the tangent at the point on the nearby a priori boundary. If sufficiently many elements are found, their locations are fit with an analytic curve such as a low-degree polynomial, and this curve becomes the representation of the boundary.

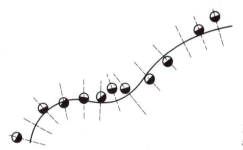

Fig. 4.2 Search orientations from an approximate boundary location.

4.2.2 Non-linear Correlation in Edge Space

In this correlation-like technique, the a priori boundary is treated as a rigid template, or piece of rigid wire along which edge operators are attached like beads. The a priori representation thus also contains relative locations at which the existence of edges will be tested (Fig. 4.3). An edge element returned by the edge-operator application "matches" the a priori boundary if its contour is tangent to the template and its magnitude exceeds some threshold. The template is to be moved around the image, and for each location, the number of matches is computed. If the number of matches exceeds a threshold, the boundary location is declared to

Fig. 4.3 A template for edge-operator application.

be the current template location. If not, the template is moved to a different image point and the process is repeated. Either the boundary will be located or there will eventually be no more image points to try.

4.2.3 Divide-and-Conquer Boundary Detection

This is a technique that is useful in the case that a low-curvature boundary is known to exist between two edge elements and the noise levels in the image are low (Algorithm 8.1). In this case, to find a boundary point in between the two known points, search along the perpendiculars of the line joining the two points. The point of maximum magnitude (if it is over some threshold) becomes a break point on the boundary and the technique is applied recursively to the two line segments formed between the three known boundary points. (Some fix must be applied if the maximum is not unique.) Figure 4.4 shows one step in this process. Divide-and-conquer boundary detection has been used to outline kidney boundaries on computed tomograms (these images were described in Section 2.3.4) [Selfridge et al. 1979].

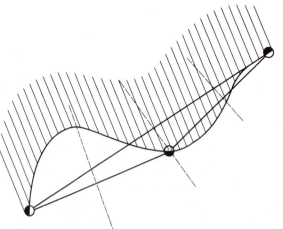

Fig. 4.4 Divide and conquer technique.

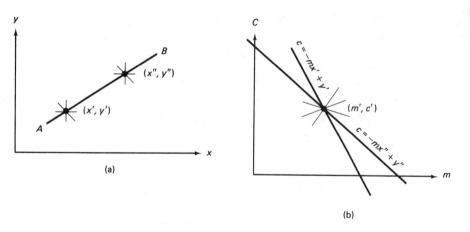

Fig. 4.5 A line (a) in image space; (b) in parameter space.

4.3 THE HOUGH METHOD FOR CURVE DETECTION

The classical Hough technique for curve detection is applicable if little is known about the location of a boundary, but its shape can be described as a parametric curve (e.g., a straight line or conic). Its main advantages are that it is relatively unaffected by gaps in curves and by noise.

To introduce the method [Duda and Hart 1972], consider the problem of detecting straight lines in images. Assume that by some process image points have been selected that have a high likelihood of being on linear boundaries. The Hough technique organizes these points into straight lines, basically by considering all possible straight lines at once and rating each on how well it explains the data.

Consider the point x' in Fig. 4.5a, and the equation for a line $y = mx + c$. What are the lines that could pass through x'? The answer is simply all the lines with m and c satisfying $y' = mx' + c$. Regarding (x', y') as fixed, the last equation is that of a line in $m-c$ space, or parameter space. Repeating this reasoning, a second point (x'', y'') will also have an associated line in parameter space and, furthermore, these lines will intersect at the point (m', c') which corresponds to the line AB connecting these points. In fact, all points on the line AB will yield lines in parameter space which intersect at the point (m', c'), as shown in Fig. 4.5b.

This relation between image space x and parameter space suggests the following algorithm for detecting lines:

Algorithm 4.1: Line Detection with the Hough Algorithm

1. Quantize parameter space between appropriate maximum and minimum values for c and m.

2. Form an accumulator array $A(c, m)$ whose elements are initially zero.

3. For each point (x,y) in a *gradient* image such that the strength of the gradient

exceeds some threshold, increment all points in the accumulator array along the appropriate line, i.e.,

$$A(c, m) := A(c, m) + 1$$

for m and c satisfying $c = -mx + y$ within the limits of the digitization.

4. Local maxima in the accumulator array now correspond to collinear points in the image array. The values of the accumulator array provide a measure of the number of points on the line.

This technique is generally known as the Hough technique [Hough 1962].

Since m may be infinite in the slope-intercept equation, a better parameterization of the line is $x \cos \theta + y \sin \theta = r$. This produces a sinusoidal curve in (r, θ) space for fixed x, y, but otherwise the procedure is unchanged.

The generalization of this technique to other curves is straightforward and this method works for any curve $f(\mathbf{x}, \mathbf{a}) = 0$, where $\mathbf{a}$ is a parameter vector. (In this chapter we often use the symbol f as various general functions unrelated to the image gray-level function.) In the case of a circle parameterized by

$$(x - a)^2 + (y - b)^2 = r^2 \qquad (4.1)$$

for fixed $\mathbf{x}$, the modified algorithm 4.1 increments values of a, b, r lying on the surface of a cone. Unfortunately, the computation and the size of the accumulator array increase exponentially as the number of parameters, making this technique practical only for curves with a small number of parameters.

The Hough method is an efficient implementation of a generalized matched filtering strategy (i.e., a template-matching paradigm). For instance, in the case of a circle, imagine a template composed of a circle of 1's (at a fixed radius r) and 0's everywhere else. If this template is convolved with the gradient image, the result is the portion of the accumulator array $A(a, b, r)$.

In its usual form, the technique yields a set of parameters for a curve that best explains the data. The parameters may specify an infinite curve (e.g., a line or parabola). Thus, if a finite curve segment is desired, some further processing is necessary to establish end points.

4.3.1 Use of the Gradient

Dramatic reductions in the amount of computation can be achieved if the gradient direction is integrated into the algorithm [Kimme et al. 1975]. For example, consider the problem of detecting a circle of fixed radius R.

Without gradient information, all values a, b lying on the circle given by (4.1) are incremented. With the gradient information, fewer points need be incremented (Fig. 4.6). Only increment points on an arc centered at (a, b),

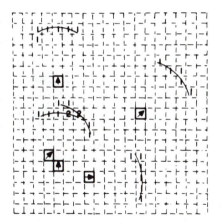

Contents of accumulator array

Gradient direction information for artifact $\Delta\phi$ = 45

□ Denotes a pixel in $f(\underline{x})$ superimposed on accumulator array

↗ Denotes the gradient direction

Fig 4.6 Reduction in computation with gradient information

$$a = x + r \cos \phi \qquad (4.2)$$
$$b = y + r \sin \phi$$

where $\phi(x)$ is the gradient angle returned by an edge operator. Implicit in these equations is the assumption that the circle is the boundary of a disk that has gray levels greater than its surroundings. These equations may also be derived by differentiating (4.1), recognizing that $dy/dx = \tan\phi$, and solving for a and b between the resultant equation and (4.1). Similar methods can be applied to other conics. In each case, the use of the gradient saves one dimension in the accumulator array.

The gradient magnitude can also be used as a heuristic in the incrementing procedure. Instead of incrementing by unity, the accumulator array location may be incremented by a function of the gradient magnitude. This heuristic can balance the magnitude of brightness change across a boundary with the boundary length, but it can lead to detection of phantom lines indicated by a few bright points, or to missing dim but coherent boundaries.

4.3.2 Some Examples

The Hough technique has been used successfully in a variety of domains. Some examples include the detection of human hemoglobin fingerprints [Ballard et al. 1975], the detection of tumors in chest films [Kimme et al. 1975], the detection of storage tanks in aerial images [Lantz et al. 1978], and the detection of ribs in chest radiographs [Wechsler and Sklansky 1977]. Figure 4.7 shows the tumor-detection application. A section of the chest film (Fig. 4.7b) is searched for disks of radius 3 units. In Fig. 4.7c, the resultant accumulator array A [a, b, 3] is shown in a pictoral fashion, by interpreting the array values as gray levels. This process is repeated for various radii and then a set of likely circles is chosen by setting a radius-dependent threshold for the accumulator array contents. This result is shown in Fig. 4.7d. The

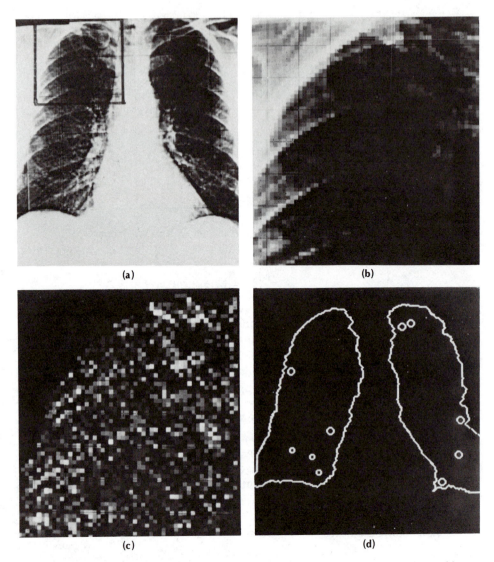

(a)

(b)

(c)

(d)

Fig. 4.7 Using the Hough technique for circular shapes. (a) Radiograph. (b) Window. (c) Accumulator array for $r = 3$. (d) Results of maxima detection.

circular boundaries detected by the Hough technique are overlaid on the original image.

4.3.3 Trading Off Work in Parameter Space for Work in Image Space

Consider the example of detecting ellipses that are known to be oriented so that a principal axis is parallel to the x axis. These can be specified by four parameters. Using the equation for the ellipse together with its derivative, and substituting for the known gradient as before, one can solve for two parameters. In the equation

$$\frac{(x - x_0)^2}{a^2} + \frac{(y - y_0)^2}{b^2} = 1 \tag{4.3}$$

$\mathbf{x}$ is an edge point and x_0, y_0, a, and b are parameters. The equation for its derivative is

$$\frac{(x - x_0)}{a^2} + \frac{(y - y_0)}{b^2} \frac{dy}{dx} = 0 \tag{4.4}$$

where $dy/dx = \tan \phi (x)$. The Hough algorithm becomes:

Algorithm 4.2: Hough technique applied to ellipses

For each discrete value of x and y, increment the point in parameter space given by a, b, x_0, y_0, where

$$x = x_0 \pm \frac{a}{(1 + b^2/a^2 \tan^2\phi)^{\frac{1}{2}}} \tag{4.5}$$

$$y = y_0 \pm \frac{b}{(1 + a^2 \tan^2 \phi/b^2)^{\frac{1}{2}}} \tag{4.6}$$

that is,

$$A (a, b, x_0, y_0) := A (a, b, x_0, y_0) + 1$$

For a and b each having m values the computational cost is proportional to m^2.

Now suppose that we consider all pairwise combinations of edge elements. This introduces two additional equations like (4.3) and (4.4), and now the four-parameter point can be determined exactly. That is, the following equations can be solved for a unique x_0, y_0, a, b.

$$\frac{(x_1 - x_0)^2}{a^2} + \frac{(y_1 - y_0)^2}{b^2} = 1 \tag{4.7a}$$

$$\frac{(x_2 - x_0)^2}{a^2} + \frac{(y_2 - y_0)^2}{b^2} = 1 \tag{4.7b}$$

$$\frac{x_1 - x_0}{a^2} + \frac{y_1 - y_0}{b^2} \frac{dy}{dx} = 0 \tag{4.7c}$$

$$\frac{x_2 - x_0}{a^2} + \frac{y_2 - y_0}{b^2} \frac{dy}{dx} = 0 \tag{4.7d}$$

$$\frac{dy}{dx} = \tan \phi \quad (\frac{dy}{dx} \text{ is known from the edge operator})$$

Their solution is left as an exercise. The amount of effort in the former case was proportional to the product of the number of discrete values of a and b, whereas this case involves effort proportional to the square of the number of edge elements.

4.3.4 Generalizing the Hough Transform

Consider the case where the object being sought has no simple analytic form, but has a particular silhouette. Since the Hough technique is so closely related to template matching, and template matching can handle this case, it is not surprising that the Hough technique can be generalized to handle this case also. Suppose for the moment that the object appears in the image with known shape, orientation, and scale. (If orientation and scale are unknown, they can be handled in the same way that additional parameters were handled earlier.) Now pick a reference point in the silhouette and draw a line to the boundary. At the boundary point compute the gradient direction and store the reference point as a function of this direction. Thus it is possible to precompute the location of the reference point from boundary points given the gradient angle. The set of all such locations, indexed by gradient angle, comprises a table termed the R-table [Ballard 1981]. Remember that the basic strategy of the Hough technique is to compute the possible loci of reference points in parameter space from edge point data in image space and increment the parameter points in an accumulator array. Figure 4.8 shows the relevant geometry and Table 4.1 shows the form of the R-table. For the moment, the reference point coordinates (x_c, y_c) are the only parameters (assuming that rotation and scaling have been fixed). Thus an edge point (x, y) with gradient orientation ϕ constrains the possible reference points to be at $\{x + r_1 (\phi) \cos [\alpha_1 (\phi)], y + r_1(\phi) \sin [\alpha_1 (\phi)]\}$ and so on.

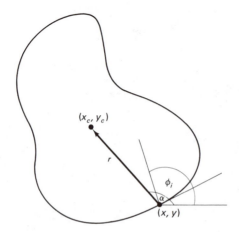

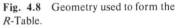
Fig. 4.8 Geometry used to form the R-Table.

Table 4.1

INCREMENTATION IN THE GENERALIZED HOUGH CASE

Angle measured from figure boundary to reference point	Set of radii $\{\mathbf{r}^k\}$ where $\mathbf{r} = (r, \alpha)$
ϕ_1	$\mathbf{r}_1^1, \mathbf{r}_2^1, \ldots, \mathbf{r}_{n_1}^1$
ϕ_2	$\mathbf{r}_1^2, \mathbf{r}_2^2, \ldots, \mathbf{r}_{n_2}^2$
.	.
.	.
.	.
ϕ_m	$\mathbf{r}_1^m, \mathbf{r}_2^m, \ldots, \mathbf{r}_{n_m}^m$

The generalized Hough algorithm may be described as follows:

Algorithm 4.3: Generalized Hough

Step 0. Make a table (like Table 4.1) for the shape to be located.

Step 1. Form an accumulator array of possible reference points $A(x_{c\min} : x_{c\max}, y_{c\min} : y_{c\max})$ initialized to zero.

Step 2. For each edge point do the following:

Step 2.1. Compute $\phi(\mathbf{x})$

Step 2.2a. Calculate the possible centers; that is, for each table entry for ϕ, compute

$$x_c := x + r\,\phi\ \cos[\alpha(\phi)]$$

$$y_c := y + r\,\phi\ \sin[\alpha(\phi)]$$

Step 2.2b. Increment the accumulator array

$$A(x_c, y_c) := A(x_c, y_c) + 1$$

Step 3. Possible locations for the shape are given by maxima in array A.

The results of using this transform to detect a shape are shown in Fig. 4.9. Figure 4.9a shows an image of shapes. The R-table has been made for the middle shape. Figure 4.9b shows the Hough transform for the shape, that is, $A(x_c, y_c)$ displayed as an image. Figure 4.9c shows the shape given by the maxima of

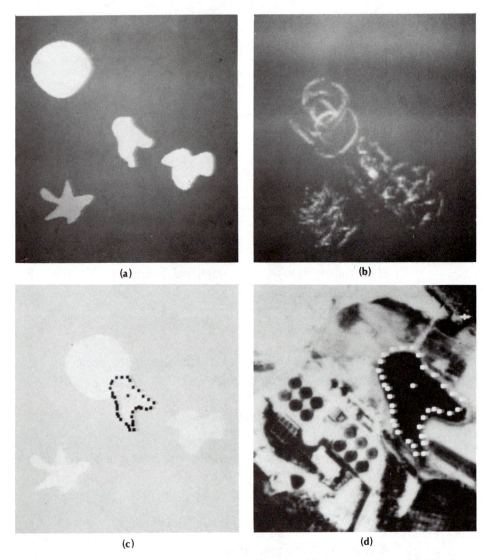

Fig. 4.9 Applying the Generalized Hough technique. (a) Synthetic image. (b) Hough Transform $A(x_c, y_c)$ for middle shape. (c) Detected shape. (d) Same shape in an aerial image setting.

$A(x_c, y_c)$ overlaid on top of the image. Finally, Fig. 4.9d shows the Hough transform used to detect a pond of the same shape in an aerial image.

What about the parameters of scale and rotation, S and θ? These are readily accommodated by expanding the accumulator array and doing more work in the incrementation step. Thus in step 1 the accumulator array is changed to

$$(x_{c\min} : x_{c\max}, \; y_{c\min} : y_{c\max}, \; S_{\min} : S_{\max}, \; \theta_{\min} : \theta_{\max})$$

and step 2.2a is changed to

$$\text{for each table entry for } \phi \text{ do}$$

$$\text{for each } S \text{ and } \theta$$

$$x_c := x + r(\phi) S \cos[\alpha(\phi) + \theta]$$

$$y_c := y + r(\phi) S \sin[\alpha(\phi) + \theta]$$

Finally, step 2.2b is now

$$A(x_c, y_c, S, \theta) := A(x_c, y_c, S, \theta) + 1$$

4.4 EDGE FOLLOWING AS GRAPH SEARCHING

A graph is a general object that consists of a set of nodes $\{n_i\}$ and arcs between nodes $<n_i, n_j>$. In this section we consider graphs whose arcs may have numerical weights or *costs* associated with them. The search for the boundary of an object is cast as a search for the lowest-cost path between two nodes of a weighted graph.

Assume that a gradient operator is applied to the gray-level image, creating the magnitude image $s(\mathbf{x})$ and direction image $\phi(\mathbf{x})$. Now interpret the elements of the direction image $\phi(\mathbf{x})$ as nodes in a graph, each with a weighting factor $s(\mathbf{x})$. Nodes $\mathbf{x}_i$, $\mathbf{x}_j$ have arcs between them if the contour directions $\phi(\mathbf{x}_i)$, $\phi(\mathbf{x}_j)$ are appropriately aligned with the arc directed in the same sense as the contour direction. Figure 4.10 shows the interpretation. To generate Fig. 4.10b impose the following restrictions. For an arc to connect from $\mathbf{x}_i$ to $\mathbf{x}_j$, $\mathbf{x}_j$ must be one of the three possible eight-neighbors in front of the contour direction $\phi(\mathbf{x}_i)$ and, furthermore, $s(\mathbf{x}_i)$

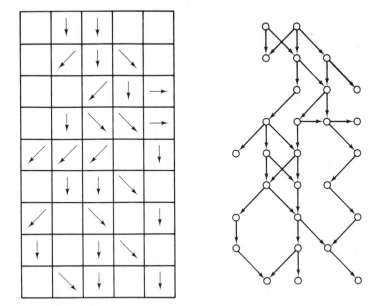

Fig. 4.10 Interpreting a gradient image as a graph (see text).

$> T$; $s(\mathbf{x}_j) > T$, where T is a chosen constant, and $|\{ [\phi(\mathbf{x}_i) - \phi(\mathbf{x}_j)] \bmod 2\pi \}| < \pi/2$. (Any or all of these restrictions may be modified to suit the requirements of a particular problem.)

To generate a path in a graph from $\mathbf{x}_A$ to $\mathbf{x}_B$ one can apply the well-known technique of heuristic search [Nilsson 1971, 1980]. The specific use of heuristic search to follow edges in images was first proposed by [Martelli 1972]. Suppose:

1. That the path should follow contours that are directed from $\mathbf{x}_A$ to $\mathbf{x}_B$

2. That we have a method for generating the successor nodes of a given node (such as the heuristic described above)

3. That we have an evaluation function $f(\mathbf{x}_j)$ which is an estimate of the optimal cost path from $\mathbf{x}_A$ to $\mathbf{x}_B$ constrained to go through $\mathbf{x}_j$

Nilsson expresses $f(\mathbf{x}_i)$ as the sum of two components: $s(\mathbf{x}_i)$, the estimated cost of journeying from the *start node* $\mathbf{x}_A$ to $\mathbf{x}_i$, and $h(\mathbf{x}_i)$, the estimated cost of the path from $\mathbf{x}_i$ to $\mathbf{x}_B$, the *goal node*.

With the foregoing preliminaries, the heuristic search algorithm (called the A algorithm by Nilsson) can be stated as:

Algorithm 4.4:   Heuristic Search (the A Algorithm)

1. "Expand" the start node (put the successors on a list called OPEN with pointers back to the start node).

2. Remove the node $\mathbf{x}_i$ of minimum f from OPEN. If $\mathbf{x}_i = \mathbf{x}_B$, then stop. Trace back through pointers to find optimal path. If OPEN is empty, fail.

3. Else expand node $\mathbf{x}_i$, putting successors on OPEN with pointers back to $\mathbf{x}_i$. Go to step 2.

The component $h(\mathbf{x}_i)$ plays an important role in the performance of the algorithm; if $h(\mathbf{x}_i) = 0$ for all i, the algorithm is a *minimum-cost search* as opposed to a *heuristic search*. If $h(\mathbf{x}_i) > h^*(\mathbf{x}_i)$ (the actual optimal cost), the algorithm may run faster, but may miss the minimum-cost path. If $h(\mathbf{x}_i) \leqslant h^*(\mathbf{x}_i)$, the search will always produce a minimum-cost path, provided that h also satisfies the following consistency condition:

> If for any two nodes $\mathbf{x}_i$ and $\mathbf{x}_j$, $k(\mathbf{x}_i, \mathbf{x}_j)$ is the minimum cost of getting from $\mathbf{x}_i$ to $\mathbf{x}_j$ (if possible), then

$$k(\mathbf{x}_i, \mathbf{x}_j) \geqslant h^*(\mathbf{x}_i) - h^*(\mathbf{x}_j)$$

With our edge elements, there is no guarantee that a path can be found since there may be insurmountable gaps between $\mathbf{x}_A$ and $\mathbf{x}_B$. If finding the edge is crucial, steps should be taken to interpolate edge elements prior to the search, or gaps may be crossed by using the edge element definition of [Martelli 1972]. He defines edges on the image grid structure so that an edge can have a direction even though there is no local gray-level change. This definition is depicted in Fig. 4.11a.

4.4.1 Good Evaluation Functions

A good evaluation function has components specific to the particular task as well as components that are relatively task-independent. The latter components are discussed here.

1. *Edge strength.* If edge strength is a factor, the cost of adding a particular edge element at **x** can be included as

$$M - s(\mathbf{x}) \qquad \text{where } M = \max_{\mathbf{x}} s(\mathbf{x})$$

2. *Curvature.* If low-curvature boundaries are desirable, curvature can be measured as some monotonically increasing function of

$$diff[\phi(\mathbf{x}_i) - \phi(\mathbf{x}_j)]$$

where diff measures the angle between the edge elements at $\mathbf{x}_j$ and $\mathbf{x}_i$.

3. *Proximity to an approximation.* If an approximate boundary is known, boundaries near this approximation can be favored by adding:

$$d = dist(\mathbf{x}_i, B)$$

to the cost measure. The dist operator measures the minimum distance of the new point $\mathbf{x}_i$ to the approximate boundary B.

4. *Estimates of the distance to the goal.* If the curve is reasonably linear, points near the goal may be favored by estimating h as $d(\mathbf{x}_i, \mathbf{x}_{\text{goal}})$, where d is a distance measure.

Specific implementations of these measures appear in [Ashkar and Modestino 1978; Lester et al. 1978].

4.4.2 Finding All the Boundaries

What if the objective is to find *all* boundaries in the image using heuristic search? In one system [Ramer 1975] Hueckel's operator (Chapter 3) is used to obtain

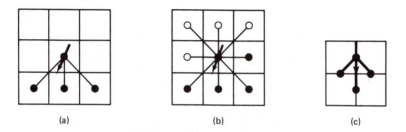

(a)   (b)   (c)

Fig. 4.11 Successor conventions in heuristic search (see text).

strokes, another name for the magnitude and direction of the local gray-level changes. Then these strokes are combined by heuristic search to form sequences of edge elements called *streaks*. Streaks are an intermediate organization which are used to assure a slightly broader coherence than is provided by the individual Hueckel edges. A bidirectional search is used with four eight-neighbors defined in front of the edge and four eight-neighbors behind the edge, as shown in Fig. 4.11b. The search algorithm is as follows:

1. Scan the stroke (edge) array for the most prominent edge.
2. Search in front of the edge until no more successors exist (i.e., a gap is encountered).
3. Search behind the edge until no more predecessors exist.
4. If the bidirectional search generates a path of 3 or more strokes, the path is a streak. Store it in a streak list and go to step 1.

Strokes that are part of a streak cannot be reused; they are marked when used and subsequently skipped.

There are other heuristic procedures for pruning the streaks to retain only *prime streaks*. These are shown in Fig. 4.12. They are essentially similar to the re-

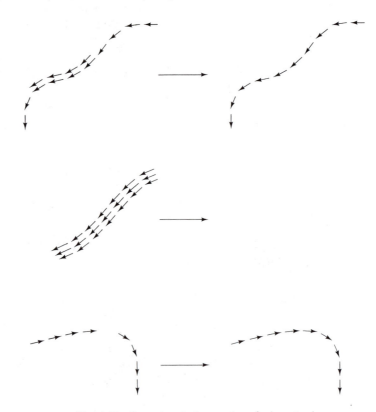

Fig. 4.12 Operations in the creation of prime streaks.

(a) Image

(b) Strokes

(c) Step in application
of heuristics

(d) Step in application
of heuristics

(e) Step in application
of heuristics

(f) Prime streaks

Fig. 4.13 Ramer's results.

laxation operations described in Section 3.3.5. The resultant streaks must still be analyzed to determine the objects they represent. Nevertheless, this method represents a cogent attempt to organize bottom-up edge following in an image. Fig. 4.13 shows an example of Ramer's technique.

4.4.3 Alternatives to the A Algorithm

The primary disadvantage with the heuristic search method is that the algorithm must keep track of a set of current best paths (nodes), and this set may become very large. These nodes represent tip nodes for the portion of the tree of possible paths that has been already examined. Also, since all the costs are nonnegative, a good path may eventually look expensive compared to tip nodes near the start node. Thus, paths from these newer nodes will be extended by the algorithm even though, from a practical standpoint, they are unlikely. Because of these disadvantages, other less rigorous search procedures have proven to be more practical, five of which are described below.

Pruning the Tree of Alternatives

At various points in the algorithm the tip nodes on the OPEN list can be pruned in some way. For example, paths that are short or have a high cost per unit length can be discriminated against. This pruning operation can be carried out whenever the number of alternative tip nodes exceeds some bound.

Modified Depth-First Search

Depth-first search is a meaningful concept if the search space is structured as a tree. Depth-first search means always evaluating the most recent expanded son. This type of search is performed if the OPEN list is structured as a stack in the A algorithm and the top node is always evaluated next. Modifications to this method use an evaluation function f to rate the successor nodes and expand the best of these. Practical examples can be seen in [Ballard and Sklansky 1976; Wechsler and Sklansky 1977; Persoon 1976].

Least Maximum Cost

In this elegant idea [Lester 1978], only the maximum-cost arc of each path is kept as an estimate of g. This is like finding a mountain pass at minimum altitude. The advantage is that g does not build up continuously with depth in the search tree, so that good paths may be followed for a long time. This technique has been applied to finding the boundaries of blood cells in optical microscope images. Some results are shown in Fig. 4.14.

Branch and Bound

The crux of this method is to have some upper bound on the cost of the path [Chien and Fu 1974]. This may be known beforehand or may be computed by actually generating a path between the desired end points. Also, the evaluation function must be monotonically increasing with the length of the path. With these conditions we start generating paths, excluding partial paths when they exceed the current bound.

Modified Heuristic Search

Sometimes an evaluation function that assigns negative costs leads to good results. Thus good paths keep getting better with respect to the evaluation function, avoiding the problem of having to look at all paths near the starting point.

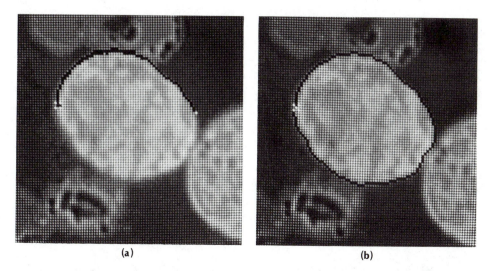

(a)   (b)

Fig. 4.14 Using least maximum cost in heuristic search to find cell boundaries in microscope images. (a) A stage in the search process. (b) The completed boundary.

However, the price paid is the sacrifice of the mathematical guarantee of finding the least-cost path. This could be reflected in unsatisfactory boundaries. This method has been used in cineangiograms with satisfactory results [Ashkar and Modestino 1978].

4.5 EDGE FOLLOWING AS DYNAMIC PROGRAMMING

4.5.1 Dynamic Programming

Dynamic programming [Bellman and Dreyfus 1962] is a technique for solving optimization problems when not all variables in the evaluation function are interrelated simultaneously. Consider the problem

$$\max_{x_i} h(x_1, x_2, x_3, x_4) \tag{4.8}$$

If nothing is known about h, the only technique that guarantees a global maximum is exhaustive enumeration of all combinations of discrete values of $x_1, \ldots, x_4$. Suppose that

$$h(\cdot) = h_1(x_1, x_2) + h_2(x_2, x_3) + h_3(x_3, x_4) \tag{4.9}$$

x_1 only depends on x_2 in h_1. Maximize over x_1 in h_1 and tabulate the best value of $h_1(x_1, x_2)$ for each x_2:

$$f_1(x_2) = \max_{x_1} h_1(x_1, x_2) \tag{4.10}$$

Since the values of h_2 and h_3 do not depend on x_1, they need not be considered at this point. Continue in this manner and eliminate x_2 by computing $f_2(x_3)$ as

$$f_2(x_3) = \max_{x_2}[f_1(x_2) + h_2(x_2, x_3)] \tag{4.11}$$

and

$$f_3(x_4) = \max_{x_3}[f_2(x_3) + h_3(x_3, x_4)] \tag{4.12}$$

so that finally

$$\max_{x_i} h = \max_{x_4} f_3(x_4) \tag{4.13}$$

Generalizing the example to N variables, where $f_0(x_1) = 0$,

$$f_{n-1}(x_n) = \max_{x_{n-1}}[f_{n-2}(x_{n-1}) + h_{n-1}(x_{n-1}, x_n)] \tag{4.14}$$

$$\max_{x_i} h(x_i, \ldots, x_N) = \max_{x_N} f_{N-1}(x_N)$$

If each x_i took on 20 discrete values, then to compute $f_N(x_{N+1})$ one must evaluate the maximand for 20 different combinations of x_N and x_{N+1}, so that the resultant computational effort involves $(N - 1)20^2 + 20$ such evaluations. This is a striking improvement over exhaustive evaluation, which would involve 20^N evaluations of h!

Consider the artificial example summarized in Table 4.2. In this example, each **x** can take on one of three discrete values. The h_i are completely described by their respective tables. For example, the value of $h_1(0, 1) = 5$. The solution steps are summarized in Table 4.3. In step 1, for each x_2 the value of x_1 that maximizes $h_1(x_1, x_2)$ is computed. This is the largest entry in each of the columns of h. Store the function value as $f_1(x_2)$ and the optimizing value of x_1 also as a function of x_2. In step 2, add $f_1(x_2)$ to $h_2(x_2, x_3)$. This is done by adding f_1 to each row of h_2, thus computing the quantity inside the braces of (4.11). Now to complete step 2, for each x_3, compute the x_2 that maximizes $h_2 + f_1$ by selecting the largest entry in each row of the appropriate table. The rest of the steps are straightforward once these are understood. The solution is found by tracing back through the tables. For example, for $x_4 = 2$ we see that the best x_3 is -1, and therefore the best x_2 is 3 and x_1 is 1. This step is denoted by arrows.

Table 4.2

DEFINITION OF h

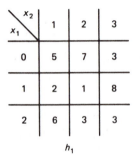

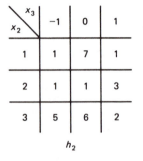

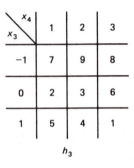

Ch. 4   Boundary Detection

Table 4.3

METHOD OF SOLUTION USING DYNAMIC PROGRAMMING

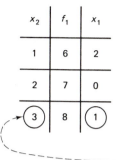

Step 1

x_2	f_1	x_1
1	6	2
2	7	0
3	8	1

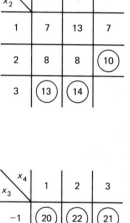

Step 2

$x_2 \backslash x_3$	-1	0	1
1	7	13	7
2	8	8	10
3	13	14	

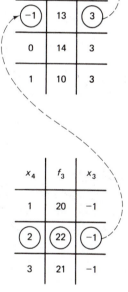

x_3	f_2	x_2
-1	13	3
0	14	3
1	10	3

Step 3

$x_3 \backslash x_4$	1	2	3
-1	20	22	21
0	16	17	20
1	15	14	11

x_4	f_3	x_3
1	20	-1
2	22	-1
3	21	-1

Step 4: Optimal x_i's are found by examing tables
(dashed line shows the order in which they
are recovered).

Solution: $h^* = 22$
$x_1^* = 1, x_2^* = 3, x_3^* = -1, x_4^* = 2$

4.5.2 Dynamic Programming for Images

To formulate the boundary-following procedure as dynamic programming, one
must define an evaluation function that embodies a notion of the "best boundary"
[Montanari 1971; Ballard 1976]. Suppose that a local edge detection operator is ap-

plied to a gray-level picture to produce edge magnitude and direction information. Then one possible criterion for a "good boundary" is a weighted sum of high cumulative edge strength and low cumulative curvature; that is, for an n-segment curve,

$$h(\mathbf{x}_1, \ldots, \mathbf{x}_n) = \sum_{k=1}^{n} s(\mathbf{x}_k) + \alpha \sum_{k=1}^{n-1} q(\mathbf{x}_k, \mathbf{x}_{k+1}) \tag{4.16}$$

where the implicit constraint is that consecutive $\mathbf{x}_k$'s must be grid neighbors:

$$\|\mathbf{x}_k - \mathbf{x}_{k+1}\| \leq \sqrt{2} \tag{4.17}$$

$$q(\mathbf{x}_k, \mathbf{x}_{k+1}) = diff[\phi(\mathbf{x}_k), \phi(\mathbf{x}_{k+1})] \tag{4.18}$$

where α is negative. Notice that this evaluation function is in the form of (4.9) and can be optimized in stages:

$$f_0(\mathbf{x}_1) \equiv 0 \tag{4.19}$$

$$f_1(\mathbf{x}_2) = \max_{x_1} [s(\mathbf{x}_1) + \alpha q(\mathbf{x}_1, \mathbf{x}_2) + f_0(\mathbf{x}_1)] \tag{4.20}$$

$$f_k(\mathbf{x}_{k+1}) = \max_{x_k} [s(\mathbf{x}_k) + \alpha q(\mathbf{x}_k, \mathbf{x}_{k+1}) + f_{k-1}(\mathbf{x}_k)] \tag{4.21}$$

These equations can be put into the following steps:

Algorithm 4.5: Dynamic Programming for Edge Finding

1. Set $k = 1$.
2. Consider only $\mathbf{x}$ such that $s(\mathbf{x}) \geq T$. For each of these $\mathbf{x}$, define low-curvature pixels "in front of" the contour direction.
3. Each of these pixels may have a curve emanating from it. For $k = 1$, the curve is one pixel in length. Join the curve to $\mathbf{x}$ that optimizes the left-hand side of the recursion equation.
4. If $k = N$, pick the best f_{N-1} and stop. Otherwise, set $k = k + 1$ and go to step 2.

This algorithm can be generalized to the case of picking a *curve* emanating from $\mathbf{x}$ (that we have already generated): Find the end of that curve, and join the best of three curves emanating from the end of that curve. Figure 4.15 shows this process. The equations for the general case are

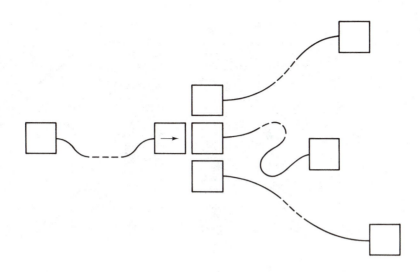

Fig. 4.15 DP optimization for boundary tracing.

$$f_0(\mathbf{x}_1) \equiv 0$$

$$f_l(\mathbf{x}_{k+1}) = \max_{x_k}[s(\mathbf{x}_k) + \alpha q(\mathbf{x}_k, t(\mathbf{x}_{k+1}))$$

$$+ f_{l-1}(\mathbf{x}_k)] \tag{4.22}$$

where the curve length n is related to α by a building sequence $n(l)$ such that $n(1) = 1$, $n(L) = N$, and $n(l) - n(l-1)$ is a member of $\{n(k)|k = 1, ..., l - 1\}$. Also, $t(\mathbf{x}_k)$ is a function that extracts the tail pixel of the curve headed by $\mathbf{x}_k$. Further details may be found in [Ballard 1976].

Results from the area of tumor detection in radiographs give a sense of this method's performance. Here it is known that the boundary inscribes an approximately circular tumor, so that circular cues can be used to assist the search. In Fig. 4.16, (a) shows the image containing the tumor, (b) shows the cues, and (c) shows the boundary found by dynamic programming overlaid on the image.

Another application of dynamic programming may be found in the pseudo-parallel road finder of Barrow [Barrow 1976].

4.5.3 Lower Resolution Evaluation Functions

In the dynamic programming formulation just developed, the components $s(\mathbf{x}_k)$ and $q(\mathbf{x}_k, \mathbf{x}_{k+1})$ in the evaluation function are very localized; the variables $\mathbf{x}$ for successive s and q are in fact constrained to be grid neighbors. This need not be the case: The $\mathbf{x}$ can be very distant from each other without altering the basic technique. Furthermore, the functions s and q need not be local gradient and absolute curvature, respectively, but can be any functions defined on permissible $\mathbf{x}$. This general formulation of the problem for images was first described by [Fischler and

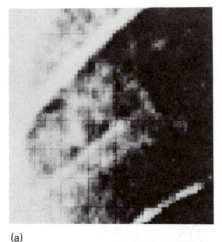

(a)

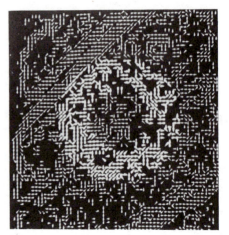

(b)

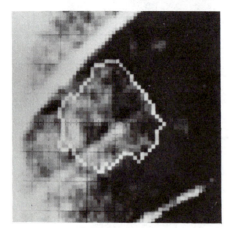

(c)

Fig. 4.16 Results of DP in boundary tracing. (a) Image containing tumor. (b) Contour cues. (c) Resultant boundary.

Elschlager 1973]. The Fischler and Elschlager formulation models an object as a set of parts and relations between parts, represented as a graph. Template functions, denoted by $g(\mathbf{x})$, measure how well a part of the model matches a part of the image at the point $\mathbf{x}$. (These local functions may be defined in any manner whatsoever.) "Relational functions," denoted by $q_{kj}(\mathbf{x}, \mathbf{y})$, measure how well the position of the match of the kth part at $(\mathbf{x})$ agrees with the position of the match of the jth part at $(\mathbf{y})$.

The basic notions are shown by a technique simplified from [Chien and Fu 1974] to find the boundaries of lungs in chest films. The lung boundaries are modeled with a polygonal approximation defined by the five key points. These points are the top of the lung, the two clavicle-lung junctions, and the two lower corners. To locate these points, local functions $g(\mathbf{x}_k)$ are defined which should be maximized when the corresponding point $\mathbf{x}_k$ is correctly determined. Similarly, $q(\mathbf{x}_k, \mathbf{x}_j)$ is a function relating points $\mathbf{x}_k$ and $\mathbf{x}_j$. In their case, Chien and Fu used the following functions:

$$T(\mathbf{x}) \equiv \text{template centered at } \mathbf{x} \text{ computed as}$$
$$\text{an aggregate of a set of chest radiographs}$$

$$g(\mathbf{x}_k) = \sum_{\mathbf{x}} \frac{T(\mathbf{x} - \mathbf{x}_k)f(\mathbf{x})}{|T||f|}$$

and

$$\theta(\mathbf{x}_k, \mathbf{x}_j) = \text{expected angular orientation of } \mathbf{x}_k \text{ from } \mathbf{x}_j$$

$$q(\mathbf{x}_k \mathbf{x}_j) = \left[\theta(\mathbf{x}_k, \mathbf{x}_j) - \arctan \frac{y_k - y_j}{x_k - x_j} \right]$$

With this formulation no further modifications are necessary and the solution may be obtained by solving Eqs. (4.19) through (4.21), as before. For purposes of comparison, this method was formalized using a lower-resolution objective function. Figure 4.17 shows Chien and Fu's results using this method with five template functions.

4.5.4 Theoretical Questions about Dynamic Programming

The Interaction Graph

This graph describes the interdependence of variables in the objective function. In the examples the interaction graph was simple: Each variable depended on only two others, resulting in the graph of Fig. 4.18a. A more complicated case is the one in 4.18b, which describes an objective function of the following form:

$$h( ) = h_1(x_1, x_2) + h_2(x_2, x_3, x_4) + h_3(x_3, x_4, x_5, x_6)$$

For these cases the dynamic programming technique still applies, but the computational effort increases exponentially with the number of interdependencies. For example, to eliminate x_2 in h_2, all possible combinations of x_3 and x_4 must be considered. To eliminate x_3 in h_3, all possible combinations of x_4, x_5, and x_6, and so forth.

Dynamic Programming versus Heuristic Search

It has been shown [Martelli 1976] that for finding a path in a graph between two points, which is an abstraction of the work we are doing here, heuristic search methods can be more efficient than dynamic programming methods. However, the point to remember about dynamic programming is that it efficiently builds paths from multiple starting points. If this is required by a particular task, then dynamic programming would be the method of choice, unless a very powerful heuristic were available.

4.6 CONTOUR FOLLOWING

If nothing is known about the boundary shape, but regions have been found in the image, the boundary is recovered by one of the simplest edge-following operations: "blob finding" in images. The ideas are easiest to present for binary images:

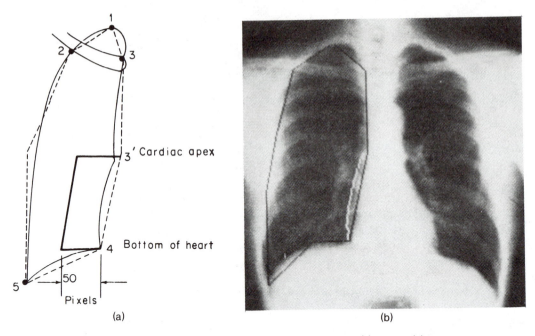

Fig. 4.17 Results of using local templates and global relations. (a) Model. (b) Results.

Given a binary image, the goal is find the boundaries of all distinct regions in the image.

This can be done simply by a procedure that functions like Papert's turtle [Papert 1973; Duda and Hart 1973]:

1. Scan the image until a region pixel is encountered.
2. If it is a region pixel, turn left and step; else, turn right and step.
3. Terminate upon return to the starting pixel.

Figure 4.19 shows the path traced out by the procedure. This procedure requires the region to be four-connected for a consistent boundary. Parts of an eight-connected region can be missed. Also, some bookkeeping is necessary to generate an exact sequence of boundary pixels without duplications.

A slightly more elaborate algorithm due to [Rosenfeld 1968] generates the boundary pixels exactly. It works by first finding a four-connected background pixel from a known boundary pixel. The next boundary pixel is the first pixel encountered when the eight neighbors are examined in a counter clockwise order from the background pixel. Many details have to be introduced into algorithms that follow contours of irregular eight-connected figures. A good exposition of these is given in [Rosenfeld and Kak 1976].

4.6.1 Extension to Gray-Level Images

The main idea behind contour following is to start with a point that is believed to be on the boundary and to keep extending the boundary by adding points in the contour directions. The details of these operations vary from task to task. The gen-

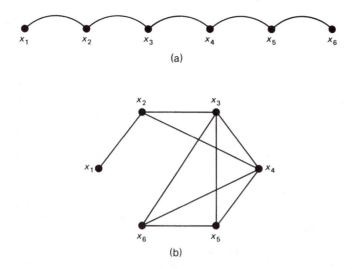

(a)

(b)

Fig. 4.18 Interaction graphs for DP (see text).

eralization of the contour follower to gray-level images uses local gradients with a magnitude $s(\mathbf{x})$ and direction $\phi(\mathbf{x})$ associated with each point $\mathbf{x}$. ϕ points in the direction of maximum change. If $\mathbf{x}$ is on the boundary of an image object, neighboring points on the boundary should be in the general direction of the contour directions, $\phi(\mathbf{x}) \pm \pi/2$, as shown by Fig. 4.20. A representative procedure is adapted from [Martelli 1976]:

1. Assume that an edge has been detected up to a point $\mathbf{x}_i$. Move to the point $\mathbf{x}_j$ adjacent to $\mathbf{x}_i$ in the direction perpendicular to the gradient of $\mathbf{x}_i$. Apply the gradient operator to $\mathbf{x}_j$; if its magnitude is greater than (some) threshold, this point is added to the edge.

2. Otherwise, compute the average gray level of the 3×3 array centered on $\mathbf{x}_j$, compare it with a suitably chosen threshold, and determine whether $\mathbf{x}_j$ is inside or outside the object.

3. Make another attempt with a point $\mathbf{x}_k$ adjacent to $\mathbf{x}_i$ in the direction perpendicular to the gradient at $\mathbf{x}_i$ plus or minus $(\pi/4)$, according to the outcome of the previous test.

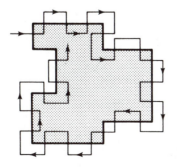

Fig. 4.19 Finding the boundary in a binary image.

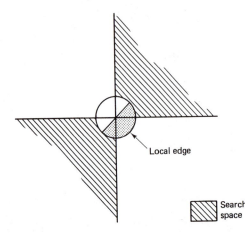

Local edge

☒ Search
space

Fig. 4.20 Angular orientations for contour following.

4.6.2 Generalization to Higher-Dimensional Image Data

The generalization of contour following to higher-dimensional spaces is straightforward [Liu 1977; Herman and Liu 1978]. The search involved is, in fact, slightly more complex than contour following and is more like the graph searching methods described in Section 4.4. Higher-dimensional image spaces arise when the image has more than two spatial dimensions, is time-varying, or both. In these images the notion of a gradient is the same (a vector describing the maximum gray-level change and its corresponding direction), but the intuitive interpretation of the corresponding edge element may be difficult. In three dimensions, edge elements are primitive surface elements, separating volumes of differing gray level. The objective of contour following is to link together neighboring surface elements with high gradient modulus values and similar orientations into larger boundaries. In four dimensions, ''edge elements'' are primitive volumes; contour following links neighboring volumes with similar gradients.

The contour following approach works well when there is little noise present and no ''spurious'' boundaries. Unfortunately, if either of these conditions is present, the contour-following algorithms are generally unsatisfactory; they are easily thwarted by gaps in the data produced by noise, and readily follow spurious boundaries. The methods described earlier in this chapter attempt to overcome these difficulties through more elaborate models of the boundary structure.

EXERCISES

4.1 Specify a heuristic search algorithm that will work with ''crack'' edges such as those in Fig. 3.12.

4.2 Describe a modification of Algorithm 4.2 to detect parabolae in gray-level images.

4.3 Suppose that a relation $h(x_1, x_6)$ is added to the model described by Fig. 4.18a so that now the interaction graph is cyclical. Show formally how this changes the optimization steps described by Eqs. (4.11) through (4.13).

4.4 Show formally that the Hough technique without gradient direction information is equivalent to template matching (Chapter 3).

4.5 Extend the Hough technique for ellipses described by Eqs. (4.7a) through (4.7d) to ellipses oriented at an arbitrary angle θ to the x axis.

4.6 Show how to use the generalized Hough technique to detect hexagons.

REFERENCES

ASHKAR, G. P. and J. W. MODESTINO. "The contour extraction problem with biomedical applications." *CGIP 7*, 1978, 331–355.

BALLARD, D. H. *Hierarchic detection of tumors in chest radiographs.* Basel: Birkhäuser-Verlag (ISR-16), January 1976.

BALLARD, D. H. "Generalizing the Hough transform to detect arbitrary shapes." *Pattern Recognition 13*, 2, 1981, 111–122.

BALLARD, D. H. and J. SKLANSKY. "A ladder-structured decision tree for recognizing tumors in chest radiographs." *IEEE Trans. Computers 25*, 1976, 503-513.

BALLARD, D. H., M. MARINUCCI, F. PROIETTI-ORLANDI, A. ROSSI-MARI, and L. TENTARI. "Automatic analysis of human haemoglobin fingerprints." *Proc.*, 3rd Meeting, International Society of Haemotology, London, August 1975.

BARROW, H. G. "Interactive aids for cartography and photo interpretation." Semi-Annual Technical Report, AI Center, SRI International, December 1976.

BELLMAN, R. and S. DREYFUS. *Applied Dynamic Programming.* Princeton, NJ: Princeton University Press, 1962.

BOLLES, R. "Verification vision for programmable assembly." *Proc.*, 5th IJCAI, August 1977, 569-575.

CHIEN, Y. P. and K. S. FU. "A decision function method for boundary detection." *CGIP 3*, 2, June 1974, 125-140.

DUDA, R. O. and P. E. HART. "Use of the Hough transformation to detect lines and curves in pictures." *Commun. ACM 15*, 1, January 1972, 11–15.

DUDA, R. O. and P. E. HART. *Pattern Recognition and Scene Analysis.* New York: Wiley, 1973.

FISCHLER, M. A. and R. A. ELSCHLAGER. "The representation and matching of pictoral patterns." *IEEE Trans. Computers 22*, January 1973.

HERMAN, G. T. and H. K. LIU. "Dynamic boundary surface detection." *CGIP 7*, 1978, 130-138.

HOUGH, P. V. C. "Method and means for recognizing complex patterns." U.S. Patent 3,069,654; 1962.

KELLY, M.D. "Edge detection by computer using planning." In *MI6*, 1971.

KIMME, C., D. BALLARD, and J. SKLANSKY. "Finding circles by an array of accumulators." *Commun. ACM 18*, 2, 1975, 120-122.

LANTZ, K. A., C. M. BROWN and D. H. BALLARD. "Model-driven vision using procedure decription: motivation and application to photointerpretation and medical diagnosis." *Proc.*, 22nd International Symp., Society of Photo-optical Instrumentation Engineers, San Diego, CA, August 1978.

LESTER, J. M., H. A. WILLIAMS, B. A. WEINTRAUB, and J. F. BRENNER, "Two graph searching techniques for boundary finding in white blood cell images." *Computers in Biology and Medicine 8*, 1978, 293-308.

LIU, H. K. "Two- and three-dimensional boundary detection." *CGIP 6*, 2, April 1977, 123-134.

MARR, D. "Analyzing natural images; a computational theory of texture vision." Technical Report 334, AI Lab, MIT, June 1975.

MARTELLI, A. "Edge detection using heuristic search methods." *CGIP 1*, 2, August 1972, 169-182.

MARTELLI, A. "An application of heuristic search methods to edge and contour detection." *Commun. ACM 19*, 2, February 1976, 73–83.

MONTANARI, U. "On the optimal detection of curves in noisy pictures." *Commun. ACM 14*, 5, May 1971, 335-345.

NILSSON, N. J. *Problem-Solving Methods in Artificial Intelligence.* New York: McGraw-Hill, 1971.

NILSSON, N. J. *Principles of Artificial Intelligence.* Palo Alto, CA: Tioga, 1980.

PAPERT, S. "Uses of technology to enhance education." Technical Report 298, AI Lab, MIT, 1973.

PERSOON, E. "A new edge detection algorithm and its applications in picture processing." *CGIP 5*, 4, December 1976, 425-446.

RAMER, U. "Extraction of line structures from photographs of curved objects." *CGIP 4*, 2, June 1975, 81-103.

ROSENFELD, A. *Picture Processing by Computer.* New York: Academic Press, 1968.

ROSENFELD, A. and A. C. KAK. *Digital Picture Processing.* New York: Academic Press, 1976.

SELFRIDGE, P. G., J. M. S. PREWITT, C. R. DYER, and S. RANADE. "Segmentation algorithms for abdominal computerized tomography scans." *Proc.*, 3rd COMPSAC, November 1979, 571-577.

WECHSLER, H. and J. SKLANSKY. "Finding the rib cage in chest radiographs." *Pattern Recognition 9*, 1977, 21-30.

Region
Growing   5

5.1 REGIONS

Chapter 4 concentrated on the linear features (discontinuities of image gray level) that often correspond to object boundaries, interesting surface detail, and so on. The "dual" problem to finding edges around regions of differing gray level is to find the regions themselves. The goal of region growing is to use image characteristics to map individual pixels in an input image to sets of pixels called *regions*. An image region might correspond to a world object or a meaningful part of one.

Of course, very simple procedures will derive a boundary from a connected region of pixels, and conversely can fill a boundary to obtain a region. There are several reasons why both region growing and line finding survive as basic segmentation techniques despite their redundant-seeming nature. Although perfect regions and boundaries are interconvertible, the processing to find them initially differs in character and applicability; besides, perfect edges or regions are not always required for an application. Region-finding and line-finding techniques can cooperate to produce a more reliable segmentation.

The geometric characteristics of regions depend on the domain. Usually, they are considered to be connected two-dimensional areas. Whether regions can be disconnected, non-simply connected (have holes), should have smooth boundaries, and so forth depends on the region-growing technique and the goals of the work. Ultimately, it is often the segmentation goal to partition the entire image into quasi-disjoint regions. That is, regions have no two-dimensional overlaps, and no pixel belongs to the interior of more than one region. However, there is no single definition of region—they may be allowed to overlap, the whole image may not be partitioned, and so forth.

Our discussion of region growers will begin with the most simple kinds and progress to the more complex. The most primitive region growers use only aggregates of properties of local groups of pixels to determine regions. More sophisticated techniques "grow" regions by *merging* more primitive regions. To do this in a structured way requires sophisticated representations of the regions and boundaries. Also, the merging *decisions* can be complex, and can depend on descriptions of the boundary structure separating regions in addition to the region semantics. A good survey of early techniques is [Zucker 1976].

The techniques we consider are:

1. *Local techniques.* Pixels are placed in a region on the basis of their properties or the properties of their close neighbors.

2. *Global techniques.* Pixels are grouped into regions on the basis of the properties of large numbers of pixels distributed throughout the image.

3. *Splitting and merging techniques.* The foregoing techniques are related to individual pixels or sets of pixels. State space techniques merge or split regions using graph structures to represent the regions and boundaries. Both local and global merging and splitting criteria can be used.

The effectiveness of region growing algorithms depends heavily on the application area and input image. If the image is sufficiently simple, say a dark blob on a light background, simple local techniques can be surprisingly effective. However, on very difficult scenes, such as outdoor scenes, even the most sophisticated techniques still may not produce a satisfactory segmentation. In this event, region growing is sometimes used conservatively to preprocess the image for more knowledgeable processes [Hanson and Riseman 1978].

In discussing the specific algorithms, the following definitions will be helpful. Regions R_k are considered to be sets of points with the following properties:

$$\mathbf{x}_i \text{ in a region } R \text{ is } connected \text{ to } \mathbf{x}_j \text{ iff there}$$
$$\text{is a sequence } \{\mathbf{x}_i, \ldots, \mathbf{x}_j\} \text{ such that } \mathbf{x}_k \text{ and } \mathbf{x}_{k+1} \tag{5.1}$$
$$\text{are connected and all the points are in } R.$$

$$R \text{ is a } connected \ region \text{ if the set of points } \mathbf{x} \text{ in } R \text{ has the} \tag{5.2}$$
$$\text{property that every pair of points is connected.}$$

$$I, \text{ the entire image} = \bigcup_{k=1}^{m} R_k \tag{5.3}$$

$$R_i \cap R_j = \phi, \qquad i \neq j \tag{5.4}$$

A set of regions satisfying (5.2) through (5.4) is known as a *partition*. In segmentation algorithms, each region often is a unique, homogeneous area. That is, for some Boolean function $H(R)$ that measures region homogeneity,

$$H(R_k) = \text{true for all } k \tag{5.5}$$

$$H(R_i \cup R_j) = \text{false for } i \neq j \tag{5.6}$$

Note that R_i does not have to be connected. A weaker but still useful criterion is that neighboring regions not be homogeneous.

5.2 A LOCAL TECHNIQUE: BLOB COLORING

The counterpart to the edge tracker for binary images is the blob-coloring algorithm. Given a binary image containing four-connected blobs of 1's on a background of 0's, the objective is to "color each blob"; that is, assign each blob a different label. To do this, scan the image from left to right and top to bottom with a special L-shaped template shown in Fig. 5.1. The coloring algorithm is as follows.

Algorithm 5.1: Blob Coloring

Let the initial color, $k = 1$. Scan the image from left to right and top to bottom.

If $f(\mathbf{x}_C) = 0$ then continue
else
    begin

        if $(f(\mathbf{x}_U) = 1$ and $f(\mathbf{x}_L) = 0)$
        then color $(\mathbf{x}_C) :=$ color $(\mathbf{x}_U)$

        if $(f(\mathbf{x}_L) = 1$ and $f(\mathbf{x}_U) = 0)$
        then color $(\mathbf{x}_C) :=$ color $(\mathbf{x}_L)$

        if $(f(\mathbf{x}_L) = 1$ and $f(\mathbf{x}_U) = 1)$
        then begin
            color $(\mathbf{x}_C) :=$ color $(\mathbf{x}_L)$
            color $(\mathbf{x}_L)$ is equivalent to color $(\mathbf{x}_U)$
            end

    comment: two colors are equivalent.

        if $(f(\mathbf{x}_L) = 0$ and $f(\mathbf{x}_U) = 0)$
        then color $(\mathbf{x}_C) := k; k := k + 1$

    comment: new color

        end

After one complete scan of the image the color equivalences can be used to assure that each object has only one color. This binary image algorithm can be used as a simple region-grower for gray-level images with the following modifications. If in a

Fig. 5.1 L-shaped template for blob coloring.

gray-level image $f(\mathbf{x}_C)$ is approximately equal to $f(\mathbf{x}_U)$, assign $\mathbf{x}_C$ to the same region (blob) as $\mathbf{x}_U$. This is equivalent to the condition $f(\mathbf{x}_C) = f(\mathbf{x}_U) = 1$ in Algorithm 5.1. The modifications to the steps in the algorithm are straightforward.

5.3 GLOBAL TECHNIQUES: REGION GROWING VIA THRESHOLDING

This approach assumes an object-background image and picks a threshold that divides the image pixels into either object or background:

$\mathbf{x}$ is part of the Object iff $f(\mathbf{x}) > T$
Otherwise it is part of the Background

The best way to pick the threshold T is to search the histogram of gray levels, assuming it is bimodal, and find the minimum separating the two peaks, as in Fig. 5.2. Finding the right valley between the peaks of a histogram can be difficult when the histogram is not a smooth function. Smoothing the histogram can help but does not guarantee that the correct minimum can be found. An elegant method for treating bimodal images assumes that the histogram is the sum of two composite normal functions and determines the valley location from the normal parameters [Chow and Kaneko 1972].

The single-threshold method is useful in simple situations, but primitive. For example, the region pixels may not be connected, and further processing such as that described in Chapter 2 may be necessary to smooth region boundaries and remove noise. A common problem with this technique occurs when the image has a background of varying gray level, or when collections we would like to call regions vary smoothly in gray level by more than the threshold. Two modifications of the threshold approach to ameliorate the difficulty are: (1) high-pass filter the image to deemphasize the low-frequency background variation and then try the original technique; and (2) use a spatially varying threshold method such as that of [Chow and Kaneko 1972].

The Chow-Kaneko technique divides the image up into rectangular subimages and computes a threshold for each subimage. A subimage can fail to have a threshold if its gray-level histogram is not bimodal. Such subimages receive inter-

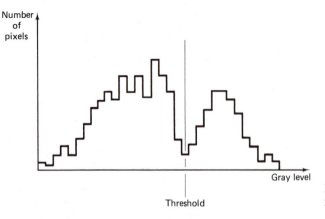

Fig. 5.2 Threshold determination from gray-level histogram.

polated thresholds from neighboring subimages that are bimodal, and finally the entire picture is thresholded by using the separate thresholds for each subimage.

5.3.1 Thresholding in Multidimensional Space

An interesting variation to the basic thresholding paradigm uses color images; the basic digital picture function is vector-valued with red, blue, and green components. This vector is augmented with possibly nonlinear combinations of these values so that the augmented picture vector has a number of components. The idea is to re-represent the color solid redundantly and hope to find color parameters for which thresholding does the desired segmentation. One implementation of this idea used the red, green, and blue color components; the intensity, saturation, and hue components; and the N.T.S.C. Y, I, Q components (Chapter 2) [Ohlander et al. 1979].

The idea of thresholding the components of a picture vector is used in a primitive form for multispectral LANDSAT imagery [Robertson et al. 1973]. The novel extension in this algorithm is the recursive application of this technique to nonrectangular subregions.

The region partitioning is then as follows:

Algorithm 5.2:   Region Growing via Recursive Splitting

1. Consider the entire image as a region and compute histograms for each of the picture vector components.

2. Apply a peak-finding test to each histogram. If at least one component passes the test, pick the component with the most significant peak and determine two thresholds, one either side of the peak (Fig. 5.3). Use these thresholds to divide the region into subregions.

3. Each subregion may have a "noisy" boundary, so the binary representation of the image achieved by thresholding is smoothed so that only a single connected subregion remains. For binary smoothing see ch. 8 and [Rosenfeld and Kak 1976].

4. Repeat steps 1 through 3 for each subregion until no new subregions are created (no histograms have significant peaks).

A refinement of step 2 of this scheme is to create histograms in higher-dimensional space [Hanson and Riseman 1978]. Multiple regions are often in the same histogram peak when a single measurement is used. The advantage of the multimeasurement histograms is that these different regions are often separated into individual peaks, and hence the segmentation is improved. Figure 5.4 shows some results using a three-dimensional RGB color space.

The figure shows the clear separation of peaks in the three-dimensional histogram that is not evident in either of the one-dimensional histograms. How many

(a)

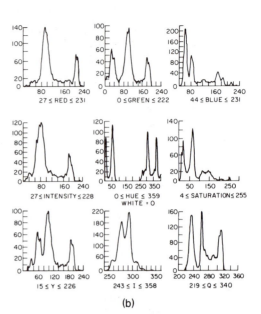

(b)

(c)

Fig. 5.3 Peak detection and threshold determination. (a) Original image. (b) Histograms. (c) Image segments resulting from first histogram peak.

(d)

Fig. 5.3 (d) Final segments.

dimensions should be used? Obviously, there is a trade-off here: As the dimensionality becomes larger, the discrimination improves, but the histograms are more expensive to compute and noise effects may be more pronounced.

5.3.2 Hierarchical Refinement

This technique uses a pyramidal image representation (Section 3.7) [Harlow and Eisenbeis 1973]. Region growing is applied to a coarse resolution image. When the algorithm has terminated at one resolution level, the pixels near the boundaries of regions are disassociated with their regions. The region-growing process is then repeated for just these pixels at a higher-resolution level. Figure 5.5 shows this structure.

5.4 SPLITTING AND MERGING

Given a set of regions R_k, $k = 1, \ldots, m$, a low-level segmentation might require the basic properties described in Section 5.1 to hold. The important properties from the standpoint of segmentation are Eqs. (5.5) and (5.6).

If Eq. (5.5) is not satisfied for some k, it means that that region is inhomogeneous and should be split into subregions. If Eq. (5.6) is not satisfied for some i and j, then regions i and j are collectively homogeneous and should be merged into a single region.

In our previous discussions we used

$$H(R) = \begin{cases} \text{true} & \text{if all neighboring pairs of points} \\ & \text{in } R \text{ are such that } f(\mathbf{x}) - f(\mathbf{y}) < T \\ \text{false} & \text{otherwise} \end{cases} \qquad (5.7)$$

and

$$H(R) = \begin{cases} \text{true} & \text{if the points in } R \text{ pass a} \\ & \text{bimodality or peak test} \\ \text{false} & \text{otherwise} \end{cases} \qquad (5.8)$$

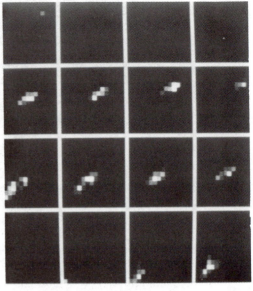

(a)

(b)

Fig. 5.4 Multi-dimensional histograms in segmentation. (a) Image. (b) RGB histogram showing successive planes through a $16 \times 16 \times 16$ color space. (c) Segments. (See color inserts.)

(c)

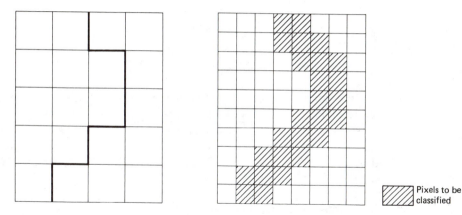

Fig. 5.5 Hierarchical region refinement.

A way of working toward the satisfaction of these homogeneity criteria is the split-and-merge algorithm [Horowitz and Pavlidis 1974]. To use the algorithm it is necessary to organize the image pixels into a pyramidal grid structure of regions. In this grid structure, regions are organized into groups of four. Any region can be split into four subregions (except a region consisting of only one pixel), and the appropriate groups of four can be merged into a single larger region. This structure is incorporated into the following region-growing algorithm.

Algorithm 5.3: Region Growing via Split and Merge [Horowitz and Pavlidis 1974]

1. Pick any grid structure, and homogeneity property H. If for any region R in that structure, $H(R)$ = false, split that region into four subregions. If for any four appropriate regions R_{k1} ,..., R_{k4}, $H(R_{k1} \bigcup R_{k2} \bigcup R_{k3} \bigcup R_{k4})$ = true, merge them into a single region. When no regions can be further split or merged, stop.

2. If there are any neighboring regions R_i and R_j (perhaps of different sizes) such that $H(R_i \bigcup R_j)$ = true, merge these regions.

5.4.1 State-Space Approach to Region Growing

The "classical" state-space approach of artificial intelligence [Nilsson 1971, 1980] was first applied to region growing in [Brice and Fennema 1970] and significantly extended in [Feldman and Yakimovsky 1974]. This approach regards the initial two-dimensional image as a discrete state, where every sample point is a separate region. Changes of state occur when a boundary between regions is either removed or inserted. The problem then becomes one of searching allowable changes in state to find the best partition.

```
·  +  ·  +  ·  +  ·  +  ·
+  O  +  O  +  O  +  O  +
·  +  ·  +  ·  +  ·  +  ·
+  O  +  O  +  O  +  O  +
·  +  ·  +  ·  +  ·  +  ·
+  O  +  O  +  O  +  O  +
```

· Unassigned
+ Edge data
O Grey level data

Fig. 5.6 Grid structure for region representation [Brice and Fennema 1970].

An important part of the state-space approach is the use of data structures to allow regions and boundaries to be manipulated as units. This moves away from earlier techniques, which labeled each individual pixel according to its region. The high-level data structures do away with this expensive practice by representing regions with their boundaries and then keeping track of what happens to these boundaries during split-and merge-operations.

5.4.2 Low-level Boundary Data Structures

A useful representation for boundaries allows the splitting and merging of regions to proceed in a simple manner [Brice and Fennema 1970]. This representation introduces the notion of a supergrid S to the image grid G. These grids are shown in Fig. 5.6, where · and + correspond to supergrid and O to the subgrid. The representation is assumed to be four-connected (i.e., $\mathbf{x}1$ is a neighbor of $\mathbf{x}2$ if $\|\mathbf{x}1 - \mathbf{x}2\| \leqslant 1$).

With this notation boundaries of regions are directed crack edges (see Sec. 3.1) at the points marked $+$. That is, if point $\mathbf{x}_k$ is a neighbor of $\mathbf{x}_j$ and $\mathbf{x}_k$ is in a different region than $\mathbf{x}_j$, insert two edges for the boundaries of the regions containing $\mathbf{x}_j$ and $\mathbf{x}_k$ at the point $+$ separating them, such that each edge traverses its associated region in a counterclockwise sense. This makes merge operations very simple: To merge regions R_k and R_l, remove edges of the opposite sense from the boundary as shown in Fig. 5.7a. Similarly, to split a region along a line, insert edges of the opposite sense in nearby points, as shown in Fig. 5.7b.

The method of [Brice and Fennema 1970] uses three criteria for merging regions, reflecting a transition from local measurements to global measurements. These criteria use measures of boundary strength s_{ij} and w_{ij} defined as

$$s_{ij} = |f(\mathbf{x}_i) - f(\mathbf{x}_j)| \tag{5.9}$$

$$w_{ij} = \begin{cases} 1 & \text{if } s_{ij} < T_1 \\ 0 & \text{otherwise} \end{cases} \tag{5.10}$$

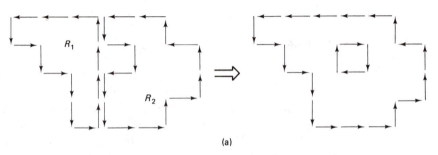

(a)

Fig. 5.7 Region operations on the grid structure of Fig. 5.6.

Ch. 5 Region Growing

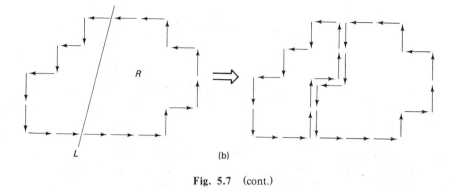

Fig. 5.7 (cont.)

where $\mathbf{x}_i$ and $\mathbf{x}_j$ are assumed to be on either side of a crack edge (Chapter 3). The three criteria are applied sequentially in the following algorithm:

Algorithm 5.4: Region Growing via Boundary Melting (T_k, $k = 1, 2, 3$ are preset thresholds)

1. For all neighboring pairs of points, remove the boundary between $\mathbf{x}_i$ and $\mathbf{x}_j$ if $i \neq j$ and $w_{ij} = 1$. When no more boundaries can be removed, go to step 2.

2. Remove the boundary between R_i and R_j if

$$\frac{W}{\min [p_i, p_j]} \geqslant T_2 \qquad (5.11)$$

where W is the sum of the w_{ij} on the common boundary between R_i and R_j, that have perimeters p_i and p_j respectively. When no more boundaries can be removed, go to step 3.

3. Remove the boundary between R_i and R_j if

$$W \geqslant T_3 \qquad (5.12)$$

5.4.3 Graph-Oriented Region Structures

The Brice–Fennema data structure stores boundaries explicitly but does not provide for explicit representation of regions. This is a drawback when regions must be referred to as units. An adjunct scheme of region representation can be developed using graph theory. This scheme represents both regions and their boundaries explicitly, and this facilitates the storing and indexing of their semantic properties.

The scheme is based on a special graph called the *region adjacency graph*, and its "dual graph." In the region adjacency graph, nodes are regions and arcs exist between neighboring regions. This scheme is useful as a way of keeping track of regions, even when they are inscribed on arbitrary nonplanar surfaces (Chapter 9).

Consider the regions of an image shown in Fig. 5.8a. The region adjacency graph has a node in each region and an arc crossing each separate boundary segment. To allow a uniform treatment of these structures, define an artificial region that surrounds the image. This node is shown in Fig. 5.8b. For regions on a plane, the region adjacency graph is *planar* (can lie in a plane with no arcs intersecting) and its edges are undirected. The "dual" of this graph is also of interest. To constuct the dual of the adjacency graph, simply place nodes in each separate region and connect them with arcs wherever the regions are separated by an arc in the adjacency graph. Figure 5.8c shows that the dual of the region adjacency graph is like the original region boundary map; in Fig. 5.8b each arc may be associated with a specific boundary segment and each node with a junction between three or more boundary segments. By maintaining both the region adjacency graph and its dual, one can merge regions using the following algorithm:

Algorithm 5.5:   Merging Using the Region-Adjacency Graph and Its Dual

Task: Merge neighboring regions R_i and R_j.

Phase 1. Update the region-adjacency graph.

1.   Place edges between R_i and all neighboring regions of R_j (excluding, of course, R_i) that do not already have edges between themselves and R_i.

2.   Delete R_j and all its associated edges.

Phase 2. Take care of the dual.

1.   Delete the edges in the dual corresponding to the borders between R_i and R_j.

2.   For each of the nodes associated with these edges:

    (a)   if the resultant degree of the node is less than or equal to 2, delete the node and join the two dangling edges into a single edge.

    (b)   otherwise, update the labels of the edges that were associated with j to reflect the new region label i.

Figure 5.9 shows these operations.

5.5 INCORPORATION OF SEMANTICS

Up to this point in our treatment of region growers, domain-dependent "semantics" has not explicitly appeared. In other words, region-merging decisions were based on raw image data and rather weak heuristics of general applicability about the likely shape of boundaries. As in early processing, the use of domain-dependent knowledge can affect region finding. Possible interpretations of regions can affect the splitting and merging process. For example, in an outdoor scene possible region interpretations might be sky, grass, or car. This kind of knowledge is quite separate from but related to measurable region properties such as intensity

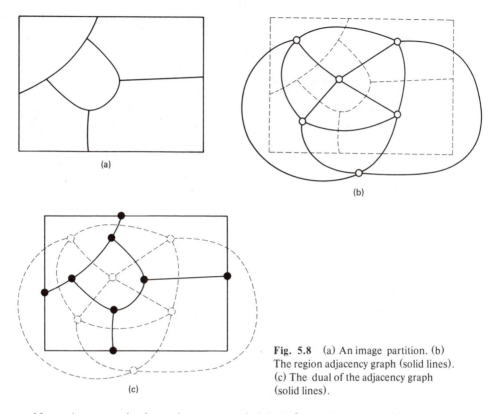

(a)

(b)

(c)

Fig. 5.8 (a) An image partition. (b) The region adjacency graph (solid lines). (c) The dual of the adjacency graph (solid lines).

and hue. An example shows how semantic labels for regions can guide the merging process. This approach was originally developed in [Feldman and Yakimovsky 1974]. it has found application in several complex vision systems [Barrow and Tenenbaum 1977; Hanson and Riseman 1978].

Early steps in the Feldman–Yakimovsky region grower used essentially the same steps as Brice–Fennema. Once regions attain significant size, semantic cri-

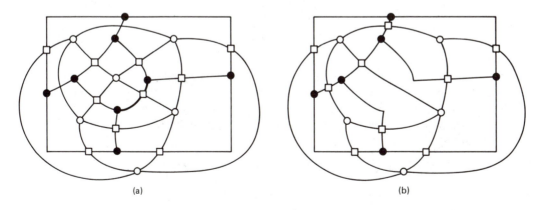

(a)

(b)

Fig. 5.9 Merging operations using the region adjacency graph and its dual. (a) Before merging regions separated by dark boundary line. (b) After merging.

teria are used. The region growing consists of four steps, as summed up in the following algorithm:

Algorithm 5.6 Semantic Region Growing

Nonsemantic Criteria
T_1 and T_2 are preset thresholds

1. Merge regions i, j as long as they have one weak separating edge until no two regions pass this test.
2. Merge regions i, j where $S(i, j) \leqslant T_2$ where

$$S(i, j) = \frac{c_1 + \alpha_{ij}}{c_2 + \alpha_{ij}}$$

where c_1 and c_2 are constants,

$$\alpha ij = \frac{(\text{area}_i)^{\frac{1}{2}} + (\text{area}_j)^{\frac{1}{2}}}{\text{perimeter}_i \cdot \text{perimeter}_j}$$

until no two regions pass this test. (This is a similar criterion to Algorithm 5.4, step 2.)

Semantic Criteria

3. Let B_{ij} be the boundary between R_i and R_j. Evaluate each B_{ij} with a Bayesian decision function that measures the (conditional) probability that B_{ij} separates two regions R_i and R_j of the *same interpretation*. Merge R_i and R_j if this conditional probability is less than some threshold. Repeat step 3 until no regions pass the threshold test.
4. Evaluate the interpretation of each region R_i with a Bayesian decision function that measures the (conditional) probability that an interpretation is the correct one for that region. Assign the interpretation to the region with the highest confidence of correct interpretation. Update the conditional probabilities for different interpretations of neighbors. Repeat the entire process until all regions have interpretation assignments.

The semantic portion of algorithm 5.6 had the goal of maximizing an evaluation function measuring the probability of a correct interpretation (labeled partition), given the measurements on the boundaries and regions of the partition. An expression for the evaluation function is (for a given partition and interpretations X and Y):

$$\max_{X, Y} \prod_{i, j} \{P[B_{ij} \text{ is a boundary between } X \text{ and } Y | \text{measurements on } B_{ij}]\}$$

$$\times \prod_i \{P[R_i \text{ is an } X | \text{measurements on } R_i]\}$$

$$\times \prod_j \{P[R_j \text{ is an } Y | \text{measurements on } R_j]\}$$

where P stands for probability and Π is the product operator.

How are these terms to be computed? Ideally, each conditional probability function should be known to a reasonable degree of accuracy; then the terms can be obtained by lookup.

However, the straightforward computation and representation of the conditional probability functions requires a massive amount of work and storage. An approximation used in [Feldman and Yakimovsky 1974] is to quantize the measurements and represent them in terms of a classification tree. The conditional probabilities can then be computed from data at the leaves of the tree. Figure 5.10 shows a hypothetical tree for the region measurements of intensity and hue, and interpretations ROAD, SKY, and CAR. Figure 5.11 shows the equivalent tree for two boundary measurements m and n and the same interpretations. These two figures indicate that $P[R_i$ is a CAR$|0 \leqslant i < I, 0 \leqslant h < H_1] = \frac{1}{6}$, and $P[B_{ij}$ divides two car regions $| M_k \leqslant m < M_{k+1}, N_l < n \leqslant N_{l+1}] = \frac{2}{13}$. These trees were created by laborious trials with correct segmentations of test images.

Now, finally, consider again step 3 of Algorithm 5.6. The probability that a boundary B_{ij} between regions R_i and R_j is false is given by

$$P_{\text{false}} = \frac{P_f}{P_t + P_f} \tag{5.13}$$

where

$$P_f = \sum \{P[B_{ij} \text{ is between two subregions } X \mid B_{ij}\text{'s measurements}]\} \tag{5.14a}$$

$$\times \{P[R_i \text{ is } X \mid \text{meas}]\} \times \{P[R_j \text{ is } X \mid \text{meas}]\}$$

$$P_t = \sum_{x,y} \{P[B_{ij} \text{ is between } X \text{ and } Y \mid \text{meas}]\} \tag{5.14b}$$

$$\times \{P[R_i \text{ is } X \mid \text{meas }]\} \times \{P[R_j \text{ is } Y \mid \text{meas}]\}$$

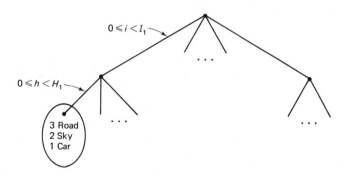

Fig. 5.10 Hypothetical classification tree for region measurements showing a particular branch for specific ranges of intensity and hue.

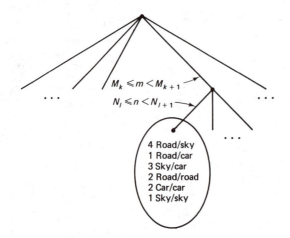

$$M_k \leqslant m < M_{k+1}$$
$$N_l \leqslant n < N_{l+1}$$

4 Road/sky
1 Road/car
3 Sky/car
2 Road/road
2 Car/car
1 Sky/sky

Fig. 5.11 Hypothetical classification tree for boundary measurements showing a specific branch for specific ranges of two measurements m and n.

And for step 4 of the algorithm,

$$\text{Confidence}_i = \frac{P[R_i \text{ is } X1 \mid \text{meas}]}{P[R_i \text{ is } X2 \mid \text{meas}]} \tag{5.15}$$

where $X1$, $X2$ are the first and second most likely interpretations, respectively. After the region is assigned interpretation $X1$, the neighbors are updated using

$$P[R_i \text{ is } X \mid \text{meas}] := Prob [Rj \text{ is } X \mid \text{meas}] \tag{5.16}$$
$$\times P[B_{ij} \text{ is between } X \text{ and } X1 \mid \text{meas}]$$

EXERCISES

5.1 In Algorithm 5.1, show how one can handle the case where colors are equivalent. Do you need more than one pass over the image?

5.2 Show for the heuristic of Eq. (5.11) that

(a) $I/T_2 \geqslant W/T_2 > P_m = P_j$

(b) $P_m < P_i + I(1/T_2 - 2)$

where P_m is the perimeter of $R_i \bigcup R_j$, I is the perimeter common to both i and j and $P_m = \min (P_i, P_j)$. What does part (b) imply about the relation between T_2 and P_m?

5.3 Write a "histogram-peak" finder; that is, detect satisfying valleys in histograms separating intuitive hills or peaks.

5.4 Suppose that regions are represented by a neighbor list structure. Each region has an associated list of neighboring regions. Design a region-merging algorithm based on this structure.

5.5 Why do junctions of regions in segmented images tend to be trihedral?

5.6 Regions, boundaries, and junctions are the structures behind the region-adjacency graph and its dual. Generalize these structures to three dimensions. Is another structure needed?

5.7 Generalize the graph of Figure 5.8 to three dimensions and develop the merging algorithm analogous to Algorithm 5.5. (Hint: see Exercise 5.6.)

REFERENCES

BARROW, H. G. and J. M. TENENBAUM. "Experiments in model-driven scene segmentation." *Artificial Intelligence 8*, 3, June 1977, 241–274.

BRICE, C. and C. FENNEMA. "Scene analysis using regions." *Artificial Intelligence 1*, 3, Fall 1970, 205–226.

CHOW, C. K. and T. KANEKO. "Automatic boundary detection of the left ventricle from cineangiograms." *Computers and Biomedical Research 5*, 4, August 1972, 388–410.

FELDMAN, J. A. and Y. YAKIMOVSKY. "Decision theory and artificial intelligence: I. A semantics-based region analyzer." *Artificial Intelligence 5*, 4, 1974, 349–371.

HANSON, A. R. and E. M. RISEMAN. "Segmentation of natural scenes." In *CVS*, 1978.

HARLOW, C. A. and S. A. EISENBEIS. "The analysis of radiographic images." *IEEE Trans. Computers 22*, 1973, 678–688.

HOROWITZ, S. L. and T. PAVLIDIS. "Picture segmentation by a directed split-and-merge procedure." *Proc.*, 2nd IJCPR, August 1974, 424–433.

NILSSON, N. J. *Principles of Artificial Intelligence.* Palo Alto, CA: Tioga, 1980.

NILSSON, N. J. *Problem-Solving Methods in Artificial Intelligence.* New York: McGraw-Hill, 1971.

OHLANDER, R., K. PRICE, and D. R. REDDY. "Picture segmentation using a recursive region splitting method." *CGIP 8*, 3, December 1979.

ROBERTSON, T. V., P. H. SWAIN, and K. S. FU. "Multispectral image partitioning." TR-EE 73-26 (LARS Information Note 071373), School of Electrical Engineering, Purdue Univ., August 1973.

ROSENFELD, A. and A. C. KAK. *Digital Picture Processing.* New York: Academic Press, 1976.

ZUCKER, S. W. "Region growing: Childhood and adolescence." *CGIP 5*, 3, September 1976, 382–399.

Texture   6

The notion of texture admits to no rigid description, but a dictionary definition of texture as "something composed of closely interwoven elements" is fairly apt. The description of interwoven elements is intimately tied to the idea of texture resolution, which one might think of as the average amount of pixels for each discernable texture element. If this number is large, we can attempt to describe the individual elements in some detail. However, as this number nears unity it becomes increasingly difficult to characterize these elements individually and they merge into less distinct spatial patterns. To see this variability, we examine some textures.

Figure 6.1 shows "cane," "paper," "coffee beans," "brickwall," "coins," and "wire braid" after Brodatz's well-known book [Brodatz 1966]. Five of these examples are high-resolution textures: they show repeated primitive elements that exhibit some kind of variation. "Coffee beans," "brick wall" and "coins" all have obvious primitives (even if it is not so obvious how to extract these from image data). Two more examples further illustrate that one sometimes has to be creative in defining primitives. In "cane" the easiest primitives to deal with seem to be the physical holes in the texture, whereas in "wire braid" it might be better to model the physical relations of a loose weave of metallic wires. However, the paper texture does not fit nicely into this mold. This is not to say that there are not possibilities for primitive elements. One is regions of lightness and darkness formed by the ridges in the paper. A second possibility is to use the reflectance models described in Section 3.5 to compute "pits" and "bumps." However, the elements seem to be "just beyond our perceptual resolving power" [Laws 1980], or in our terms, the elements are very close in size to individual pixels.

Fig. 6.1 Six examples of texture. (a) Cane. (b) Paper. (c) Coffee beans. (d) Brick wall. (e) Coins. (f) Wire braid.

The exposition of texture takes place under four main headings:

1. Texture primitives
2. Structural models
3. Statistical models
4. Texture gradients

We have already described texture as being composed of elements of *texture primitives*. The main point of additional discussion on texture primitives is to refine the idea of a primitive and its relation to image resolution.

The main work that is unique to texture is that which describes how primitives are related to the aim of recognizing or classifying the texture. Two broad classes of techniques have emerged and we shall study each in turn. The *structural* model regards the primitives as forming a repeating pattern and describes such patterns in terms of rules for generating them. Formally, these rules can be termed a grammar. This model is best for describing textures where there is much regularity in the placement of primitive elements and the texture is imaged at high resolution. The "reptile" texture in Fig. 6.9 is an example that can be handled by the structured approach. The *statistical* model usually describes texture by statistical rules governing the distribution and relation of gray levels. This works well for many natural textures which have barely discernible primitives. The "paper" texture is such an example. As we shall see, we cannot be too rigid about this division since statistical models can describe pattern-like textures and vice versa, but in general the dichotomy is helpful.

The examples suggest that texture is almost always a property of *surfaces*. Indeed, as the example of Fig. 6.2 shows, human beings tend to relate texture elements of varying size to a plausible surface in three dimensions [Gibson 1950; Stevens 1979]. Techniques for determining surface orientation in this fashion are termed texture *gradient* techniques. The gradient is given both in terms of the direction of greatest change in size of primitives and in terms of the spatial placement of primitives. The notion of a gradient is very useful. For example, if the texture is embedded on a flat surface, the gradient points toward a vanishing point in the image. The chapter concludes with algorithms for computing this gradient. The gradient may be computed directly or indirectly via the computation of the vanishing point.

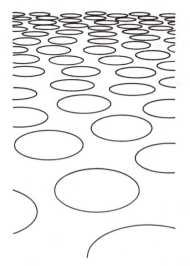

Fig. 6.2 Texture as a surface property.

6.2 TEXTURE PRIMITIVES

The notion of a primitive is central to texture. To highlight its importance, we shall use the appelation *texel* (for texture element) [Kender 1978]. A texel is (loosely) a visual primitive with certain invariant properties which occurs repeatedly in different positions, deformations, and orientations inside a given area. One basic invariant property of such a unit might be that its pixels have a constant gray level, but more elaborate properties related to shape are possible. (A detailed discussion of planar shapes is deferred until Chapter 8.) Figure 6.3 shows examples of two kinds of texels: (a) ellipses of approximately constant gray level and (b) linear edge segments. Interestingly, these are nearly the two features selected as texture primitives by [Julesz, 1981], who has performed extensive studies of human texture perception.

For textures that can be described in two dimensions, image-based descriptions are sufficient. Texture primitives may be pixels, or aggregates of pixels such as curve segments or regions. The "coffee beans" texture can be described by an image-based model: repeated dark ellipses on a lighter background. These models describe equally well an image of texture or an image of a picture of texture. The methods for creating these aggregates were discussed in Chapters 4 and 5. As with all image-based models, three-dimensional phenomena such as occlusion must be handled indirectly. In contrast, structural approaches to texture sometimes require knowledge of the three-dimensional world producing the texture image. One example of this is Brodatz's "coins" shown in Fig. 6.1. A three-dimensional model of the way coins can be stacked is needed to understand this texture fully.

An important part of the texel definition is that primitives must occur repeatedly inside a given area. The question is: How many times? This can be answered qualitatively by imagining a window that corresponds approximately to our field of view superimposed on a very large textured area. As this window is made smaller, corresponding to moving the viewpoint closer to the texture, fewer and fewer texels are contained in it. At some distance, the image in the window no longer

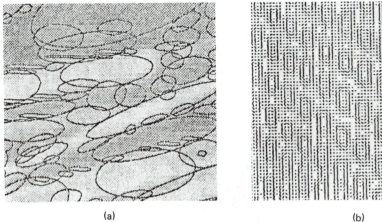

(a)   (b)

Fig. 6.3 Examples of texels. (a) Ellipses. (b) Linear segments.

appears textured, or if it does, translation of the window changes the perceived texture drastically. At this point we no longer have a texture. A similar effect occurs if the window is made increasingly larger, corresponding to moving the field of view farther away from the image. At some distance textural details are blurred into continuous tones and repeated elements are no longer visible as the window is translated. (This is the basis for halftone images, which are highly textured patterns meant to be viewed from enough distance to blur the texture.) Thus the idea of an appropriate *resolution*, or the number of texels in a subimage, is an implicit part of our qualitative definition of texture. If the resolution is appropriate, the texture will be apparent and will "look the same" as the field of view is translated across the textured area. Most often the appropriate resolution is not known but must be computed. Often this computation is simpler to carry out than detailed computations characterizing the primitives and hence has been used as a precursor to the latter computations. Figure 6.4 shows such a resolution-like computation, which examines the image for repeating peaks [Connors 1979].

Textures can be hierarchical, the hierarchies corresponding to different resolutions. The "brick wall" texture shows such a hierarchy. At one resolution, the highly structured pattern made by collections of bricks is in evidence; at higher resolution, the variations of the texture of each brick are visible.

6.3 STRUCTURAL MODELS OF TEXEL PLACEMENT

Highly patterned textures tesselate the plane in an ordered way, and thus we must understand the different ways in which this can be done. In a regular tesselation the

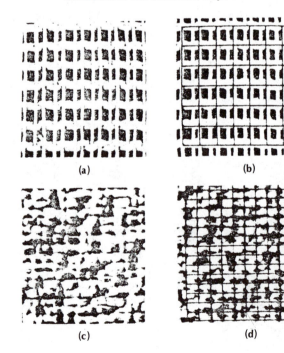

(a)

(b)

(c)

(d)

Fig. 6.4 Computing texture resolutions. (a) French canvas. (b) Resolution grid for canvas. (c) Raffia. (d) Grid for raffia.

polygons surrounding a vertex all have the same number of sides. Semiregular tesselations have two kinds of polygons (differing in number of sides) surrounding a vertex. Figure 2.11 depicts the regular tesselations of the plane. There are eight semiregular tesselations of the plane, as shown in Fig. 6.5. These tesselations are conveniently described by listing in order the number of sides of the polygons sur-

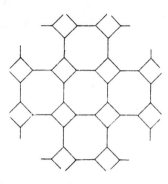

(4, 8, 8)

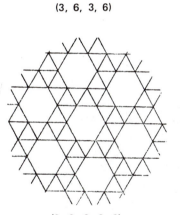

(3, 6, 3, 6)

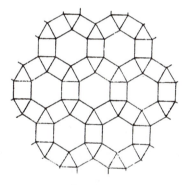

(3, 4, 6, 4)

(3, 3, 3, 3, 6)

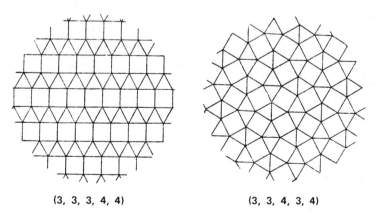

(3, 3, 3, 4, 4)

(3, 3, 4, 3, 4)

Fig. 6.5 Semiregular tesselations.

rounding each vertex. Thus a hexagonal tesselation is described by (6,6,6) and every vertex in the tesselation of Fig. 6.5 can be denoted by the list (3,12,12). It is important to note that the tesselations of interest are those which describe the *placement* of primitives rather than the primitives themselves. When the primitives define a tesselation, the tesselation describing the primitive placement will be the dual of this graph in the sense of Section 5.4. Figure 6.6 shows these relationships.

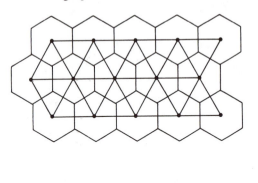

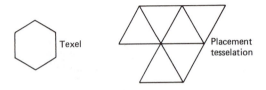

Texel   Placement
tesselation

Fig. 6.6 The primitive placement tesselation as the dual of the primitive tesselation.

6.3.1 Grammatical Models

A powerful way of describing the rules that govern textural structure is through a grammar. A grammar describes how to generate patterns by applying *rewriting rules* to a small number of *symbols*. Through a small number of rules and symbols, the grammar can generate complex textural patterns. Of course, the symbols turn out to be related to texels. The mapping between the stored model prototype texture and an image of texture with real-world variations may be incorporated into the grammar by attaching probabilities to different rules. Grammars with such rules are termed *stochastic* [Fu 1974].

There is no unique grammar for a given texture; in fact, there are usually infinitely many choices for rules and symbols. Thus texture grammars are described as *syntactically ambiguous*. Figure 6.7 shows a syntactically ambiguous texture and two of the possible choices for primitives. This texture is also *semantically ambiguous* [Zucker 1976] in that alternate ridges may be thought of in three dimensions as coming out of or going into the page.

There are many variants of the basic idea of formal grammars and we shall examine three of them: shape grammars, tree grammars, and array grammars. For a basic reference, see [Hopcroft and Ullman 1979]. Shape grammars are distinguished from the other two by having high-level primitives that closely correspond to the shapes in the texture. In the examples of tree grammars and array grammars that we examine, texels are defined as pixels and this makes the

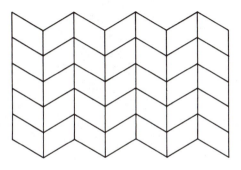

Two choices for primitives:

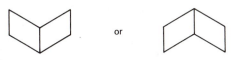

or

Fig. 6.7 Ambiguous texture.

grammars correspondingly more complicated. A particular texture that can be described in eight rules in a shape grammar requires 85 rules in a tree grammar [Lu and Fu 1978]. The compensating trade-off is that pixels are gratis with the image; considerable processing must be done to derive the more complex primitives used by the shape grammar.

6.3.2 Shape Grammars

A shape grammar [Stiny and Gips 1972] is defined as a four-tuple $< V_t, V_m, R, S>$ where:

1. V_t is a finite set of shapes
2. V_m is a finite set of shapes such that $V_t \cap V_m = \phi$
3. R is a finite set of ordered pairs (u, v) such that u is a shape consisting of elements of V_m^+ and v is a shape consisting of an element of V_t^* combined with an element of V_m^*
4. S is a shape consisting of an element of V_t^* combined with an element of V_m^*.

Elements of the set V_t are called terminal shape elements (or terminals). Elements of the set V_m are called nonterminal shape elements (or markers). The sets V_t and V_m must be disjoint. Elements of the set V_t^+ are formed by the finite arrangement of one or more elements of V_t in which any elements and/or their mirror images may be used a multiple number of times in any location, orientation, or scale. The set $V_t^* = V_t^+ \cup \{\Lambda\}$, where Λ is the empty shape. The sets V_m^+ and V_m^* are defined similarly. Elements (u, v) of R are called shape rules and are written $u\,v$. u is called the left side of the rule; v the right side of the rule. u and v usually are enclosed in identical dashed rectangles to show the correspondence between the two shapes. S is called the initial shape and normally contains a u such that there is a (u, v) which is an element of R.

A texture is generated from a shape grammar by beginning with the initial shape and repeatedly applying the shape rules. The result of applying a shape rule R to a given shape S is another shape, consisting of S with the right side of R substituted in S for an occurrence of the left side of R. Rule application to a shape proceeds as follows:

1. Find part of the shape that is geometrically similar to the left side of a rule in terms of both terminal elements and nonterminal elements (markers). There must be a one-to-one correspondence between the terminals and markers in the left side of the rule and the terminals and markers in the part of the shape to which the rule is to be applied.

2. Find the geometric transformations (scale, translation, rotation, mirror image) which make the left side of the rule identical to the corresponding part in the shape.

3. Apply those transformations to the right side of the rule.

4. Substitute the transformed right side of the rule for the part of the shape that corresponds to the left side of the rule.

The generation process is terminated when no rule in the grammar can be applied.

As a simple example, one of the many ways of specifying a hexagonal texture $\{V_t, V_m, R, S\}$ is

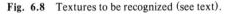

$$V_t = \{ \hexagon \}$$
$$V_m = \{ \cdot \} \tag{6.1}$$
$$R: \hexagon \rightarrow \hexagon\hexagon \ ; \ \hexagon\hexagon \ ; \text{etc.}$$
$$S = \{ \hexagon \}$$

Hexagonal textures can be *generated* by the repeated application of the single rule in R. They can be *recognized* by the application of the rule in the opposite direction to a given texture until the initial shape, I, is produced. Of course, the rule will generate only hexagonal textures. Similarly, the hexagonal texture in Fig. 6.8a will be recognized but the variants in Fig. 6.8b will not.

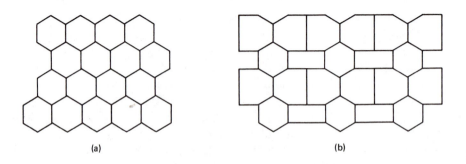

(a)   (b)

Fig. 6.8 Textures to be recognized (see text).

A more difficult example is given by the "reptile" texture. Except for the occasional new rows, a $(3, 6, 3, 6)$ tesselation of primitives would model this texture exactly. As shown in Fig. 6.9, the new row is introduced when a seven-sided polygon splits into a six-sided polygon and a five-sided polygon. To capture this with a shape grammar, we examine the dual of this graph, which is the primitive placement graph, Fig. 6.9b. This graph provides a simple explanation of how the extra row is created; that is, the diamond pattern splits into two. Notice that the dual graph is composed solely of four-sided polygons but that some vertices are $(4, 4, 4)$ and some are $(4, 4, 4, 4, 4, 4)$. A shape grammar for the dual is shown in Fig. 6.10. The image texture can be obtained by forming the dual of this graph. One further refinement should be added to rules (6) and (7); so that rule (7) is used less often, the appropriate probabilities should be associated with each rule. This would make the grammar stochastic.

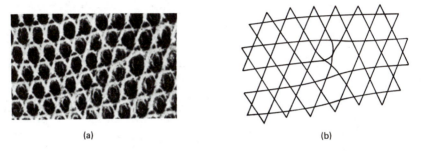

(a)   (b)

Fig. 6.9 (a) The reptile texture. (b) The reptile texture as a $(3, 6, 3, 6)$ semiregular tesselation with local deformations.

6.3.3 Tree Grammars

The symbolic form of a tree grammar is very similar to that of a shape grammar. A grammar

$$G_t = (V_t, \ V_m, \ r, \ R, \ S)$$

is a tree grammar if

V_t is a set of terminal symbols
V_m is a set of symbols such that
$V_m \cap V_t = \phi$
$r : V_t \rightarrow N$ (where N is the set of nonnegative integers)
   is the rank associated with symbols in V_t
S is the start symbol
R is the set of rules of the form

$$X_0 \rightarrow x \qquad \text{or} \quad X_0 \rightarrow x$$

$$X_0 \dots X_{r(x)}$$
with x in V_t and $X_0 \dots X_{r(x)}$ in V_m

For a tree grammar to generate arrays of pixels, it is necessary to choose some way of embedding the tree in the array. Figure 6.11 shows two such embeddings.

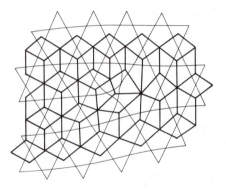

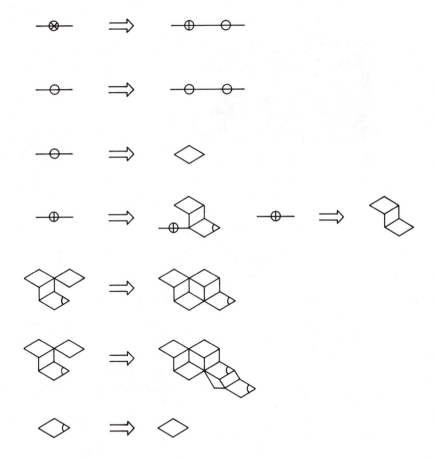

Fig. 6.10 Shape grammar for the reptile texture.

In the application to texture [Lu and Fu 1978], the notion of pyramids or hierarchical levels of resolution in texture is used. One level describes the placement of repeating patterns in texture windows—a rectangular texel placement tesselation—and another level describes texels in terms of pixels. We shall illus-

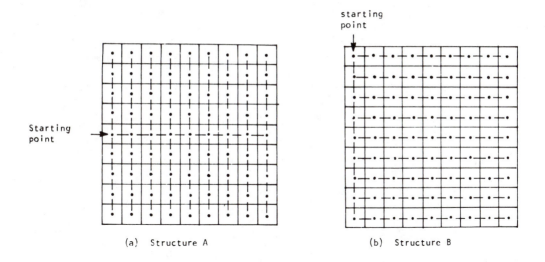

starting
point

Starting
point

(a) Structure A

(b) Structure B

Fig. 6.11 Two ways of embedding a tree structure in an array.

trate these ideas with Lu and Fu's grammar for "wire braid." The texture windows are shown in Fig. 6.12a. Each of these can be described by a "sentence" in a second tree grammar. The grammar is given by:

$$G_w = (V_t, V_m, r, R, X)$$

where

$$V_t = \{A_1, C_1\}$$
$$V_m = \{X, Y, Z\}$$
$$r = \{0, 1, 2\}$$

$$R : X \rightarrow \underset{X \quad Y}{A_1} \quad \text{or} \quad \underset{Y}{A_1}$$

$$Y \rightarrow \underset{Z}{C_1} \quad \text{or} \quad C_1$$

$$Z \rightarrow \underset{Y}{A_1} \quad \text{or} \quad A_1$$

(6.2)

and the first embedding in Fig. 6.11 is used. The pattern inside each of these windows is specified by another grammatical level:

$$G = (V_t, V_m, r, R, S)$$

where

$$V_t = \{1, 0\}$$
$$V_m = \{A_1, A_2, A_3, A_4, A_5, A_6, A_7, C_1, C_2, C_3, C_4, C_5, C_6, C_7,$$
$$N_0, N_1, N_2, N_3, N_4\}$$
$$r = \{0, 1, 2\}$$
$$S = \{A_1, C_1\}$$

R:

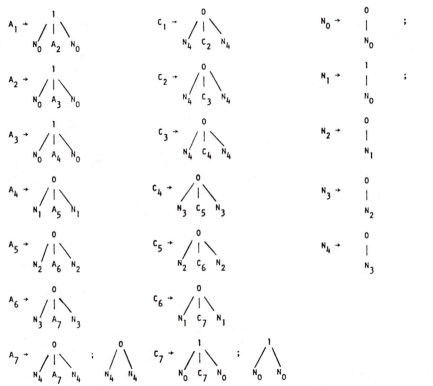

The application of these rules generates the two different patterns of pixels shown in Fig. 6.13.

6.3.4 Array Grammars

Like tree grammars, array grammars use hierarchical levels of resolution [Milgram and Rosenfeld 1971; Rosenfeld 1971]. Array grammars are different from tree grammars in that they do not use the tree-array embedding. Instead, prodigious use of a blank or null symbol is used to make sure the rules are applied in appropriate contexts. A simple array grammar for generating a checkerboard pattern is

$$G = \{V_t, V_n, R\}$$

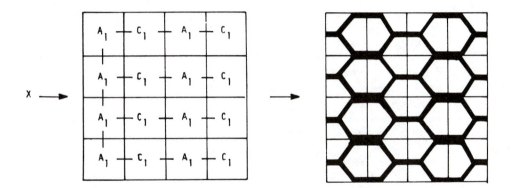

Fig. 6.12 Texture window and grammar (see text).

where

$$V_t = \{0, 1\} \text{ (corresponding to black and white pixels, respectively)}$$
$$V_n = \{b, S\}$$

b is a "blank" symbol used to provide context for the application of the rules. Another notational convenience is to use a subscript to denote the orientation of symbols. For example, when describing the rules R we use

$$0_x b \rightarrow 0_x 1 \qquad \text{where } x \text{ is one of } \{U, D, L, R\}$$

to summarize the four rules

$$\frac{0}{b} \rightarrow \frac{0}{1}, \qquad \frac{b}{0} \rightarrow \frac{1}{0}, \qquad 0b \rightarrow 01, \qquad b0 \rightarrow 10$$

Thus the checkerboard rule set is given by

$$R: S \rightarrow 0 \text{ or } 1$$
$$0_x b \rightarrow 0_x 1 \qquad x \text{ in } \{U, D, L, R\}$$
$$1_x b \rightarrow 1_x 0$$

A compact encoding of textural patterns [Jayaramamurthy 1979] uses levels of array grammars defined on a pyramid. The terminal symbols of one layer are the start symbols of the next grammatical layer defined lower down in the pyramid. This corresponds nicely to the idea of having one grammar to generate primitives and another to generate the primitive placement tesselations.

As another example, consider the herringbone pattern in Fig. 6.14a, which is composed of 4×3 arrays of a particular placement pattern as shown in Fig. 6.14b. The following grammar is sufficient to generate the placement pattern.

$$G_w = \{V_t, V_m, R, S\}$$

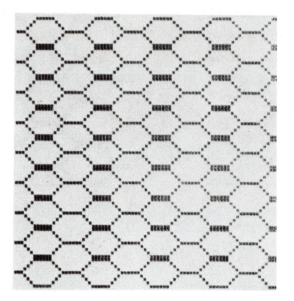

Fig. 6.13 Texture generated by tree grammar.

where

$$V_t = \{a\}$$

$$V_n = \{b, S\}$$

$$R : S \rightarrow a$$

$$a_x b \rightarrow a_x a \qquad x \text{ in } \{U, D, L, R\}$$

We have not been precise in specifying how the terminal symbol is projected onto the lower level. Assume without loss of generality that it is placed in the upper left-hand corner, the rest of the subarray being initially blank symbols. Thus a simple grammar for the primitive is

$$G_t = \{V_t, V_n, R, S\}$$

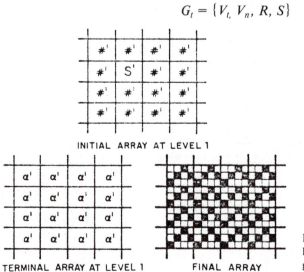

Fig. 6.14 Steps in generating a herringbone texture with an array grammar.

where

$$V_t = \{0, 1\}$$

$$V_n = \{a, b\}$$

$$R: \begin{matrix} a & b & b & b \\ b & b & b & b \\ b & b & b & b \end{matrix} \rightarrow \begin{matrix} 0 & 0 & 1 & 0 \\ 0 & 1 & 0 & 1 \\ 1 & 0 & 0 & 0 \end{matrix}$$

6.4 TEXTURE AS A PATTERN RECOGNITION PROBLEM

Many textures do not have the nice geometrical regularity of "reptile" or "wire braid"; instead, they exhibit variations that are not satisfactorily described by shapes, but are best described by statistical models. *Statistical pattern recognition* is a paradigm that can classify statistical variations in patterns. (There are other statistical methods of describing texture [Pratt et al. 1981], but we will focus on statistical pattern recognition since it is the most widely used for computer vision purposes.) There is a voluminous literature on pattern recognition, including several excellent texts (e.g., [Fu 1968; Tou and Gonzalez 1974; Fukunaga 1972], and the ideas have much wider application than their use here, but they seem particularly appropriate for low-resolution textures, such as those seen in aerial images [Weszka et al. 1976]. The pattern recognition approach to the problem is to classify instances of a texture in an image into a set of classes. For example, given the textures in Fig. 6.15, the choice might be between the classes "orchard," "field," "residential," "water."

The basic notion of pattern recognition is the *feature vector*. The feature vector **v** is a set of measurements $\{v_1 \cdots v_m\}$ which is supposed to condense the description of relevant properties of the textured image into a small, Euclidean *feature space* of m dimensions. Each point in feature space represents a value for the feature vector applied to a different image (or subimage) of texture. The measurement values for a feature should be correlated with its class membership. Figure 6.16a shows a two-dimensional space in which the features exhibit the desired correlation property. Feature vector values cluster according to the texture from which they were derived. Figure 6.16b shows a bad choice of features (measurements) which does not separate the different classes.

The pattern recognition paradigm divides the problem into two phases: training and test. Usually, during a training phase, feature vectors from known samples are used to partition feature space into regions representing the different classes. However, self teaching can be done; the classifier derives its own partitions. Feature selection can be based on parametric or nonparametric models of the distributions of points in feature space. In the former case, analytic solutions are sometimes available. In the latter, feature vectors are *clustered* into groups which are taken to indicate partitions. During a test phase the feature-space partitions are used to classify feature vectors from unknown samples. Figure 6.17 shows this process.

Given that the data are reasonably well behaved, there are many methods for clustering feature vectors [Fukunaga 1972; Tou and Gonzales 1974; Fu 1974].

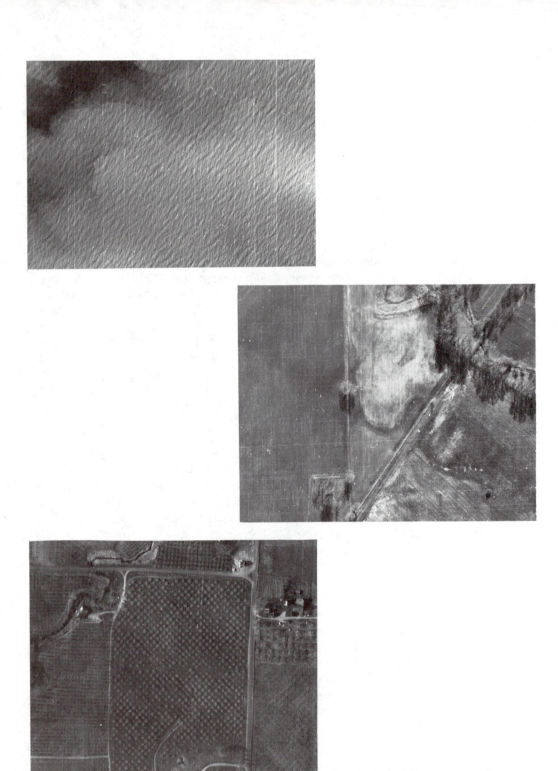

Fig. 6.15 Aerial image textures for discrimination.

Fig. 6.15 (cont.)

One popular way of doing this is to use prototype points for each class and a *nearest-neighbor* rule [Cover 1968]:

assign v to class w_i if i minimizes
$$\min_i d(\mathbf{v}, \mathbf{v}_{w_i})$$

where $\mathbf{v}_{w_i}$ is the prototype point for class w_i.

Parametric techniques assume information about the feature vector probability distributions to find rules that maximize the likelihood of correct classification:

assign $\mathbf{v}$ to class w_i if i maximizes
$$\max_i p(w_i|\mathbf{v})$$

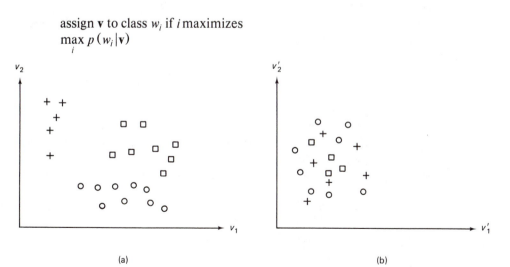

(a)   (b)

Fig. 6.16 Feature space for texture discrimination. (a) effective features (b) ineffective features.

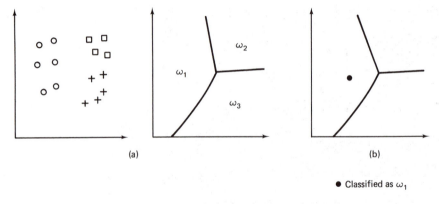

● Classified as ω_1

Fig. 6.17   Pattern recognition paradigm.

The distributions may also be used to formulate rules that minimize errors.

Picking good features is the essence of pattern recognition. No elaborate formalism will work well for bad features such as those of Fig.6.16b. On the other hand, almost any method will work for very good features. For this reason, texture is a good domain for pattern recognition: it is fairly easy to define features that (1) cluster in feature space according to different classes, and (2) can separate texture classes.

The ensuing subsections describe features that have worked well. These subsections are in reverse order from those of Section 6.2 in that we begin with features defined on pixels—Fourier subspaces, gray-level dependencies—and conclude with features defined on higher-level texels such as regions. However, the lesson is the same as with the grammatical approach: hard work spent in obtaining high-level primitives can both improve and simplify the texture model. Space does not permit a discussion of many texture features; instead, we limit ourselves to a few representative samples. For further reading, see [Haralick 1978].

6.4.1 Texture Energy

Fourier Domain Basis

If a texture is at all spatially periodic or directional, its power spectrum will tend to have peaks for corresponding spatial frequencies. These peaks can form the basis of features of a pattern recognition discriminator. One way to define features is to search Fourier space directly [Bajcsy and Lieberman 1976]. Another is to partition Fourier space into bins. Two kinds of bins, radial and angular, are commonly used, as shown in Fig. 6.18. These bins, together with the Fourier power spectrum are used to define features. If F is the Fourier transform, the Fourier power spectrum is given by $|F|^2$.

Radial features are given by

$$V_{r_1 r_2} = \int\int |F(u, v)|^2 \, du \, dv \tag{6.5}$$

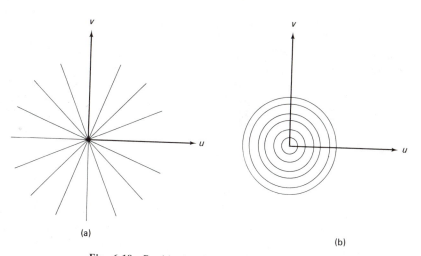

(a)

(b)

Fig. 6.18 Partitioning the Fourier domain into bins.

where the limits of integration are defined by

$$r_1^2 \leqslant u^2 + v^2 < r_2^2$$

$$0 \leqslant u, v < n-1$$

where $[r_1, r_2]$ is one of the radial bins and $\mathbf{v}$ is the vector (not related to v) defined by different values of r_1 and r_2. Radial features are correlated with texture coarseness. A smooth texture will have high values of $V_{r_1 r_2}$ for small radii, whereas a coarse, grainy texture will tend to have relatively higher values for larger radii.

Features that measure angular orientation are given by

$$V_{\theta_1 \theta_2} = \iint |F(u, v)|^2 \, du \, dv \tag{6.6}$$

where the limits of integration are defined by

$$\theta_1 \leqslant \tan^{-1} \left[\frac{v}{u} \right] \leqslant \theta_2$$

$$0 < u, v \leqslant n-1$$

where $[\theta_1, \theta_2)$ is one of the sectors and $\mathbf{v}$ is defined by different values of θ_1 and θ_2. These features exploit the sensitivity of the power spectrum to the directionality of the texture. If a texture has many lines or edges in a given direction θ, $|F|^2$ will tend to have high values clustered around the direction in frequency space $\theta + \pi/2$.

Texture Energy in the Spatial Domain

From Section 2.2.4 we know that the Fourier approach could also be carried out in the image domain. This is the approach taken in [Laws 1980]. The advantage of this approach is that the basis is not the Fourier basis but a variant that is more matched to intuition about texture features. Figure 6.19 shows the most important of Laws' 12 basis functions.

The image is first histogram-equalized (Section 3.2). Then 12 new images are made by convolving the original image with each of the basis functions (i.e., $f'_k = f * h_k$ for basis functions $h_1, ..., h_{12}$). Then each of these images is transformed into an "energy" image by the following transformation: Each pixel in the convolved image is replaced by an average of the absolute values in a local window of 15×15 pixels centered over the pixel:

$$f''_k(x, y) = \sum_{x', y' \text{ in window}} (|f'_k(x', y')|) \qquad (6.7)$$

The transformation $f \to f''_k$, $k = 1, ... 12$ is termed a "texture energy transform" by Laws and is analogous to the Fourier power spectrum. The f''_k, $k = 1, ... 12$ form a set of features for each point in the image which are used in a nearest-neighbor classifier. Classification details may be found in [Laws 1980]. Our interest is in the particular choice of basis functions used.

Figure 6.20 shows a composite of natural textures [Brodatz 1966] used in Laws's experiments. Each texture is digitized into a 128×128 pixel subimage. The texture energy transforms were applied to this composite image and each pixel was classified into one of the eight categories. The average classification accuracy was about 87% for interior regions of the subimages. This is a very good result for textures that are similar.

6.4.2 Spatial Gray-Level Dependence

Spatial gray-level dependence (SGLD) matrices are one of the most popular sources of features [Kruger et al. 1974; Hall et al. 1971; Haralick et al. 1973]. The SGLD approach computes an intermediate matrix of measures from the digitized image data, and then defines features as functions on this intermediate matrix. Given an image f with a set of discrete gray levels I, we define for each of a set of discrete values of d and θ the intermediate matrix $S(d, \theta)$ as follows:

S$(i, j|d, \theta)$, an entry in the matrix, is the number of times gray level i is oriented with respect to gray level j such that where

$$f(\mathbf{x}) = i \quad \text{and} \quad f(\mathbf{y}) = j \quad \text{then}$$

$$\mathbf{y} = \mathbf{x} + (d\cos\theta, \ d\sin\theta)$$

$$\begin{bmatrix} -1 & -4 & -6 & -4 & -1 \\ -2 & -8 & -12 & -8 & -2 \\ 0 & 0 & 0 & 0 & 0 \\ 2 & 8 & 12 & 8 & 2 \\ 1 & 4 & 6 & 4 & 1 \end{bmatrix} \qquad \begin{bmatrix} 1 & -4 & 6 & -4 & 1 \\ -4 & 16 & -24 & 16 & -4 \\ 6 & -24 & 36 & -24 & 6 \\ -4 & 16 & -24 & 16 & -4 \\ 1 & -4 & 6 & -4 & 1 \end{bmatrix}$$

$$\begin{bmatrix} -1 & 0 & 2 & 0 & -1 \\ -2 & 0 & 4 & 0 & -2 \\ 0 & 0 & 0 & 0 & 0 \\ 2 & 0 & -4 & 0 & 2 \\ 1 & 0 & -2 & 0 & 1 \end{bmatrix} \qquad \begin{bmatrix} -1 & 0 & 2 & 0 & -1 \\ -4 & 0 & 8 & 0 & -4 \\ -6 & 0 & 12 & 0 & -6 \\ -4 & 0 & 8 & 0 & -4 \\ -1 & 0 & 2 & 0 & -1 \end{bmatrix}$$

Fig. 6.19 Laws' basis functions (these are the low-order four of twelve actually used).

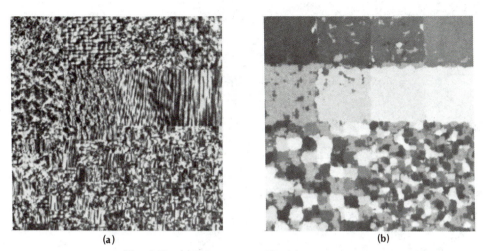

(a)　　　　　　　　　　　　　　　　　　　(b)

Fig. 6.20 (a) Texture composite. (b) Classification.

Note that the gray-level values appear as indices of the matrix S, implying that they are taken from some well-ordered discrete set $0, \ldots, K$. Since

$$S(d, \theta) = S(d, \theta + \pi).$$

common practice is to restrict θ to multiples of $\pi/4$. Furthermore, information is not usually retained at both θ and $\theta + \pi$. The reasoning for the latter step is that for most texture discrimination tasks, the information is redundant. Thus we define

$$S(d, \theta) = \tfrac{1}{2}\,[S(d, \theta) + S(d, \theta + \pi)]$$

The intermediate matrices S yield potential features. Commonly used features are:

1. *Energy*

$$E(d, \theta) = \sum_{i=0}^{K} \sum_{j=0}^{K} [S(i, j\,|\,d, \theta)]^2 \tag{6.8}$$

2. *Entropy*

$$H(d, \theta) = \sum_{i=0}^{K} \sum_{j=0}^{K} S(i, j\,|\,d, \theta) \, \log \, S(i, j\,|\,d, \theta) \tag{6.9}$$

3. *Correlation*

$$C(d, \theta) = \frac{\sum_{i=0}^{K} \sum_{j=0}^{K} (i - \mu_x)(j - \mu_y) S(i, j\,|\,d, \theta)}{\sigma_x \sigma_y} \tag{6.10}$$

4. *Inertia*

$$I(d, \theta) = \sum_{i=0}^{K} \sum_{j=0}^{K} (i - j)^2 \, S(i, j\,|\,d, \theta) \tag{6.11}$$

5. *Local Homogeneity*

$$L(d, \theta) = \sum_{i=0}^{K} \sum_{j=0}^{K} \frac{1}{1 + (i-j)^2} S(i, j|d, \theta) \tag{6.12}$$

where $S(i, j|d, \theta)$ is the (i, j) the element of $S(d, \theta)$, and

$$\mu_x = \sum_{i=0}^{K} i \sum_{j=0}^{K} S(i, j|d, \theta) \tag{6.13a}$$

$$\mu_y = \sum_{i=0}^{K} j \sum_{j=0}^{K} S(i, j|d, \theta) \tag{6.13b}$$

$$\sigma_x^2 = \sum_{i=0}^{K} (i - \mu_x)^2 \sum_{j=0}^{K} S(i, j|d, \theta) \tag{6.13c}$$

and

$$\sigma_y^2 = \sum_{i=0}^{K} (j - \mu_y)^2 \sum_{i=0}^{K} S(i, j|d, \theta) \tag{6.13d}$$

One important aspect of this approach is that the features chosen do not have psychological correlates [Tamura et al. 1978]. For example, none of the measures described would take on specific values corresponding to our notions of "rough" or "smooth." Also, the texture gradient is difficult to define in terms of SGLD feature values [Bajcsy and Lieberman 1976].

6.4.3 Region Texels

Region texels are an image-based way of defining primitives above the level of pixels. Rather than defining features directly as functions of pixels, a region segmentation of the image is created first. Features can then be defined in terms of the shape of the resultant regions, which are often more intuitive than the pixel-related features. Naturally, the approach of using edge elements is also possible. We shall discuss this in the context of texture gradients.

The idea of using regions as texture primitives was pursued in [Maleson et al. 1977]. In that implementation, all regions are ultimately modeled as ellipses and a corresponding five-parameter shape description is computed for each region. These parameters only define gross region shape, but the five-parameter primitives seem to work well for many domains. The texture image is segmented into regions in two steps. Initially, the modified version of Algorithm 5.1 that works for gray-level images is used. Figure 6.21 shows this example of the segmentation applied to a sample of "straw" texture. Next, parameters of the region grower are controlled so as to encourage convex regions which are fit with ellipses. Figure 6.22 shows the resultant ellipses for the "straw" texture. One set of ellipse parameters is $\mathbf{x}_0$, a, b, θ where $\mathbf{x}_0$ is the origin, a and b are the major and minor axis lengths and θ is the orientation of the major axis (Appendix 1). Besides these shape parameters, elliptical texels are also described by their average gray level. Figure 6.23 gives a qualitative indication of how ranges on feature values reflect different texels.

(a) Image

(b) With Region Boundaries

Fig. 6.21 Region segmentation for straw texture.

6.5 THE TEXTURE GRADIENT

The importance of texture in determining surface orientation was described by Gibson [Gibson 1950]. There are three ways in which this can be done. These methods are depicted in Fig. 6.24. All these methods assume that the texture is embedded on a planar surface.

First, if the texture image has been segmented into primitives, the maximum rate of change of the projected size of these primitives constrains the orientation of

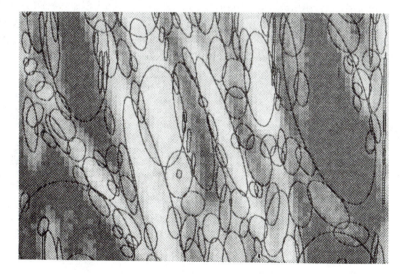

Fig. 6.22 Ellipses for straw texture.

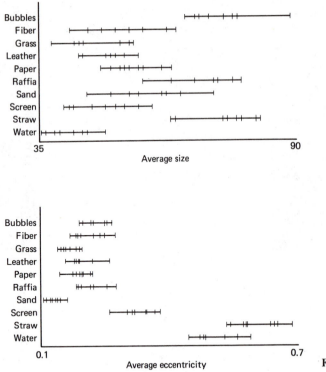

Average size

Average eccentricity

Fig. 6.23 Features defined on ellipses.

the plane in the following manner. The direction of maximum rate of change of projected primitive size is the direction of the *texture gradient*. The orientation of this direction with respect to the image coordinate frame determines how much the plane is rotated about the camera line of sight. The magnitude of the gradient can help determine how much the plane is tilted with respect to the camera, but knowledge about the camera geometry is also required. We have seen these ideas before in the form of gradient space; the rotation and tilt characterization is a polar coordinate representation of gradients.

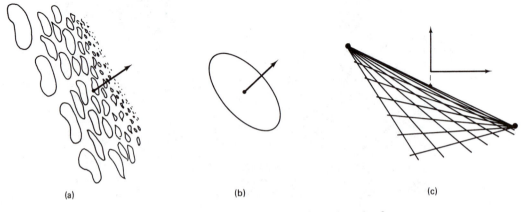

Fig. 6.24 Methods for calculating surface orientation from texture.

The second way to measure surface orientation is by knowing the shape of the texel itself. For example, a texture composed of circles appears as ellipses on the tilted surface. The orientation of the principal axes defines rotation with respect to the camera, and the ratio of minor to major axes defines tilt [Stevens 1979].

Finally, if the texture is composed of a regular grid of texels, we can compute vanishing points. For a perspective image, vanishing points on a plane P are the projection onto the image plane of the points at infinity in a given direction. In the examples here, the texels themselves are (conveniently) small line segments on a plane that are oriented in two orthogonal directions in the physical world. The general method applies whenever the placement tesselation defines lines of texels. Two vanishing points that arise from texels on the same surface can be used to determine orientation as follows. The line joining the vanishing points provides the orientation of the surface and the vertical position of the plane with respect to the z axis (i.e., the intersection of the line joining the vanishing points with $x = 0$) determines the tilt of the plane.

Line segment textures indicate vanishing points [Kender 1978]. As shown in Fig. 6.25, these segments could arise quite naturally from an urban image of the windows of a building which has been processed with an edge operator.

As discussed in Chapter 4, lines in images can be detected by detecting their parameters with a Hough algorithm. For example, by using the line parameterization

$$x \cos\theta + y \sin\theta = r$$

and by knowing the orientation of the line in terms of its gradient $\mathbf{g} = (\Delta x, \Delta y)$, a line segment $(x, y, \Delta x, \Delta y)$ can be mapped into r, θ space by using the relations

$$r = \frac{\Delta x x + \Delta y y}{\sqrt{\Delta x^2 + \Delta y^2}} \tag{6.14}$$

$$\theta = \tan^{-1}\left(\frac{\Delta y}{\Delta x}\right) \tag{6.15}$$

These relationships can be derived by using Fig. 6.26 and some geometry. The Cartesian coordinates of the $r-\theta$ space vector are given by

$$\mathbf{a} = \left(\frac{\mathbf{g} \cdot \mathbf{x}}{\|\mathbf{g}\|^2}\right) \mathbf{g} \tag{6.16}$$

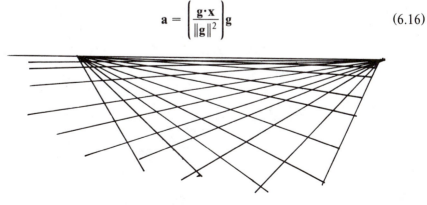

Fig. 6.25 Orthogonal line segments comprising a texture.

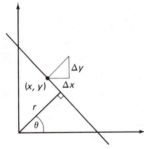

Fig. 6.26 r-θ transform.

Using this transformation, the set of line segments L_1 shown in Fig. 6.27 are all mapped into a single point in r-θ space. Furthermore, the set of lines L_2 which have the same vanishing point (x_v, y_v) project onto a circle in r-θ space with the line segment $((0, 0), (x_v, y_v))$ as a diameter. This scheme has two drawbacks: (1) vanishing points at infinity are projected into infinity, and (2) circles require some effort to detect. Hence we are motivated to use the transform $(x, y, \Delta x, \Delta y) \rightarrow \left[\dfrac{k}{r}, \theta\right]$ for some constant k. Now vanishing points at infinity are projected into the origin and the locus of the set of points L_2 is now a line. This line is perpendicular to the vector x_v and $\dfrac{k}{\|\mathbf{x}_v\|}$ units from the origin, as shown in Fig. 6.28. It can be detected by a second stage of the Hough transform; each point $\mathbf{a}$ is mapped into an r'-θ' space. For every $\mathbf{a}$, compute all the r', θ' such that

$$a \cos\theta' + b \sin\theta' = r' \tag{6.17}$$

and increment that location in the appropriate r', θ' accumulator array. In this second space a vanishing point is detected as

$$r' = \frac{k}{\|\mathbf{x}_v\|} \tag{6.18}$$

$$\theta' = \tan^{-1}\left(\frac{y_v}{x_v}\right) \tag{6.19}$$

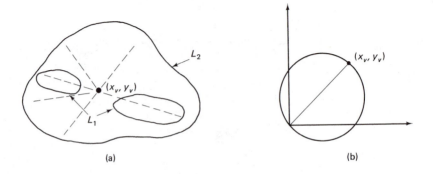

(a) (b)

Fig. 6.27 Detecting the vanishing point with the Hough transform.

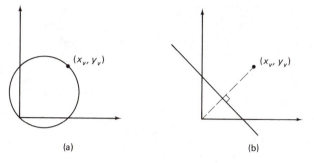

Fig. 6.28 Vanishing point loci.

In Kender's application the texels and their placement tesselation are similar in that the primitives are parallel to arcs in the placement tesselation graph. In a more general application the tesselation could be computed by connecting the centers of primitives.

EXERCISES

6.1 Devise a computer algorithm that, given a set of texels from each of a set of different "windows" of the textured image, checks to see of the resolution is appropriate. In other words, try to formalize the discussion of resolution in Section 6.2.

6.2 Are any of the grammars in Section 6.3 suitable for a parallel implementation (i.e., parallel application of rules)? Discuss, illustrating your arguments with examples or counterexamples from each of the three main grammatical types (shape, tree, and array grammars).

6.3 Are shape, array, and tree grammars context free or context-sensitive as defined? Can such grammars be translated into "traditional" (string) grammars? If not, how are they different; and if so, why are they useful?

6.4 Show how the generalized Hough transform (Section 4.3) could be applied to texel detection.

6.5 In an outdoors scene, there is the problem of different scales. For example, consider the grass. Grass that is close to an observer will appear "sharp" and composed of primitive elements, yet grass distant from an observer will be much more "fuzzy" and homogeneous. Describe how one might handle this problem.

6.6 The texture energy transform (Section 6.4.1) is equivalent to a set of Fourier-domain operations. How do the texture energy features compare with the ring and sector features?

6.7 The texture gradient is presumably a gradient in some aspect of texture. What aspect is it, and how might it be quantified so that texture descriptions can be made gradient independent?

6.8 Write a texture region grower and apply it to natural scenes.

REFERENCES

Bajcsy, R. and L. Lieberman. "Texture gradient as a depth cue." *CGIP 5*, 1, March 1976, 52–67.

Brodatz, P. *Textures: A Photographic Album for Artists and Designers.* Toronto: Dover Publishing Co., 1966.

CONNORS, R. "Towards a set of statistical features which measure visually perceivable qualities of textures." *Proc.*, PRIP, August 1979, 382–390.

COVER, T. M. "Estimation by the nearest neighbor rule." *IEEE Trans. Information Theory 14*, January 1968, 50–55.

FU, K. S. *Sequential Methods in Pattern Recognition and Machine Learning*. New York: Academic Press, 1968.

FU, K. S. *Syntactic Methods in Pattern Recognition*. New York: Academic Press, 1974.

FUKUNAGA, K. *Introduction to Statistical Pattern Recognition*. New York, Academic Press, 1972.

GIBSON, J. J. *The Perception of the Visual World*. Cambridge, MA: Riverside Press, 1950.

HALL, E. L, R. P. KRUGER, S. J. DWYER III, D. L. HALL, R. W. McLAREN, and G. S. LODWICK. "A survey of preprocessing and feature extraction techniques for radiographic images." *IEEE Trans. Computers 20*, September 1971.

HARALICK, R. M. "Statistical and structural approaches to texture." *Proc.*, 4th IJCPR, November 1978, 45–60.

HARALICK, R. M., R. SHANMUGAM, and I. DINSTEIN. "Textural features for image classification." *IEEE Trans. SMC 3*, November 1973, 610–621.

HOPCROFT, J. E. and J. D. ULLMAN. *Introduction to Automata Theory, Languages and Computation*. Reading, MA: Addison-Wesley, 1979.

JAYARAMAMURTHY, S. N. "Multilevel array grammars for generating texture scenes." *Proc.*, PRIP, August 1979, 391–398.

JULESZ, B. "Textons, the elements of texture perception, and their interactions." *Nature 290*, March 1981, 91–97.

KENDER, J. R. "Shape from texture: a brief overview and a new aggregation transform." *Proc.*, DARPA IU Workshop, November 1978, 79–84.

KRUGER, R. P., W. B. THOMPSON, and A. F. TURNER. "Computer diagnosis of pneumoconiosis." *IEEE Trans. SMC 45*, 1974, 40–49.

LAWS, K. I. "Textured image segmentation." Ph.D. dissertation, Dept. of Engineering, Univ. Southern California, 1980.

LU, S. Y. and K. S. FU. "A syntactic approach to texture analysis." *CGIP 7*, 3, June 1978, 303–330.

MALESON, J. T., C. M. BROWN, and J. A. FELDMAN. "Understanding natural texture." *Proc.*, DARPA IU Workshop, October 1977, 19–27.

MILGRAM, D. L. and A. ROSENFELD. "Array automata and array grammars." *Proc.*, IFIP Congress 71, Booklet TA-2. Amsterdam: North-Holland, 1971, 166–173.

PRATT, W. K., O. D. FAUGERAS, and A. GAGALOWICZ. "Applications of Stochastic Texture Field Models to Image Processing." *Proc. of the IEEE*. Vol.69, No. 5, May 1981

ROSENFELD, A. "Isotonic grammars, parallel grammars and picture grammars." In *MI6*, 1971.

STEVENS, K.A. "Representing and analyzing surface orientation." In *Artificial Intelligence: An MIT Perspective*, Vol. 2, P. H. Winston and R. H. Brown (Eds.). Cambridge, MA: MIT Press, 1979.

STINY, G. and J. GIPS. *Algorithmic Aesthetics: Computer Models for Criticism and Design in the Arts*. Berkeley, CA: University of California Press, 1972.

TAMURA, H., S. MORI, and T. YAMAWAKI. "Textural features corresponding to visual perception." *IEEE Trans. SMC 8*, 1978, 460–473.

TOU, J. T. and R. C. GONZALEZ. *Pattern Recognition Principles*. Reading, MA: Addison-Wesley, 1974.

WESZKA, J. S., C. R. DYER, and A. ROSENFELD. "A comparative study of texture measures for terrain classification." *IEEE Trans. SMC 6*, 4, April 1976, 269–285.

ZUCKER, S. W. "Toward a model of texture." *CGIP 5*, 2, June 1976, 190–202.

Motion 7

7.1 MOTION UNDERSTANDING

Motion imagery presents many interesting challenges to computer vision, but static scene analysis received more attention in the 1960's and 1970's. In part, this may have been due to a technical problem: With most types of input media and domains, motion vision input is much more voluminous than static vision input. However, we believe that a more basic problem has been the assumption that motion vision could best be understood (or implemented) as many static frames analyzed very quickly, with results linked up in temporal sequence. This characterization of motion vision is extreme but perhaps illuminating. First, it assumes that vision involves processing static scenes. Second, it acknowledges that massive amounts of data may be required. Third, in it motion understanding degenerates to a postprocessing step which is mostly a matching operation—the differences or similarities between (understood) frames are analyzed and recorded. The extreme "static is basic" view is that motion is an unnaturally complex or difficult problem because it is ill suited to the techniques available.

A modified view is that object motion provides good image cues for segmentation, much as color might. This approach leads to the use of motion for segmentation, so that motion gets a more basic role in the understanding process. In this view, motion as such is useful for basic image understanding; a motion image sequence may actually be easier to understand than a static image, because the effects of motion can help in segmentation. Recent examples may be found in [Snyder 1981].

A further departure from the "static is basic" view is that motion understanding is qualitatively different from static vision. A logical extreme of this view is that there are many visual processing operations whose primitives are points in motion, and that in fact static vision is the puzzle, being ill-suited to the needs and mechanisms of biological systems. Serious work in computer motion understanding has begun even more recently than computer vision as a whole, and it is too early to dismiss any approach out of hand. There are domains and applications in which the "static is basic" paradigm seems natural, but it also seems very reasonable that animals have perceptual systems or subsystems for which "motion is basic."

Section 7.2 is concerned with processing and understanding the "flow" of the world image across the retina. Section 7.3 considers several techniques for understanding sequences of static images.

7.1.1 Domain Independent Understanding

Domain independent motion processing extracts information from time-varying images using the weakest possible assumptions about the world. Processing that merely transforms the input data into another image-like structure is in the province of generalized image processing. However, if the motion processing aggregates spatial information on the basis of a common feature, then the processing is a form of segmentation.

The basic visual input for domain-independent work in motion vision understanding is *optical flow*. Although Helmholtz noted the striking immediacy of three-dimensional perception mediated through motion [Helmholtz 1925], Gibson is usually credited with pioneering the theory that a primary visual stimulus for motion is the flow of elements in the optic array, or pattern of luminance in the full sphere of solid angle surrounding the observer [Gibson 1950, 1957, 1965, 1966]. Human beings undoubtedly are sensitive to optical flow, as evidenced by the "looming" reflex [Schiff 1965], the effect of flow on balance [Lee and Lishman 1975], and many other documented phenomena [Nakayama and Loomis 1974]. The basic input to an "optical flow understander" is a continuously changing visual field, which may be considered a field of vectors, each expressing the instantaneous change of position on the optic array of the image of a world point. A field of such vectors is shown in Fig. 7.1. The extraction of the vectors from the changing image is a low-level operation often posited by optical flow research; one computational mechanism was given in Chapter 3. Flow may also be approximated in an image sequence by matching and difference operations (Section 7.3.1).

Computer vision researchers have recently begun to concern themselves with both the geometry and computational mechanisms that might be useful in the understanding of optical flow [Horn and Schunck 1980; Clocksin 1980; Prager 1979; Prazdny 1979; Lawton 1981]. Many formalisms are in use. Cartesian, polar space, and spherical coordinates all have their appeal in different situations; differential vector geometry and simple analytic geometry are both used; even the geometry of the eye or camera varies from one study to another. This chapter does not contain a "unified flow theory;" instead it briefly describes several approaches, each of which uses a different aspect of optical flow.

7.1.2 Domain Dependent Understanding

The use of models, or at least stronger assumptions about the world, is complementary to domain-independent processing. The changing image, or even the field of optical flow, can be treated as input to a model-driven vision process whose goal

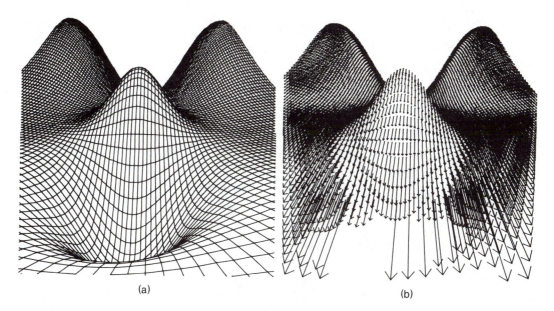

(a)	(b)

Fig. 7.1 An example of an optical flow field for an approaching "hill." (a) The hill. (b) Flow field.

is typically to segment the input into areas corresponding to meaningful world objects. The optical flow field becomes just another component of the generalized image, together with intensity, texture, or color. Motion often reveals information similar to that from range data; flow and range are discontinuous at object boundaries, surface orientation may be derived, and so forth. Object (or world) motions determine image (or retinal) motions; we shall be explicit about which motion we mean when confusion can occur.

Section 7.3 describes how knowledge of object motion phenomena can help in segmenting the flow field. One useful assumption is that the world contains rigid bodies. Tests for rigid bodies and calculations using data from them are quite useful—for example, the three-dimensional position of four points on a rigid object may be determined uniquely from three views (Section 7.3.2). A weaker object model, that they are assemblies of compound rigid pendula (linkages), is enough to accomplish successful segmentation of very sparse motion input which consists only of images of the end points of links (Section 7.3.3). Section 7.3.4 describes work with a highly specific and detailed model which is used in several ways to restrict low-level image processing and aid in three-dimensional interpretation of human motion images. Section 7.3.5 considers the processing of sequences of segmented images.

The coherence of most three-dimensional objects and their continuity through time are two general principles which, although occasionally violated, guide many segmentation and point-matching heuristics. The assumed correspondence of regions in images with objects is one example. Motion images provide another example; object coherence implies the likelihood of many "continuity" (actually similarity) conditions on the positions and velocities of neighboring image points.

Here are five heuristics for use in matching points from images separated by a small time interval [Prager 1979] (Fig. 7.2).

1. *Maximum velocity.* If a world point is known to have a maximum velocity V with respect to a stationary imaging device, then it can move at most $V\ dt$ between two images made dt time units apart. Thus given the location of the point in one image (and some assumptions about depth), this constraint limits where the point can appear on the second image.

2. *Small velocity change.* Since most visible physical objects have finite mass, this heuristic is a conseqence of physical laws and the assumption of a "small interval" between images. Of course, the definition of "small interval" depends on the definition of the velocity changes one desires to measure.

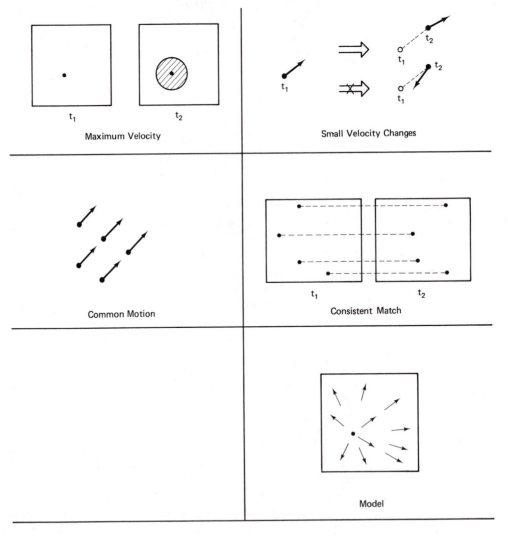

Maximum Velocity

Small Velocity Changes

Common Motion

Consistent Match

Model

Fig. 7.2 Five heuristics.

3. *Common motion.* Spatially coherent objects often appear in successive images as regions of points sharing a "common motion." It is interesting that such a weak notion as common motion (and the related "common position") actually can serve to segment very sparse scenes of a few points with very complex motion behavior if a long–enough sequence of images is used (Sections 7.3.3 and 7.3.4).

4. *Consistent match.* Two points from one image generally do not match a single point from another image (exceptions arise from occlusions). This is one of the main heuristics in the stereopsis algorithm described in Chapter 3.

5. *Known motion.* If a world model can supply information about object motions, perhaps retinal motions can be derived, predicted, and recognized.

In the discussions to follow these heuristics (and others) are often used or implicitly taken as principles. A careful catalog of the probable behavior of objects in motion is often a useful practical adjunct to a mathematical treatment. The mathematics itself must be based on a set of assumptions, and often these are closely related to the phenomenological heuristics noted above.

7.2 UNDERSTANDING OPTICAL FLOW

This section describes some more direct calculations on optical flow, using no other input information. Information may be obtained from flow that seems useful both for survival in the world and (on a less existential level) for automated image understanding. As with shape from shading research (Chapter 3), the paradigm here is often to see mathematically what information resides in the input and to use this to suggest mechanisms for doing the computation. The flow input is assumed to be known (Chapter 3 showed how to derive optical flow by local analysis of changing intensity in the image).

7.2.1 Focus of Expansion

As one moves through a world of static objects, the visual world as projected on the retina seems to flow past. In fact, for a given direction of translatory motion and direction of gaze, the world seems to be flowing out of one particular retinal point, the *focus of expansion* (FOE). Each direction of motion and gaze induces a unique FOE, which may be a point at infinity if the motion is parallel to the retinal (image) plane.

These aspects of optical flow have been studied by computing the simulated flow pattern an observer would see while moving through a "forest" of vertical cylinders [Prager 1979] or Gaussian hills and valleys [Lawton 1981]. Some sample FOEs are shown in Fig. 7.3. Figure 7.3c shows a second FOE when the field of view contains an object which is itself in motion.

Our first model of the imaging situation is a simplification of the imaging geometry given in Appendix 1. Let the viewpoint be at the origin with the view

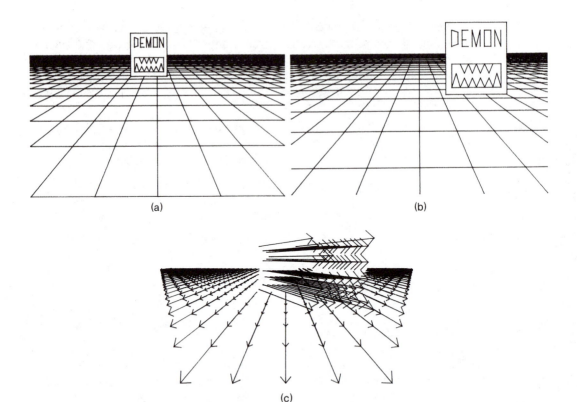

Fig. 7.3 FOE for rectilinear observer motion. (a) An image. (b) Later image. (c) Flow shows different FOEs for static floor and moving object.

direction out along the positive Z axis, and let the focal length $f = 1$. Then the perspective distortion equations simplify to

$$x' = \frac{x}{z} \tag{7.1}$$

$$y' = \frac{y}{z} \tag{7.2}$$

In the next two sections the letters u, v, and w (sometimes written as functions of t) denote world point velocity components, or the time derivatives of world coordinates (x, y, z). Observer motion with instantaneous velocity $(-dx/dt, -dy/dt, -dz/dt) = (-u, -v, -w)$, keeping the coordinate system attached to the viewpoint, gives points in a stationary world a relative velocity (u, v, w). Consider a point located at (x_0, y_0, z_0) at some initial time. After a time interval t, its image will be at

$$(x', y') = \left(\frac{x_0 + ut}{z_0 + wt}, \frac{y_0 + vt}{z_0 + wt} \right) \tag{7.3}$$

As t varies, this parametric "flow-path" equation is that of a straight line; as t goes to minus infinity, the image of the point travels back along the straight line toward a particular point on the image, namely,

$$\text{FOE} = \left(\frac{u}{w}, \frac{v}{w} \right) \tag{7.4}$$

This focus of expansion is where the optical flow originates on the image. If the observer changes direction (or objects in the world change their direction), the FOE changes as well.

7.2.2 Adjacency, Depth, and Collision

The flow path equation of a point moving with a constant velocity reveals information about its depth in z. The information is not provided directly, since all flow paths for points at a given depth do not look alike. However, there is the elegant relation

$$\frac{D(t)}{V(t)} = \frac{z(t)}{w(t)} \tag{7.5}$$

Here again w is dz/dt, and V is dD/dt. D is the distance along the straight flow path from the FOE to the image of the point. Thus the distance/velocity ratio of the point's image is the same as the distance/velocity ratio of the world point. This result is basic, but perhaps not immediately obvious.

The above relation is called the time-to-adjacency relation, because the right-hand side, z/w, is the z-distance of the point from the image plane divided by its velocity toward the plane. It is thus the time until the point passes through the image plane. This basic time interval is clearly useful when dealing with world objects; it changes when the magnitude of the world point's velocity (or the observer's) changes.

Knowing the depth of any point determines the depth of all others of the same velocity w, for it follows from the two time to adjacency equations of the points that

$$z_2(t) = \frac{z_1(t) D_2(t) V_1(t)}{V_2(t) D_1(t)} \tag{7.6}$$

The time-to-adjacency equation allows easy determination of the world coordinates of a point, scaled by its z velocity. If the observer is mobile and in control of his own velocity, and if the world is stationary, such scaled coordinates may be useful. Using the perspective distortion equations,

$$z(t) = \frac{w(t) D(t)}{V(t)} \tag{7.7}$$

$$y(t) = \frac{y'(t) w(t) D(t)}{V(t)} \tag{7.8}$$

$$x(t) = \frac{x'(t) w(t) D(t)}{V(t)} \tag{7.9}$$

As a last example, let us relate optical flow to the sensing of impending collisions with world objects. The focal point of the imaging system, or origin of coordinates, is at any instant headed "toward the focus of expansion," whose image coordinates are $(u/w, v/w)$. It is thus traveling in the direction

$$\mathbf{O} = (\frac{u}{w}, \frac{v}{w}, 1) \qquad (7.10)$$

and is following at any instant a path in the environment instantaneously defined by the parametric equation

$$(x, y, z) = t\mathbf{O} = t(\frac{u}{w}, \frac{v}{w}, 1) \qquad (7.11)$$

where t acts like a real scalar measure of time. Given this vector expression for the path of the observer, one can apply well-known vector formulas from analytic solid geometry to derive useful information about the relation of this path to world points, which are also vectors.

For example, the position $\mathbf{P}$ along the observer's path at which a world point approaches closest is given by

$$\mathbf{P} = \mathbf{x} - \frac{\mathbf{O}(\mathbf{O} \cdot \mathbf{x})}{(\mathbf{O} \cdot \mathbf{O})} \qquad (7.12)$$

where $\mathbf{O}$ is the direction of observer motion and $\mathbf{x}$ the position of the world point. Here the period (.) is the dot product operator. The squared distance Q^2 between the observer and the world point at closest approach is then

$$Q^2 = (\mathbf{x} \cdot \mathbf{x}) - (\mathbf{x} \cdot \mathbf{O})^2 / (\mathbf{O} \cdot \mathbf{O}) \qquad (7.13)$$

7.2.3 Surface Orientation and Edge Detection

It is possible to derive surface orientation and to characterize certain types of surface discontinuities (edges) by their motion. A formalism, computer program, and biologically motivated computational mechanism for these calculations was developed in [Clocksin 1980].

This section outlines mainly the surface orientation aspect of this work. As usual, the model is for a monocular observer, whose focal point is the origin of coordinates. An unusual feature of the model is that the observer has a spherical retina. The world is thus projected onto an "image unit sphere" instead of an image plane. World points and surface orientation are represented in an observer-centered Cartesian coordinate system. The image sphere has a spherical coordinate system which may be considered as "longitude" θ and "latitude" ϕ. These coordinates bear no relation to the orientation of the retina. World points are then determined by their image coordinates and a range r. An observer-centered Cartesian coordinate system is also useful; it is related to the sphere as shown in Fig. 7.4, and by the transformations given in Appendix 1.

The flow of the image of a freely moving world point may be found through the following derivation. As before, let the world velocity of the point (possibly induced by observer motion) $(dx/dt, dy/dt, dz/dt)$ be written (u, v, w). Similarly,

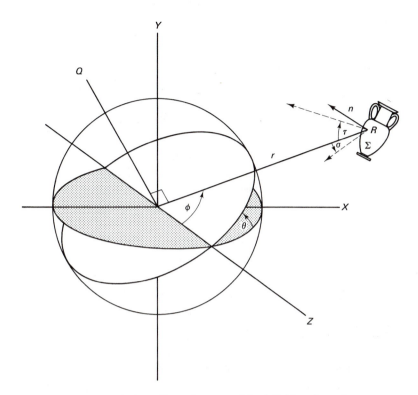

Fig. 7.4 Spherical coordinate system, and the definition of σ and τ.

write the angular velocities of the image point in the θ and ϕ directions as

$$\delta = \frac{d\theta}{dt} \tag{7.14}$$

$$\epsilon = \frac{d\phi}{dt} \tag{7.15}$$

Then from the coordinate transformation equations of Appendix 1,

$$y = x \tan\theta \tag{7.16}$$

Differentiating and solving for $d\theta/dt$ (written as δ) gives

$$\delta = \frac{v - u \tan\theta}{x \, sec^2\theta} \tag{7.17}$$

Substituting for x its spherical coordinate expression $r \sin\phi \cos\theta$ and simplifying yields the general expression for flow in the θ direction:

$$\delta = \frac{v \cos\theta - u \sin\theta}{r \sin\phi} \tag{7.18}$$

The derivation of ϵ proceeds from the coordinate transformation equation

$$z = r \cos\phi \tag{7.19}$$

Differentiating, solving for $d\phi/dt$ (written as ϵ), and using

$$\frac{dr}{dt} = \frac{xu + yv + zw}{r} \qquad (7.20)$$

yields the general expression for flow in the ϕ direction:

$$\epsilon = \frac{(xu + yv + zw)\cos\phi - rw}{r^2 \sin\phi} \qquad (7.21)$$

As usual, general point motions are rather complicated to deal with, and more constraints are needed if the optic flow is to be "inverted" to discover much about the outside world. Let us then make the simplification that the world is stationary and the observer is traveling along the z direction at some speed S (This assumption is briefly discussed below.) Explicitly, suppose that

$$u = 0, \quad v = 0, \quad w = -S$$

Substituting these into the general flow equations (7.18) and (7.21) yields simplified flow equations:

$$\delta = 0 \qquad (7.22)$$

$$\epsilon = \frac{S \sin\phi}{r} \qquad (7.23)$$

Thus r is a function of θ and ϕ and therefore so is ϵ.

It is this simplified flow equation which forms the basis for surface orientation calculation and edge detection. The goals are to assign to any point in the flow field one of three interpretations: *edge*, *surface*, or *space* and also to derive the type of edge and the orientation of the surface.

To find surface orientation, represent the surface normal of a surface Σ by two angles σ and τ defined as in Fig. 7.4 with the two planes of σ and τ being the RZ and QR planes, respectively. The slant is measured relative to the line of sight, denoted by R in the figure. σ and τ correspond to depth changes in "depth profiles" oriented along lines of constant θ and ϕ, respectively. Thus,

$$\tan\sigma = \left[\frac{1}{r}\right]\frac{\partial r}{\partial \phi} \qquad (7.24)$$

$$\tan\tau = \left[\frac{1}{r}\right]\frac{\partial r}{\partial \theta} \qquad (7.25)$$

Surface orientation is defined by σ and τ or equivalently by their tangents. A surface perpendicular to the line of sight has $\sigma = \tau = 0$.

Equations (7.24) and (7.25) assume the range r is known. However, one can determine them without knowing r through the simplified flow equation, Eq. (7.23). The latter may be written

$$r = \frac{S \sin\phi}{\epsilon(\theta, \phi)}$$

where $\epsilon(\theta, \phi)$ gives the flow in the ϕ direction. Differentiating this with respect to θ and ϕ gives

$$\frac{\partial r}{\partial \phi} = S\ \frac{\epsilon \cos \phi - \sin \phi\ (\partial \epsilon / \partial \phi)}{\epsilon^2} \tag{7.26}$$

$$\frac{\partial r}{\partial \theta} = -\ \frac{S \sin \phi\ (\partial \epsilon / \partial \theta)}{\epsilon^2} \tag{7.27}$$

These last three equations may be substituted into Eqs. (7.24) and (7.25), and the results may then be simplified to the following surface orientation equations:

$$\tan \sigma = \cot \phi - \frac{\partial}{\partial \phi} \ln \epsilon \tag{7.28}$$

$$\tan \tau = -\frac{\partial}{\partial \theta} (\ln \epsilon) \tag{7.29}$$

These tangents are thus easily computed from optical flow. The result does not depend on velocity, and no depth scaling is required. In fact, absolute depth is not computable unless we know more, such as the observer speed.

Turning briefly to edge perception: Although physical edges are a depth phenomenon, in flow they are mirrored by ϵ, the flow measure that allows determination of orientation without depth. In particular, it is possible to demonstrate that the Laplacian of ϵ has singularities where the Laplacian of depth has singularities. An arc on the sphere projects out onto a "depth profile" in the world, along which depth may vary. If the arc is parameterized by α, relations among the depth profile, flow profile, and the singularities in flow are shown in Fig. 7.5. Thus the Laplacian of ϵ provides information about edge type but not about edge depth.

The formal derivations are at an end. Implementing them in a computer program or in a biological system requires solutions to several technical problems. More details on the implementation of this model on a computer and a possible

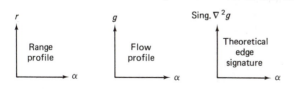

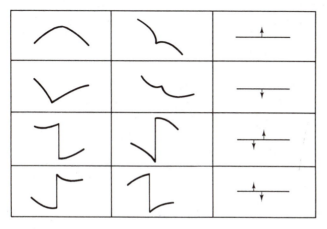

Fig. 7.5 The singularities of the second derivative of the flow profile inform about the type of edge.

implementation using low-level physiological vision primitives appear in [Clocksin 1980]. There are some data on human performance for the types of tasks attempted by the program. The assumption of a fixed environment basically implies that flow motions in the environment are likely to be interpreted as observer motions. This view is rather strikingly borne out by "swaying room" experiments [Lee and Lishman 1975], in which a subject stands in a swayable visual environment. (A large, low-mass bottomless box suspended from above may be lowered around the subject, giving him a room-like visual environment.) When the hanging "room" is made to sway, the subject inside tends to lose balance. Further, moving surfaces in the real world are quite often objects of interest, such as animals.

A survey of depth perception experiments [Braunstein 1976] points to motion as the dominant indicator of surface orientation perception. Random-dot displays of monocular flow patterns [Rogers and Graham 1979] evoke striking perceptions of solid oriented surfaces; flow may be adequate for shape and depth perception even with no other depth information. The experiments on perception of "edges," or discontinuities in flow caused by discontinuities in depth of textured surfaces, are less common. However, there have been enough to provide some confirmation of the model.

The computational model is consistent with and has correctly predicted psychological data on human thresholds for slant and edge perception in optical flow fields. (The thresholds are on the amount of slant to the surface and the depth difference of the edge sides.) The computational model can be used to determine range, but only to poor accuracy; this happens to correspond with the human trait that orientation is much more accurately determined by flow than is range. Quantitatively, the accuracy of orientation and range determinations are the same for the model and for human beings under similar conditions.

7.2.4 Egomotion

It is possible to extract information about complex observer motions from optical flow, although at considerable computational cost. In one formulation [Prazdny 1979], a model observer is allowed to follow any space curve in an environment of stationary objects, while at the same time turning its head. It is possible to derive formulae that determine the observer's instantaneous velocity vector and head rotational vector from a small number (six) of flow vectors in the image on a (standard flat) retina.

The equations that describe flow given observer motion and head rotation can be quite compactly written by using vector operators and a polar coordinate system (similar to that of the last section). The inherent elegance and power of the vector operations is well displayed in these calculations. Inverting the equations results in a system of three cubic equations of 20 terms each. Such a system can be solved by normal methods for simultaneous nonlinear equations, but the solutions tend to be relatively sensitive to noise. In the noise-free case, the method seems to perform quite adequately.

The calculation yields a method for deriving relative depth, or the ratio of the distances of points from the observer. An approximation to surface orientation may be obtained using several relative depth measurements in a small area and assuming that the surface normal varies slowly in the area.

7.3 UNDERSTANDING IMAGE SEQUENCES

An image sequence is an ordered set of images. The image sequences of interest here are samplings of four-dimensional space-time. Commonly, as in a movie, the images are two-dimensional projections of a three-dimensional physical world, sequenced through time. Sometimes the sequence consists of two-dimensional images of essentially two-dimensional slices of the three-dimensional world, sequenced through the third spatial dimension. Some of the techniques in this section are useful in interpreting the three-dimensional nature of objects from such spatial image sequences, but the main concern here is with temporal image sequences. In many practical applications, the input must be such a sequence, and continuous motion must be inferred from discrete location differences of image points. The thrust of work under these assumptions is often to extend static image understanding by making models that incorporate or explain objects in motion, extending segmentation to work across time [Thompson 1979, Tsotsos 1980].

When asked why he was listening to a metronome ticking, Ezra Pound is said to have replied that he did not listen to the ticks, but to the "spaces between them." Like Pound, we take the ticks, or images, as given, and are really interested in what goes on "between the ticks." We usually want to determine and describe how the images are related to each other. This information must be derived from the static images, and two approaches immediately present themselves: broadly, the first is to look for differences between the images, and the second is to look for similarities.

These two approaches are complementary, and are often used together. A general paradigm for object-oriented motion analysis is the following:

1. Segment (describe) the individual images. This process may be complex, yielding a relational structure or a segmentation into regions or edges. An important special case is the one in which the description (segmentation) process is null and the description is just the image itself. For example, an initial high-level static description is impossible if motion is to be used as an aid to segmentation.

2. Compute and describe the differences or similarities between the descriptions (or undescribed images).

3. Build a description of the sequence as a whole from the single-frame primitives and descriptions of difference or similarity that are relevant to the purpose at hand.

7.3.1 Calculating Flow from Discrete Images

This method is a form of disparity calculation that is not only used for flow calculations, but may also be used for stereo matching or tracking applications. The computations are implemented with "relaxation" techniques.

The flow calculations have so far assumed an underlying continuous image which was densely sampled. With those assumptions and a few more the fundamental motion equation allows the calculation of flow (Chapter 3). The approach of this section is to identify discrete points in the image that are very different from their surround. Given such discrete points from each of two images at different times, the problem becomes one of matching a point in one image with the right point (if it exists) in the other image. This matching problem is known as the *correspondence* problem [Duda and Hart 1973, Aggarwal et al 1981]. The solution to the correspondence problem in the case of motion is, of course, the optic flow.

One algorithm for matching distinct points from two different frames [Barnard and Thompson 1979] breaks the matching problem into two steps. The first is the identification of candidate match points in each of the two frames. The second is an iterative algorithm which adjusts match probabilities for pairs of match points. After successful termination of the algorithm, correct matches have high probabilities and incorrect matches have very low probabilities.

The Moravec interest operator ([Moravec 1977]; Section 3.2) produces candidate match points by measuring the distinctness of a local piece of the image from its surround. Each frame is analyzed separately so that the end result is two sets of points S_1 and S_2, one from each frame, which are candidates to be matched. Candidates in S_1 are indexed by i and those in S_2 by j.

The iterative part of the algorithm is initialized with a data structure for the possible matches that exploits the heuristic that a point in the world does not move large distances between frames. Potential matches for a given point $\mathbf{x}_i$ in S_1, the first image, are all points $\mathbf{y}_j$ in S_2 such that

$$\|\mathbf{x}_i - \mathbf{y}_j\| \leqslant v_{\max} \qquad (7.30)$$

where $v_{\max}$ is the maximum disparity allowed between points. All points that are selected by the Moravec operator have a given disparity vector $\mathbf{v}_{ij}$ and are kept as possible matches. Each disparity has an associated probability P_{ij} which changes through time as the most likely disparities are found. The information kept for each point $\mathbf{x}_i$ in S_1 looks like

$$(\mathbf{x}_i \, (\mathbf{v}_{ij_1}, \, P_{ij_1}) \, (\mathbf{v}_{ij_2}, \, P_{ij_2}) \cdots (V^*, P^*)) \qquad (7.31)$$

where V^* is a special symbol that denotes "no match," and all the j_k are members of S_2. Storing the flow vectors $\mathbf{v}$ implicitly stores the corresponding point in S_2 since $\mathbf{y}_j = \mathbf{x}_i + \mathbf{v}_{ij}$. Since the probabilities are adjusted iteratively, one final index is needed to denote the iteration value so that P_{ij} actually becomes P_{ij}^n for $n \geqslant 0$.

The initial approximation for the probabilities P_{ij}^0 takes advantage of the "common motion" heuristic: If $\mathbf{y}_j$ is the correct match point for $\mathbf{x}_i$, the image near $\mathbf{y}_j$ should look like the image near $\mathbf{x}_i$. Thus P_{ij}^0 can be defined by

$$P_{ij}^0 = \frac{1}{1 + cw_{ij}} \qquad \text{for } x \text{ in } S_1 \qquad (7.32)$$

where

$$w_{ij} = \sum_{|d\mathbf{x}| \leqslant k} [(f(\mathbf{x}_i + d\mathbf{x}, t_1) - f(\mathbf{y}_j + d\mathbf{x}, t_2)]^2 \qquad (7.33)$$

and c is constant. The updating formula is complex in form but basically is a weighted sum of neighboring match probabilities where the neighboring match is consistent (i.e., has nearly the same velocity). A neighboring match k is consistent if

$$\|\mathbf{v}_{ij} - \mathbf{v}_{kl}\| \leqslant dV_{\max} \tag{7.34}$$

The goodness of a particular match is measured by q_{ij}, where

$$q_{ij}^{n-1} = \sum_{k \text{ a neighbor of } i} \sum_{l \text{ s.t. } kl \text{ satisfies (7.34)}} P_{kl}^{n-1} \tag{7.35}$$

and the probabilities are updated by

$$\tilde{P}_{ij}^n = P_{ij}^{n-1}(A + Bq_{ij}) \tag{7.36}$$

$$P_{ij}^n = \frac{\tilde{P}_{ij}^n}{\sum_{j \text{ s.t. } ij \text{ is a match}} \tilde{P}_{ij}^n} \tag{7.37}$$

where the function of Eq. (7.36) is to renormalize the probabilities and A and B are constants.

The following simplified example makes these ideas more concrete.

Consider the situation given in Fig. 7.6, where the points in (a) are from S_1 and the points in (b) are from S_2. Using hypothetical values for P^0, an initial match data structure is, in terms of Eq. (7.31):

$$((4, 10) \quad ((5, 0), 0.7) \quad ((4, -5), 0.25) \quad ((2, -8), 0.05))$$
$$((4, 6) \quad ((5, 4), 0.5) \quad ((4, -1), 0.3) \quad ((2, -4), 0.2))$$
$$((2, 3) \quad ((7, 7), 0.3) \quad ((6, 2), 0.35) \quad ((4, -1, 0.2))$$

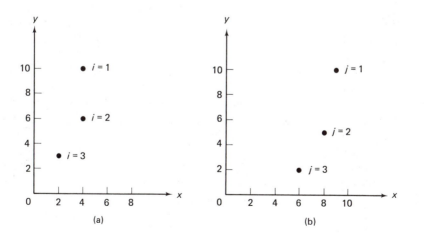

Fig. 7.6 Discrete matching: a concrete example.

Also, $Dv_{max} = 3$, using the chessboard norm. Using the updating formula (7.35), the first set of q_{ij}'s is given by

$$[q_{ij}^1] = \begin{bmatrix} 0.3 & 0.2 & 0 \\ 0 & 0.9 & 0.25 \\ 0 & 0 & 0.3 \end{bmatrix}$$

and the corresponding unnormalized probabilities, with $A = 0.3$ and $B = 3$, are

$$[\tilde{P}_{ij}^1] = \begin{bmatrix} 1.11 & 0.875 & 0.015 \\ 0.15 & 2.79 & 0.80 \\ 0.09 & 0.105 & 0.65 \end{bmatrix}$$

which are normalized to be

$$[P_{ij}^1] = \begin{bmatrix} 0.55 & 0.44 & 0.01 \\ 0.04 & 0.75 & 0.21 \\ 0.11 & 0.12 & 0.74 \end{bmatrix}$$

So after one iteration the match structure is already starting to converge to the best match of $P_{ii} = 1$, $P_{ij} = 0$ for $i \neq j$. Note that in general P_{ij} and q_{ij} are, in matrix form, sparse due to the consistency condition (7.34). To see the results for an example of a more appropriate scale, consult Fig. 7.7.

7.3.2 Rigid Bodies from Motion

The human visual system is predisposed to interpret (perceive) two-dimensional projections of moving three-dimensional rigid objects as just that—moving rigid objects. This facility is an interesting one, since it persists even when all three-dimensions information is removed from any single static view. This sort of result has been known for some time [Wallach and O'Connell 1953; Johansson 1964]. The ability to interpret points as three-dimensional objects demonstrated by Johansson means that the interpretation process does not rely solely on monitoring the changes of angles and length of lines, as suggested by Wallach and O'Connell.

Of course any change between two two-dimensional projections of points in three dimensions can be explained by any number of configurations and motions. Our visual system only accepts a few interpretations, often only one. This one is, in the world of moving objects in which we live, usually correct. This ability to reject unlikely interpretations is consistent with a "rigidity assumption" [Ullman 1979]: Any set of elements undergoing a two-dimensional transformation which has a unique interpretation as a rigid body moving in space should be so interpreted. It seems likely that something like this rigidity assumption is built into our visual system. However, saying that does not tell us much about how it could possibly work. Below we consider the problem of obtaining three-dimensional structure from sets of corresponding two-dimensional points.

One related area of work is the reconstruction of three-dimensional structure when the corresponding points in two dimensions are not known. The reconstruc-

(a)

(b)

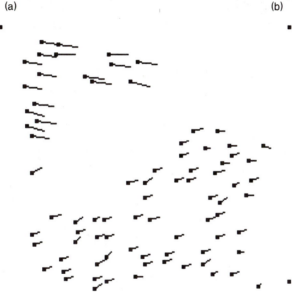

(c)

Fig. 7.7 Optical flow from feature point analyses. (a) An image. (b) Later image. (c) Optical flow found by relaxation.

tion procedure must begin by matching points in the several views. It can be shown [Shapira 1974] that general wire-frame objects of straight wires (of which the edges of polyhedra are only a special case) may be reconstructed from a finite number of perspective projections, but that for general wire-frame objects, the number of projections needed may be quite large. In fact, given any set of projections (viewpoints and viewing planes), an object may be constructed that is only ambiguously specified by those projections. Further work on reconstruction from projections is reported in [Shapira and Freeman 1978, Wesley and Markovsky 1981].

If point correspondences are known, it is possible to compute a unique three-dimensional location of four noncoplanar points from just three (orthographic) projections [Ullman 1979]. If the projections result from noncoplanar viewpoints, the recovery of three-dimensional structure is straightforward and is outlined below. If the projections are from coplanar viewpoints, the computations become more complex but still yield a unique result up to reflection. This second case is an important one; it applies if the camera is stationary and the object revolves about a single axis, for instance. Since the reconstruction is unique, the method never gets a wrong structure from accurate two-dimensional evidence about a rigid body. The probability that three views of four nonrigidly connected points can be interpreted as a rigid body is very low. Thus, the method is unlikely to report structure that is not there.

The method may be heuristically extended to multiple objects. Given the capability of describing the three-dimensional structure of four points, one can segment large collections of points by treating them in groups of four, deriving their structure and hence their motion. Groups of points that are not rigid have a very low probability of being interpreted as rigid, and the rest will presumably cluster into sets that share motions associated with rigid objects in the imaged scene. Thus the method to be described may be adaptable for image segmentation.

The calculation may be applied to coplanar points. If a unique result is derived, it is correct; otherwise, the fact that the points are coplanar is revealed. Generally, accuracy of two-dimensional positional information can be sacrificed to some degree if more points or more views are supplied. Perspective projections are more difficult to analyze. Such views can easily be treated approximately by the technique of breaking them into four element groups and treating each group as if it were orthographically projected in a direction depending on its position in the scene. Thus perspective may be dealt with globally, although each group is locally treated as an orthogonal projection. The assumption of orthographic projection implies that the method cannot recover relative depth of objects. The method does not lend itself well to "structure from receding motion" in which the motion information is largely encoded in the perspective effects which render objects larger or smaller as they advance and recede. The method does not serve well to explain human performance on moving images of a few points on nonrigid objects (such as those in Section 7.3.3).

Assume that three orthographic projections of four noncoplanar points are given, and that the correspondence between the points in the projection is known. Translational motion perpendicular to a projection plane is unrecoverable, and translation in a plane parallel to the projection plane is explicitly reproduced in the image by the projection process. The problem thus easily reduces to the case that one of the points is chosen as the origin of coordinates, and stays fixed throughout the process. This treatment follows that of [Ullman 1979].

Let the four points be 0, A, B, and C. Three orthographic views, projections on some planes Π_1, Π_2, and Π_3, are the input to the process. A coordinate system is chosen with origin at 0, and $\mathbf{a}$, $\mathbf{b}$, and $\mathbf{c}$ are vectors from 0 to A, B, and C. Then each view has a two-dimensional coordinate system with the image of 0 at its origin. Let $\mathbf{p}_i$ and $\mathbf{q}_i$ be the orthogonal unit basis vectors of the coordinate systems of the Π_i. Let the image coordinates of A, B, and C on Π_i be $(x(a_i), y(a_i))$, $(x(b_i),$

$y(b_i))$, and $(x(c_i), y(c_i))$ for $i = 1, 2, 3$. The calculations produce vectors $\mathbf{u}_{ij}$, which are unit vectors along the lines of intersection of Π_i with Π_j.

The image coordinates are in fact

$$
\begin{aligned}
x(a_i) &= \mathbf{a} \cdot \mathbf{p}_i & y(a_i) &= \mathbf{a} \cdot \mathbf{q}_i \\
x(b_i) &= \mathbf{b} \cdot \mathbf{p}_i & y(b_i) &= \mathbf{b} \cdot \mathbf{q}_i \\
x(c_i) &= \mathbf{c} \cdot \mathbf{p}_i & y(c_i) &= \mathbf{c} \cdot \mathbf{q}_i
\end{aligned} \tag{7.38}
$$

The unit vector $\mathbf{u}_{ij}$ is on both Π_i and Π_j; hence for some r_{ij}, s_{ij}, t_{ij}, and v_{ij},

$$\mathbf{u}_{ij} = r_{ij} \mathbf{p}_i + s_{ij} \mathbf{q}_i \tag{7.39}$$

$$r_{ij}^2 + s_{ij}^2 = 1$$

$$\mathbf{u}_{ij} = t_{ij} \mathbf{p}_j + v_{ij} \mathbf{q}_j \tag{7.40}$$

$$t_{ij}^2 + v_{ij}^2 = 1$$

Equations (7.39) and (7.40) yield

$$r_{ij} \mathbf{p}_i + s_{ij} \mathbf{q}_i = t_{ij} \mathbf{p}_j + v_{ij} \mathbf{q}_j \tag{7.41}$$

Taking the scalar product of $\mathbf{a}$, $\mathbf{b}$, and $\mathbf{c}$ with Eq. (7.41) yields three more equations, which are linearly independent. These equations in r_{ij}, s_{ij}, t_{ij}, and v_{ij}, combined with Eqs. (7.39) and (7.40), yield two solutions differing only in sign. But this means that (up to a sign) $\mathbf{u}_{ij}$ is determined in terms of the image coordinate basis vectors $(\mathbf{p}_i, \mathbf{q}_i)$ and $(\mathbf{p}_j, \mathbf{q}_j)$. Two $\mathbf{u}$ vectors determine one of the planes of orthogonal projection. For instance, $\mathbf{u}_{13}$ and $\mathbf{u}_{23}$ lie in P_3. Given the plane equation for the Π_i, the three-dimensional locations are computed as the intersection of lines perpendicular to the Π_i and through the two-dimensional image points. Of course, because of the ambiguity in sign, the expected mirror image ambiguity of structure exists.

The extension to the case that $\mathbf{u}_{12} = \mathbf{u}_{23} = \mathbf{u}_{31}$, where the three viewpoints are coplanar, is not difficult. It is perhaps a little surprising that coplanar viewpoints still yield a unique interpretation.

An extension of the mathematics to perspective imaging is not difficult to formulate, but the equations are nonlinear and must be solved either conventionally, say by the multidimensional Newton-Raphson technique of Appendix 1, or perhaps by cooperative algorithms of a more artificial intelligence flavor [Lawton 1981].

In geometrically underconstrained situations, plausible interpretations can sometimes be made by using other knowledge to give constraints. For example, one can minimize a second-difference approximation to the acceleration of points in order to use the "constraint" of smooth motion. Such a criterion may find a single "best" location for points. Another example is the use of position and velocity commonality over time to establish rigid members in linkages (Section 7.3.3), a first step to location determination.

To see how the equations might be set up, consider the perspective geometry of Section 7.2.1. In this simplified Cartesian system, Eqs. (7.1) and (7.2) are used as before. Since $z(x', y', 1) = (x, y, z)$, the location of any point is determined (up to a scale factor, since the focal length is not explicit) from its image coordinates and its depth coordinate, z. For $F \geqslant 1$ images and $N \geqslant 3$ points there are $FN - 1$ unknowns (the ability to scale distance allows one point to be placed arbitrarily).

To apply the rigid body constraint, enough pairwise distances between points must be specified to lock them into a rigid configuration. For three points, three distances are necessary. Each additional point requires another three distances, and so for each interframe interval $3(N - 2)$ constraints are needed, for a total of $3(F - 1)(N - 2)$ constraints. Thus, whenever

$$2FN - 6F - 3N + 7 \geqslant 0 \qquad (7.42)$$

consistent equations from the constraints can be solved [Lawton 1981]. With two views, five points are needed; with three views, four points. This is not surprising, given the preceding analysis for orthographic projections.

Consider the simple case of two points seen in two frames. If they are rigidly connected, one constraint equation holds. It is equivalent to

$$(\mathbf{x}_{11} - \mathbf{x}_{12}) \cdot (\mathbf{x}_{11} - \mathbf{x}_{12}) = (\mathbf{x}_{21} - \mathbf{x}_{22}) \cdot (\mathbf{x}_{21} - \mathbf{x}_{22}) \qquad (7.43)$$

$(\mathbf{x}_{ij}, \mathbf{x}'_{ij}$ are, respectively, the world and image coordinate vectors of point j in frame ij). Since $\mathbf{x}_{ij} = z_{ij} \mathbf{x}'_{ij}$, (recall (7.1) and (7.2)) the constraint becomes

$$z_{11}^2 (\mathbf{x}'_{11} \cdot \mathbf{x}'_{11}) + z_{12}^2 (\mathbf{x}'_{12} \cdot \mathbf{x}'_{12}) - 2 z_{11} z_{12} (\mathbf{x}'_{11} \cdot \mathbf{x}'_{12})$$
$$- z_{21}^2 (\mathbf{x}'_{21} \cdot \mathbf{x}'_{21}) - z_{22}^2 (\mathbf{x}'_{22} \cdot \mathbf{x}'_{22}) + 2 z_{21} z_{22} (\mathbf{x}'_{21} \cdot \mathbf{x}'_{22}) = 0 \qquad (7.44)$$

A further constraint that objects only move in the "ground plane," or at a constant y, has the effect of removing two unknowns through substitution in the constraint equation above. Since for arbitrary m and n,

$$y_{im} = z_{im} y'_{im} = y_{in} = z_{in} y'_{in} \qquad (7.45)$$

$$z_{in} = \frac{z_{im} y'_{im}}{y'_{in}} \qquad (7.46)$$

As a final example, a restriction to purely translational motion of the point configurations yields the constraint

$$(\mathbf{x}_{11} - \mathbf{x}_{21}) - (\mathbf{x}_{12} - \mathbf{x}_{22}) = 0 \qquad (7.47)$$

Expanding this as the product of unknown depths (z) and known image positions ($\mathbf{x}'$) yields a vector equation that may be written componentwise as three linear equations in four unknowns. Recall that a focal length must be fixed, effectively setting one unknown: setting one z_{ij} to 1 gives a system of three linear equations in the other three z_{ij}.

7.3.3 Interpretation of Moving Light Displays—A Domain-Independent Approach

One of the domains that provides the purest aspects of motion vision is moving light displays (MLDs). These are sequences of images which track only a few discrete points per frame. A typical way to produce an MLD is to attach small glass bead reflectors to a person's major joints (shoulders, elbows, wrists, hips, knees, ankles), focus a strong light on him or her, and manipulate the contrast of a video-tape recorder so as to produce on videotape a record of the movement of the reflective points on the joints. A single frame from such a record is unrecognizable by an inexperienced subject (Fig. 7.8).

However, a sequence of such frames quickly gives (typically in 0.4 second) not only a compelling perception of motion of a three-dimensional body, but allows recognition of the sequence as depicting a walking person, and a description of the type of motion (walking backward, jumping, walking left). Complicated scenes such as several independently moving bodies and couples dancing can be recognized. Sophisticated judgments can be made, such as determining the sex of a subject from an MLD, or recognizing the gait of a friend [Johannson 1964].

MLDs thus present quite a challenge to computer vision. It could be that MLDs of moving people are interpreted by specialized neural mechanisms expressly tailored to the purpose of dealing with any visual input whatever that sug-

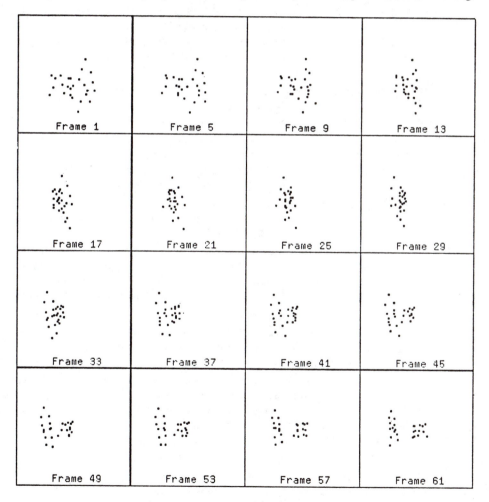

Fig. 7.8 An MLD for a man walking his dog.

gests moving people. MLDs certainly demonstrate that texture, continuous fields of flow, and especially that the interpretability of static versions of the scene are not necessary for human beings to do complex perception of certain three-dimensional objects.

This section is concerned with MLDs of moving human beings, and the interpretation we desire consists of separating images of individuals, in deriving their "connectivity" (i.e., the rigid links that connect the points), and possibly in describing the three-dimensional motion in which the subjects are engaged.

MLDs produced with perspective projection have few of the pleasant properties of the rigid orthographic projection which were used in Section 7.3.1. In particular, both translating and rotating objects are inherently ambiguous in perspective projections [Roache and Aggarwal 1979]. The approximate method outlined in Section 7.3.1, in which local groups of four points are considered rigid and orthographically projected, fails for MLDs of walking people. In many applications, digitization error will limit severely the accuracy returned. Worse, in a typical 12-point MLD of a moving person, there is never a rigid system of four noncoplanar points. The small departures from rigidity occurring in 30 ms of normal walking are enough to render the rigidity assumptions invalid [Rashid 1980].

An algorithm in [Badler 1975] extracts the trajectory of two moving points if they move in parallel paths and are viewed by spherical projection. The projection conditions are approximately met in typical moving-person MLDs, but the lack of points moving in parallel paths is enough to render the algorithm inapplicable.

A good start in the interpretation of MLDs involves solving the point-correspondence problem between frames. Knowing how points move from frame to frame gives at least a start on perceiving the continuity of the objects in the scene. Solving this problem from frame to frame may be attacked in any number of ways; the relaxation approach of Section 7.2.3 is an example.

Another is to predict the location of a point in the two-dimensional image from its velocity in the preceding frame. Velocity is computed from the differences in position of the point in the preceding two frames. Predicting where a point will be in frame 3 implies that one knew which point it was in frames 1 and 2. One way of getting the process started is to associate points in frames 1 and 2 that are nearest neighbors. Evidence suggests that human beings in fact are not infallible trackers of points in MLDs [Rashid 1980]. However, they do not let local inconsistencies in point interpretation (say, if the ankle momentarily "turns into" the knee) detract from their overall perception of a moving person. This is a good example of how inconsistent interpretations arise in human vision.

A program can be given similar resilience by having it suspend judgment on contradictory clues and use succeeding frames to resolve the problem [Rashid 1980; O'Rourke 1980]. Having established local point correspondences, the next problem is to group the points into coherent three-dimensional structures and separate individual bodies moving in the scene. When constraints on the scene are available that make analytic techniques applicable (Section 7.3.1), explicit grouping of points prior to analysis may be unnecessary. In fact, with complex MLDs such as Ullman studied (e.g. two transparent but spotty coaxial cylinders rotating in opposite directions about an axis in the viewing plane), most naive grouping strategies based on two-dimensional motion in the image will fail. Ullman's method chooses four-tuples of points from such a scene; on the average seven-eighths of such groups involve points from both cylinders, but with accurate data the algorithm can identify such nonrigid four-tuples. The remaining one-eighth of the groups have consistent interpretations as rigid rotating groups, and the groups fall into two classes, one for each cylinder.

One straightforward heuristic approach to MLD interpretation enjoys moderate success and does not use domain-dependent models [Rashid 1980]. It has the characteristic that it deals exclusively with two-dimensional motions in order to extract information about three dimensions. The approach is more heuristic than Lawton's and certainly more than Ullman's (Section 7.3.1). It is prey to many of the same pitfalls that threaten any image-based (as opposed to world-based) approach to computer vision. With sparse MLDs of nonrigid objects, clustering algorithms may be used to group points into related structures. Rashid's method computes the minimum spanning tree of points in a four-dimensional space of two-dimensional position and two-dimensional velocity. That is, each point in the MLD is represented at any time t by a four-vector

$$(x(t), \ y(t), \ u(t), \ v(t))$$

where u and v are the velocity in image x and y coordinates. Points may be clustered in this position-velocity space on the basis of a four-dimensional Euclidean metric, modified by information about distances derived from preceding frames. Perspective distortion can affect the usefulness of two-dimensional distances computed in previous frames, and data scaling is useful to establish a reasonable relation between units in the four-dimensional space. Rashid's technique is to scale the data in each dimension to have unit variance and zero mean, and to compute cumulative distances between points in a frame by a function such as

$$D_n(i, j) = d(i, j) + D_{n-1}(i, j) \times 0.95 \tag{7.48}$$

where $D_n(i, j)$ is the cumulative distance between points i and j in frame n, and $d(i, j)$ is their Euclidean distance.

This clustering method can successfully group points on the two cylinders in the rotating-cylinder sequence mentioned above after seven frames. Figure 7.9 gives the results of clustering the data for the MLD of Fig. 7.8. Clustering is stable after some 25 frames (about one-half of a step).

7.3.4 Human Motion Understanding—A Model-Directed Approach

Human motion understanding may be done with a much different approach than the heuristic clustering applied to MLDs in Section 7.3.3. A very detailed model of the domain can help restrict search, make inferences, disambiguate clues, and so forth. A program for understanding images of human motion successfully uses such an approach [O'Rourke 1980; O'Rourke and Badler 1980].

The body model accounts for such factors as relative location of body parts, joint angle ranges, joint angle acceleration limits, collision checking, and gravity. A motion simulation program drives a "bubble man" representation of a person

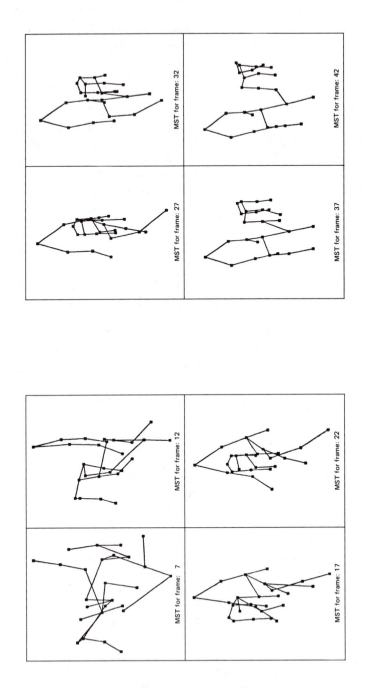

Fig. 7.9 The minimal spanning tree for the man and dog.

MST for frame: 32

MST for frame: 42

MST for frame: 27

MST for frame: 37

MST for frame: 12

MST for frame: 22

MST for frame: 7

MST for frame: 17

(Fig. 7.10a) [Badler and Smoliar 1979]. This representation is used to produce a shaded graphic rendition which serves as input to the motion understanding program (Fig. 7.10b). Knowledge of the imaging process also provides constraints on the configuration of the figure represented. For instance, perspective, the figure/ground distinction, the location of features, and occlusion all have implications for the interpretation of the scene as a configuration of the model.

The system is another example of a cooperative, constraint-satisfying system (Chapter 12), this time one that involves a high-level domain-dependent model.

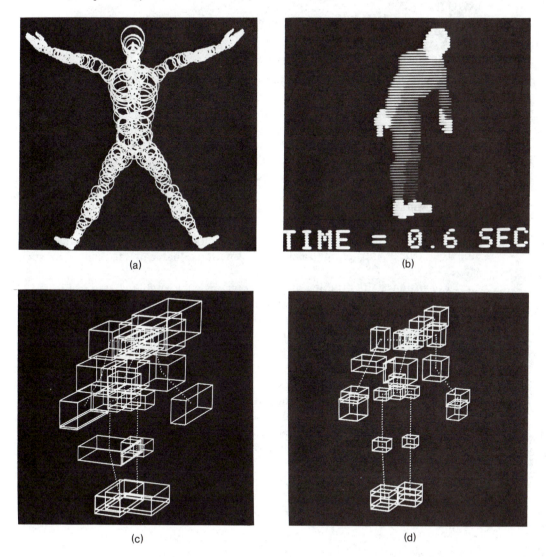

(a)

(b)

(c)

(d)

Fig. 7.10 Understanding human motion through the incorporation of many constraints. (a) Bubble Man from simulation program. (b) Input to motion understander; a bowing man. (c, d) Initial and final stages in understanding the motion of the bowing man.

The constraints imposed by the model restrict the application of low-level operators, and their results reduce uncertainty in parts of the model configuration.
Through the relations between model parts, improved estimates for part locations
are evolved and propagate throughout the model. Figure 7.10c and d show how the
image of the bowing man is understood more accurately as time passes and more
constraints are propagated through the model. It should be noted that only the
hand, foot, and head features are explicitly searched for in the image. The boxes
represent possible locations for the obvious body parts. Note how the occlusion has
been understood.

7.3.5 Segmented Images

Moving Polygons and Line Drawings

As one step along the way to motion understanding, the analysis of ideal polygonal images was popular for a time [Aggarwal and Duda 1975; Martin and Aggarwal 1978; Potter 1975]. The assumptions are usually that opaque polygons
move in parallel planes and may obscure one another (this is often called a 2.5-
dimensional situation). The viewpoint is somewhere "above" the collection of
moving shapes. The viewer (program) is presented with a sequence of frames either of line drawings or gray level images of the scene (Fig. 7.11). Polygon motion
is assumed small between frames. The goal is usually to segment the scenes into
polygons, and to extract such information as their direction and speed of motion.
The solutions to these problems usually reflect assumptions about the connectivity
of the polygons, or restrictions on their motion, and often revolve about the allowable topological and geometrical transformations that can take place in such
scenes.

For instance, in a frame with two polygons such as that shown in Fig. 7.12,
certain scene vertices belong to primitive polyhedra (they are "true" vertices),
whereas others are "false" artifacts of occlusion. The lines impinging at true vertices will not change their angle of meeting through time, but false vertices may
change angles if the polygons rotate as they move. False vertices are usually ob-
tuse.

Complex connectivity changes can arise when nonconvex polygons slide past
one another. Sorting out a coherent interpretation of a sequence of frames, especially in the presence of noisy vertex positions, is a challenging exercise.

A system was designed in [Badler 1975] which used sequences of line drawings produced by a spherical projection of a three-dimensional world to reconstruct

Fig. 7.11 Two frames from a motion image of three moving polygons.

Ch. 7 Motion

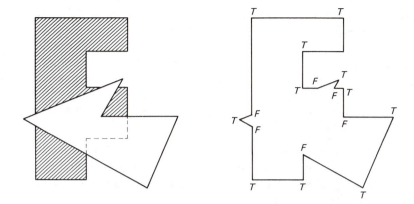

Fig. 7.12 True (T) and False (F) vertices in a scene of two overlapping polygons.

some three-dimensional aspects of the input, and to transform the pictorial input into natural language descriptions of motion.

Similarity Analysis, Then Difference Measurement

This approach is probably the most intuitive if motion perception is thought to be built up from perception of successive frames. The idea is simply to extract an object in one frame, and to search for it in the next frame. Obviously, the basic techniques here are the description-extraction process (i.e., static computer vision, the topic of most of this book) and matching (Chapter 11).

The entire range of matching techniques, from image matching to description matching, has been applied to image sequences. One characteristic of this approach in its pure form is that motion is merely a nuisance—segmentation is performed without using motion information. Usually the approach is pursued in a more pragmatic and domain-dependent fashion: for instance, the matching may be guided by knowledge about the motions.

One advanced system that uses this basic paradigm is described in [Price 1976; Price 1978; Price and Reddy 1977]. It segments and describes both images first. Using the symbolic descriptions, it matches complex scenes (such as houses or aerial images) that have been relatively rotated by large amounts (45 to 180°) and have size differences as well. It also derives the geometric transformation that produced the second image from the first.

Clearly, the major problems in systems of this sort come from generating and matching descriptions. The matching must be sophisticated, and to be successful in general it must combine symbolic and geometric components. The constraint that successive frames do not reflect violent motions eases the matching problem considerably, and iconic correlation techniques may sometimes apply.

Difference Measurement, Then Similarity Analysis

The idea behind this approach is to guide the similarity analysis with information about image differences. This seems a promising idea, because differences are easy to compute, whereas the very definition of similarity is open to question, and computing it may be arbitrarily complex.

In particular, in locating moving objects in an image sequence, one is invited to ignore the stationary background. The area of changing image can be tracked easily from image to image, and subjected to further analysis. Rather than trying to track an object from image to image, it is attractive to consider letting the object move far enough that it does not overlap between two images. Then the difference between the images will actually reflect the structure of the object.

One possible method [Nagel 1978a, 1978b; Jain and Nagel 1978] proceeds as follows:

1. Obtain two images from the motion sequence such that the object of interest will have moved far enough not to overlap in position in the two images. (One clearly needs information about the objects and the imaging parameters to assure no overlap.)

2. Segment the two images into regions.

3. Compute a dissimilarity measure between the overlapping areas of regions in the two images. One reasonable measure is the likelihood ratio for the two hypotheses that the intensities in the overlaps come from the same distribution of intensities or from different distributions.

4. In one of the images, take all regions that are most consistent with the hypothesis of different distributions and assume that they arise from the moving object (or its old vacated position). Merge these regions by a reasonable technique into one which is taken to include the moving object.

5. Take the boundary of the candidate region and use it as a template for correlation detection tracking between adjacent frames.

6. The offsets revealed by the correlation process give the velocity, and can be used to "subtract out" the motion, register the views of the object in several images, and thus obtain a more accurate characterization of the object.

This approach leads to results such as those shown in Fig. 7.13.

EXERCISES

7.1 Write a geometric explanation of the FOE phenomenon.

7.2 Devise a motion segmentation scheme for rigid bodies in translational three-dimensional motion that uses the FOE calculation.

7.3 Prove that the parametric flow path equation (7.3) indeed does produce a straight line in image coordinates.

7.4 Prove the time-to-adjacency relation (7.5). A geometrical demonstration may be made with similar triangles; an algebraic one is not very hard.

7.5 Express Eq. (7.12) as much as possible in terms of observables in the optical flow "image." What is left unspecified?

7.6 Perform Exercise 7.5 with equation (7.13).

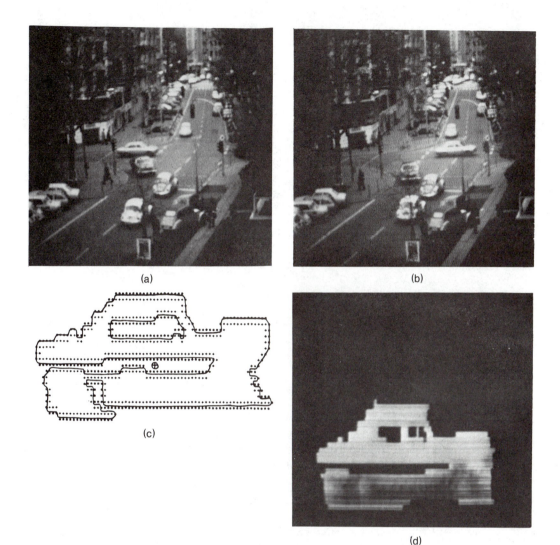

(a)

(b)

(c)

(d)

Fig. 7.13 Motion from segmented images. Initial (a) and final (b) frames from 16-frame sequence. The object of interest is the car moving left to right in the intersection. (c) Car segmentation from an intermediate frame. (d) Car reconstructed from several frames; the gray values result from aligning the values extracted from individual frames by segmentation.

7.7 Specialize the result of Exercise 7.6 to the case that the observer is moving in the direction of his direction of view [the FOE is at $(0,0)$].

7.8 Fill in the steps in the derivations of the general and special cases of δ and ϵ (Eqs. (7.18) and (7.21) through (7.23)).

7.9 Fill in the steps in the derivations of $\tan \sigma$ and $\tan \tau$ (Eqs. (7.28) and (7.29)).

7.10 Show how to compute absolute depth from flow (Section 7.2.2) if the observer speed is known.

7.11 The Laplacian of ϵ in Section 7.2.3 is the sum of the second partial derivatives of ϵ

with respect to θ and ϕ. Write it out and show that it has singularities only when the Laplacian of depth (r) does except at $\phi = 0$ or π or $r = 0$.

7.12 In Section 7.2.2, the θ, ϕ system is divorced from the retinal position. How might this coordinate system be deduced from optical flow, or how might this deduction be unnecessary?

7.13 Work out the details of the vector equation referred to in the last paragraph of Section 7.3.2.

7.14 What do flow paths look like if the observer (or the environment) only executes rotational motion? Pick a congenial coordinate system and prove your supposition.

7.15 Tighten up the "common motion" heuristic in Section 7.1.2. What domains under what sorts of world motion yield what sorts of "common" image motions for objects?

REFERENCES

AGGARWAL, J. K. and R. O. DUDA. "Computer analysis of moving polygonal images." *IEEE Trans. Computers 24*, 1975, 966–976.

AGGARWAL, J. K., L. S. DAVIS, and W. N. MARTIN. "Correspondence processes in dynamic scene analysis." *Proc. IEEE 69*, 5, May 1981, 562–571.

BADLER, N. "Temporal scene analysis: conceptual descriptions of object movements." Technical Report 80, Dept. of Computer Science, Univ. Toronto, February 1975.

BADLER, N. I. and S. W. SMOLIAR. "Digital representations of human movement." *Computing Surveys 11*, 1, 19–38, March 1979.

BARNARD, S. T. and W. B. THOMPSON. "Disparity analysis of images." Technical Report 79-1, Computer Science Dept., Univ. Minnesota, January 1979.

BRAUNSTEIN, M. L. *Depth Perception through Motion*. New York: Academic Press, 1976.

CLOCKSIN, W. F. "Computer prediction of visual thresholds for surface slant and edge detection from optical flow fields." Ph.D. dissertation, Univ. Edinburgh, 1980.

DUDA, R. O. and P. E. HART. *Pattern Recognition and Scene Analysis*. New York: Wiley, 1973.

GIBSON, J. J. *The Perception of the Visual World*. Boston: Houghton Mifflin, 1950.

GIBSON, J. J. "Continuous perspective transformations and the perception of rigid motion." *J. Experimental Psychology 54*, 1957, 129–138.

GIBSON, J. J. "Research on the visual perception of motion and change." In *Readings in the Study of Visually Perceived Movement*, Irwin M. Spigel (Ed.). New York: Harper & Row, 1965.

GIBSON, J. J. *The Ecological Approach to Visual Perception*. Ithaca, NY: Cornell University Press, 1966.

HELMHOLTZ, H. VON. *Treatise on Physiological Optics* (translated by J. P. C. Southall). New York: Dover Publications, 1925.

HORN, B. K. P and B. G. SCHUNCK. "Determining optical flow." AI Memo 572, AI Lab, MIT, April 1980.

JAIN, R. and H.-H. NAGEL. "On a motion analysis process for image sequences from real world scenes." *Proc.*, IEEE Workshop on Pattern Recognition and Artificial Intelligence, Princeton, NJ, 1978.

JOHANSSON, G. "Perception of motion and changing form." *Scandinavian J. Psychology 5*, 1964, 181–208.

LAWTON, D. T. "The processing of dynamic images and the control of robot behavior." Ph.D. dissertation, Univ. Massachusetts, 1981.

LEE, D. N. and J. R. LISHMAN. "Visual proprioceptive control of stance." *J. Human Movement Studies 1*, 1975, 87–95.

MARTIN, W. N., and J. K. AGGARWAL. "Dynamic scene analysis." *CGIP 7*, 1978, 356–374.

MORAVEC, H. P. "Towards automatic visual obstacle avoidance." *Proc.*, 5th IJCAI, August 1977, 584.

NAGEL, H.-H. "Formation of an object concept by analysis of systematic time variations in the optically perceptible environment." *CGIP 7*, 2, June 1978a, 149–194.

NAGEL, H.-H. "Analysis techniques for image sequences." *Proc.*, 4th IJCPR, November 1978b, 186–211.

NAKAYAMA, K. and J. M. LOOMIS. "Optical velocity patterns, velocity sensitive neurons, and space perception." *Perception 3*, 1974, 63–80.

O'ROURKE, J. "Image analysis of human motion." Ph.D. dissertation, The Moore School of Electrical Engineering, Univ. Pennsylvania, 1980.

O'ROURKE, J. and N. I. BADLER. "Model-based image analysis of human motion using constraint propagation." *IEEE Trans. PAMI 2*, 4, November 1980.

POTTER, J. L. "Velocity as a cue to segmentation." *IEEE Trans. SMC 5*, 1975, 390–394.

PRAGER, J. M. "Segmentation of static and dynamic scenes." COINS Technical Report 79-7, Computer and Information Science, Univ. Massachusetts, May 1979.

PRAZDNY, K. "Egomotion and relative depth map from optical flow." Computer Science Dept., Univ. Essex, March 1979.

PRICE, K. E. "Change detection and analysis in multi-spectral images." Ph.D. dissertation, Dept. of Computer Science, Carnegie-Mellon Univ., 1976.

PRICE, K. E. "Symbolic matching and analysis with substantial changes in orientation." *Proc.*, IEEE Workshop on Pattern Recognition and Artificial Intelligence, Princeton, NJ, 1978.

PRICE, K. E., and R. REDDY. "Change detection and analysis in multi-spectral images." *Proc.*, 5th IJCAI, August 1977, 619–625.

RASHID, R. F. "LIGHTS: a system for interpretation of moving light displays." Ph.D. dissertation, Computer Science Dept., Univ. Rochester, April 1980.

ROACHE, J. W. and J. K. AGGARWAL. "On the ambiguity of three-dimensional analysis of a moving object from its images." IEEE Workshop on Computer Analysis of Time-Varying Imagery, April 1979.

ROGERS, B. and M. GRAHAM. "Motion parallax as an independent cue for depth." *Perception 8*, 1979, 125–134.

SCHIFF, W. "The perception of impending collision: A study of visually directed avoidant behavior." *Psychological Monographs 79*, 1965.

SHAPIRA, R. "A technique for the reconstruction of a straight-edge, wire-frame object from two or more central projections." *CGIP 3*, 4, December 1974, 318–326.

SHAPIRA, R. and H. FREEMAN. "Computer description of bodies bounded by quadratic surfaces from a set of important projections." *IEEE Trans. Computers 27*, 9, September 1978, 841–854.

SNYDER, W. E. (ed.). "Computer analysis of time varying images," *IEEE Computer 14*, 8, August 1981.

THOMPSON, W. B. "Combining motion and contrast for segmentation." Technical Report 79-7, Computer Science Dept., Univ. Minnesota, March 1979.

TSOTSOS, J. K., J. MYLOPOULOS, H. D. COVVEY and S. W. ZUCKER. "A framework for visual motion understanding." *IEEE Trans. PAMI 2*, 6, November 1980, 563–573.

ULLMAN, S. *The Interpretation of Visual Motion* (Ph.D. dissertation). Cambridge, MA: MIT Press, 1979.

WALLACH, H. and D. N. O'CONNELL. "The kinetic depth effect." *J. Experimental Psychology 45*, 4, 1953, 205–217.

WESLEY, M. A. and G. MARKOVSKY. "Fleshing out projections," Research Rpt. RC8884, Computer Sciences Dept., IBM, T. J. Watson Research Center, April 1981.

GEOMETRICAL
STRUCTURES   III

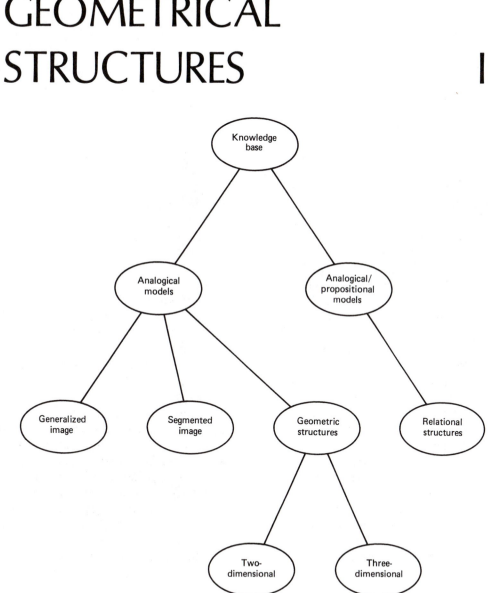

Ultimately, one of the most important things to be determined from an image is the *shape* of the objects in it. Shape is an intrinsic property of three-dimensional objects; in a sense it is the primal intrinsic property for the vision system, from which many others (surface normals, object boundaries) can be derived. It is primal in the sense that we associate the definitions of objects with shape, rather than with color or reflectivity, for example.

Webster defines shape as "that quality of an [object] which depends on the relative position of all points composing its outline or external surface." This definition emphasizes the fact that we are aware of shapes through outlines and surfaces of objects, both of which may be visually perceived. It also makes the distinction between the two-dimensional outline and the three-dimensional surface. We preserve this distinction: Chapter 8 deals with two dimensional shapes, Chapter 9 with three dimensional shapes.

If our goal is to understand flat images, why bring solids into consideration? Our simple answer is that we believe in many cases vision without a "solid basis" is a practical impossibility. Much of the recent history of computer vision demonstrates the advantages that can be gained by acknowledging the three-dimensional world of objects. The appearance of objects in images may be understood by understanding the physics of objects and the imaging process. The purest form of two-dimensional recognition, template matching, clearly does not practically extend to a world where objects appear in arbitrary positions, much less to a world of nonrigid objects. It is true that in some important image understanding tasks (interpretation of chest radiographs, ERTS images or some microscope slides), the third dimension is irrelevant. But where the three-dimensionality of objects is important, the considerable effort necessary to develop a usable three-dimensional model will always be amply repaid.

Shape recognition is doubtless one of the most important facilities of the mammalian visual system. We have seen how important shape information can be extracted from images in early processing and segmentatiou. One of the major challenges to computer vision is to represent shapes, or the important aspects of shapes, so that they may be learned, matched against, recollected, and used. This effort is hampered by several factors.

1. *Shapes are often complex.* Whereas color, motion, and intensity are relatively simply quantified by a few well-understood parameters, shape is much more subtle. Common manufactured or natural shapes are incredibly complex; they may be represented "explicitly" (say by representing their surface) only with hundreds of parameters. Worse, it is not clear what aspects of shapes are important for applications such as recognition. An explicit and complete representation may be computationally intractable for such basic uses as matching. What "shape features" can be used to ease the burden of computation with complex shapes?

2. *Introspection is no help.* Human beings seem to have a large fraction of their brains devoted to the single task of shape recognition. This important activity is largely "wired in" at a level below our conscious introspection. Why is shape recognition so easy for human beings and shape description so hard? The fact that we have no precise language for shape may argue for the inaccessibility of our shape-processing algorithms or data structures. This lack of cognitive leverage is a trifle daunting, especially when taken with the complexity of everyday shapes.

3. *There is little classical guidance.* Mathematics traditionally has not concerned itself with shape. For instance, only recently has there been a mathematical definition of "rigid solid" that accords with our intuition and of set operations on solids that preserve their solidity. The fact that such basic questions are only now being addressed indicates that computer science must do more than encode some already existing proven ideas. Thus we have the next point.

4. *The discipline is young.* Until very recently, human beings communicated about complex shapes mainly through words, gestures, and two-dimensional drawings. It was not until the advent of the digital computer that it became of interest to represent complex shapes so that they could be specified to the machine, manipulated, computed with, and represented as output graphics. No generally accepted single representation scheme is available for all shapes; several exist, each with its advantages and disadvantages. Algorithms for manipulating shapes (for example, for computing how to move a sofa up a flight of stairs, or computing the volume of a specified shape) are surprisingly complex, and are research topics. Often the representations good for one application, such as recognition, are not good for other computations.

It is the intention of this part of the book to indicate some of what is known about the representation of shape. Although the details of geometric representations may be still under development, they are an essential part of our layered computer vision organization. They are more abstract than segmented structures and are distinguished from relational structures by their preponderance of metric information.

Representation of
Two-Dimensional
Geometric Structures 8

8.1 TWO-DIMENSIONAL GEOMETRIC STRUCTURES

The structures of this chapter are the intuitive ones of well-behaved planar regions and curves. A mathematical characterization of these structures that bars "pathological" cases (such as regions of a single point and space-filling curves) is possible [Requicha 1977]. Basically the requirement is that regions be "homogeneously two-dimensional" (contain no hanging or isolated structures of different dimension—solids, lines or points). Similarly, curves should be homogeneously one-dimensional. The property of regularity is sometimes important; a regular set is one that is the closure of its interior (in the relevant one- or two-dimensional topology). Intuitively, regularizing a two-dimensional set (taking the closure of its interior) first removes any hanging one- and zero-dimensional parts, then covers the remainder with a tight skin (Fig. 8.1). In computer vision, often regions and curves are discrete, being defined on a raster of pixels or on an orthogonal grid of possible primitive edge segments. It is frequently convenient to associate a direction with a curve, hence ordering the points along it and defining portions of the plane to its left and right.

The one-dimensional closed curve that bounds a well-behaved region is an unambiguous representation of it; Section 8.2 deals with representations of curves and hence indirectly of regions. Section 8.3 deals with other unambiguous representations of regions that are not based on the boundary. Sometimes unambiguous representation is not the issue; it may be important to have qualitative description of a region (its size or shape, say). Section 8.4 presents several terse descriptive properties for regions.

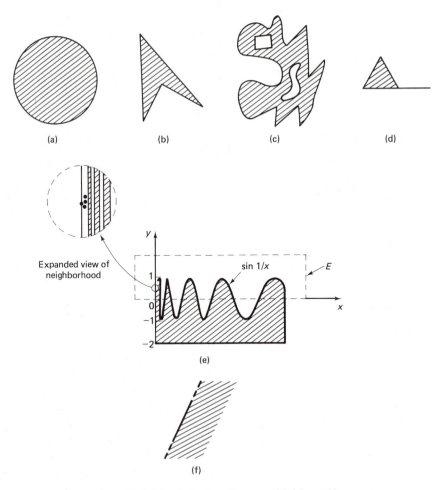

Fig. 8.1 (a, b, c) are Regions; (d) (e) and (f) are not.

8.2 BOUNDARY REPRESENTATIONS

8.2.1 Polylines

The "two-point" form of a line segment (see Appendix 1) extends easily to the *polyline*, which represents a concatenation of line segments as a list of points. Thus the point list $\mathbf{x}_1$, $\mathbf{x}_2$, $\mathbf{x}_3$ represents the concatenation of the line segments from $\mathbf{x}_1$ to $\mathbf{x}_2$ and from $\mathbf{x}_2$ to $\mathbf{x}_3$. If the first point is the same as the last, a closed boundary is represented.

Polylines can approximate most useful curves to any desired degree of accuracy. One might think there is one obvious way to approximate a boundary curve (or raw data) with a polygonal line. This is not so: many different approaches are possible. Finding a satisfying polygonal approximation to a given curve basically involves segmentation issues. The problem is to find corners or *breakpoints* that

Ch. 8   Representation of Two-Dimensional Geometric Structures

yield the "best" polyline. As with region-based segmentation schemes, the ideas here can be characterized by the concepts of *merging* and *splitting*. Splitting and merging schemes may be combined, especially if the appropriate number of linear segments is known beforehand. For details, see [Horowitz and Pavlidis 1976].

In a merging algorithm, points along a curve (possibly in image data) are considered in order and accepted into a linear segment as long as they fit sufficiently well. When they do not, a new segment is begun. The efficiency and characteristics of these schemes are quite variable, and endless variations on the general idea are possible. A few examples of "one pass" merging schemes are given here: explicit algorithms are available in [Pavlidis 1977].

If the boundary (represented on a discrete grid) is known to be piecewise linear, it is specified by its breakpoints. To find them, one can look along the boundary, monitoring the angle between two line segments. One segment is between the current point and a point several points back along the boundary; the other is between the current point and one several points forward. When the angle between these segments reaches a maximum over some threshold, a breakpoint is declared at the current point. This scheme does not adjust breakpoint positions, and so is fast [Shirai 1975] but works best for piecewise linear input curves.

Tolerance-band solutions place a point on either side of the curve at the maximum allowable error distance, and then find the longest piece of the curve that lies entirely between parallel lines through the two points [Tomek 1974]. This method proceeds without breakpoint adjustment, and may not find the most economical set of segments (Fig. 8.2).

An approximation of a curve with a polyline of minimum length in error by at most a pixel is given in [Sklansky and Kibler 1976]. Each curve pixel is considered a square and the resulting pixel structure is four-connected. The approximation describes the shape of an elastic thread placed in the pixel structure (Fig. 8.3). The

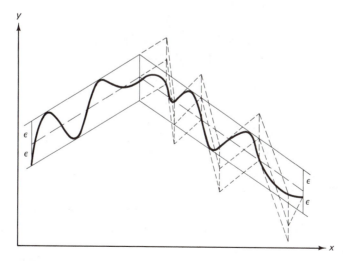

Fig. 8.2 Simple tolerance-band solution (dotted lines). Better solution (solid lines).

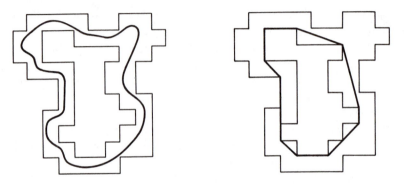

Fig. 8.3   Minimum length polyline.

method tends to have difficulties with curves that are sharp relative to the grid size.

Another scheme, [Roberts 1965] is to keep a running least-squared-error best-fit line calculation for points as they are merged into segments [Appendix 1]. When the residual (error) of a point goes over some threshold or the accumulated error for a segment exceeds a threshold, a new segment is started. Difficulties arise here because the concept of a breakpoint is nonexistent; they just occur at the intersections of the best-fit lines, and without a phase of adjusting the set of points to be fit by each line (analogous to breakpoint adjustment), they may not be intuitively appealing.

Generally, one-pass merging schemes do not produce the most satisfying polylines possible under all conditions. Part of the problem is that breakpoints are only introduced after the fit has deteriorated, usually indicating that an earlier breakpoint would have been desirable.

In a *splitting* scheme, segments are divided (usually into two parts) as long as they fail some fitting condition [Duda and Hart 1973; Turner 1974]. Algorithm 8.1 provides an example.

Algorithm 8.1:   Curve Approximation

1. Given a curve as in Fig. 8.4a, draw a straight line between its end points (Fig. 8.4b).

2. For every point on the curve, compute its perpendicular distance to the approximating (poly)line. If it is everywhere within some tolerance, exit.

3. Otherwise, pick the curve point farthest from the approximating (poly)line, make it a new breakpoint (Fig. 8.4c) and replace the relevant segment of polyline with two new line segments.

4. Recursively apply the algorithm to the two new segments (Fig. 8.4d).

A straightforward extension is needed to deal with the case of curve segments parallel to the approximating one at maximum distance (Fig. 8.4e).

Ch. 8   Representation of Two-Dimensional Geometric Structures

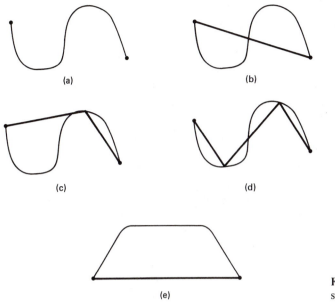

(a)

(b)

(c)

(d)

(e)

Fig. 8.4 Stages in the recursive linear segmenter (see text).

The area of a polygon may easily be computed from its polyline representation [Roberts 1965]. For a closed polyline of n points $(x(i), y(i))$, $i=0, ..., n-1$, labeled clockwise around a polygonal boundary, the area of the polygon is

$$\frac{1}{2} \sum_{i=0}^{n-1} (x_{i+1}y_i - x_i y_{i+1})$$

(8.1)

where subscript calculations are modulo n. This formula can be proved by considering it as the sum of (signed) areas of triangles, each with a vertex at the origin, or of parallelograms constructed by dropping perpendiculars from the polyline points to an axis. This method specializes to chain codes, which are a limiting case of polylines.

8.2.2 Chain Codes

Chain codes [Freeman 1974] consist of line segments that must lie on a fixed grid with a fixed set of possible orientations. This structure may be efficiently represented because of the constraints on its construction. Only a starting point is represented by its location; the other points on a curve are represented by successive displacements from grid point to grid point along the curve. Since the grid is uniform, direction is sufficient to characterize displacement. The grid is usually considered to be four- or eight- connected; directions are assigned as in Fig. 8.5, and each direction can be represented in 2 or 3 bits (it takes 18 bits to represent the starting point in a 512 × 512 image).

Chain codes may be made position-independent by ignoring the "start point." If they represent closed boundaries they may be "start point normalized" by choosing the start point so that the resulting sequence of direction codes forms an integer of minimum magnitude. These normalizations may help in matching. Periodic correlation (Section 3.2.1) can provide a measure of chain code similarity. The chain codes without their start point information are considered to be periodic functions of "arc length." (Here the arc length is just the number of steps in the chain code.) The correlation operation finds the (arc length) displacement of the functions at which they match up best as well as quantifying the goodness of the match. It can be sensitive to slight differences in the code.

The "derivative" of the chain code is useful because it is invariant under boundary rotation. The derivative (really a first difference mod 4 or 8) is simply another sequence of numbers indicating the relative direction of chain code segments; the number of left hand turns of $\pi/2$ or $\pi/4$ needed to achieve the direction of the next chain segment.

Chain codes are also well-suited for merging of regions [Brice and Fennema 1970] using the data structure described in Section 5.4.1. However, the pleasant properties for merging do not extend to union and intersection. Chain codes lend themselves to efficient calculation of certain parameters of the curves, such as area. Algorithm 8.2 computes the area enclosed by a four-neighbor chain code.

Algorithm 8.2:   Chain Code Area

Comment: For a four-neighbor code $(0: +x, 1: +y, 2: -x, 3: -y)$ surrounding a region in a counterclockwise sense, with starting point (x, y):

> *begin* Chain Area;
> 1. area := 0;
> 2. *y*position := *y*;
> 3. *For each* element of chain code
>   *case* element-direction of
>   *begin case*
>   [0] area := area-*y*position;
>   [1] *y*position := *y*position + 1;
>   [2] area := area + *y*position;
>   [3] *y*position := *y*position − 1;
>   *end case;*
> *end* Chain Area;

To merge two region boundaries is to remove any boundary they share, obtaining a boundary for the region resulting from gluing the two abutting regions together. As we saw in Chapter 5, the chain codes for neighboring regions are closely related at their common boundary, being equal and opposite in a clearly defined sense (for N-neighbor chain codes, one number is equal to the other plus $N/2$ modulo N (see Chapter 5). This property allows such sections to be identified readily, and easily scissored out to give a new merged boundary. As with polylines, it is not immediately obvious from a chain-coded boundary and a point whether the point is within the boundary or outside. Many algorithms for use with chain code representations may be found in [Freeman 1974; Gallus and Neurath 1970].

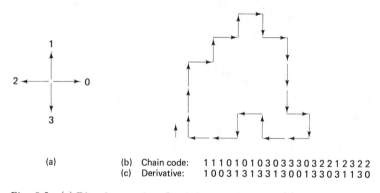

(a)	(b) Chain code:	1 1 1 0 1 0 1 0 3 0 3 3 3 0 3 2 2 1 2 3 2 2
	(c) Derivative:	1 0 0 3 1 3 1 3 3 1 3 0 0 1 3 3 0 3 1 1 3 0

Fig. 8.5 (a) Direction numbers for chain code elements. (b) Chain code for the boundary shown. (c) Derivative of (b).

8.2.3 The ψ -s Curve

The $\psi - s$ curve is like a continuous version of the chain code representation; it is the basis for several measures of shape. ψ is the angle made between a fixed line and a tangent to the boundary of a shape. It is plotted against s, the arc length of the boundary traversed. For a closed boundary, the function is periodic, with a discontinuous jump from 2π back to 0 as the tangent reattains the angle of the fixed line after traversing the boundary.

Horizontal straight lines in the $\psi - s$ curve correspond to straight lines on the boundary (ψ is not changing). Nonhorizontal straight lines correspond to segments of circles, since ψ is changing at a constant rate. Thus the $\psi - s$ curve itself may be segmented into straight lines [Ambler et al. 1975], yielding a segmentation of the boundary of the shape in terms of straight lines and circular arcs (Fig. 8.6).

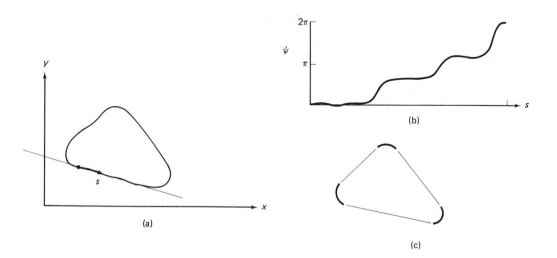

Fig. 8.6 ψ-s segmentation. (a) Triangular curve and a tangent. (b) ψ-s curve showing regions of high curvature. (c) Resultant segmentation.

8.2.4 Fourier Descriptors

Fourier descriptors represent the boundary of a region as a periodic function which can be expanded in a Fourier series. There are several possible parameterizations, summarized in [Persoon and Fu 1974]. These frequency-domain descriptions provide an increasingly accurate characterization of shape as more coefficients are included. In the infinite limit, they are unambiguous; individual coefficients are descriptive representations indicating "lobedness" of various degrees.

The boundary itself may provide the parameters for the Fourier transform as shown in Fig. 8.7. The parameterization of Fig. 8.7 gives the following series expansions:

$$\mathbf{x}(s) = \Sigma \mathbf{X}_k e^{jkw_o s} \qquad w_o = 2\pi/P, \quad P = \text{perimeter} \qquad (8.2)$$

where the discrete Fourier coefficients $\mathbf{X}_k$ are given by

$$\mathbf{X}_k = \frac{1}{P} \int_0^P \mathbf{x}(s) e^{-jkw_o s} ds \qquad (8.3)$$

A common feature for the Fourier descriptors is that typically the general shape is given rather well by a few of the low-order terms in the expansion of the boundary curve. Properly parameterized, the coefficients are independent of size, translation, and rotation of the shape to be described. The descriptors do not lend themselves well to reconstruction of the boundary; for one thing, the resulting curve may not be closed if only a finite number of coefficients is used for the reconstruction.

The $\psi-s$ curve may be used as the basis for a Fourier transform shape description [Barrow and Popplestone 1971]. $\psi(s)$ is converted to $\phi(s)$: $\phi(s) = \psi(s) - 2\pi \, s/P$. This operation subtracts out the rising component. A number of shape-indicating numbers arise from taking the root-mean-square amplitudes of the Fourier components of $\phi(s)$, discarding phase information. The shape descriptors are again indicative of the "lobedness" of the shape.

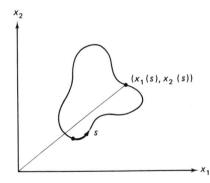

Fig. 8.7 Parameterization for Fourier Series Expansion.

Ch. 8 Representation of Two-Dimensional Geometric Structures

8.2.5 Conic Sections

Polynomials are a natural choice for curve representation, and certain polynomials of degree 2 (namely, circles and ellipses) are closed curves and hence define regions. Circles may be represented with three parameters, ellipses by five, and general conics by six. Thus the coefficients or parameters of conic sections are terse representations. Conics are often good models for physical curves such as the edges of manufactured objects.

Conics are commonly used to represent general curves approximately [Paton 1970]. Conics have some annoying properties, however; an important one is the difficulty of producing a well-behaved conic from noisy data to be fitted. Unless one is careful in defining the error measure [Turner 1974], a "least-squared error" fit of a conic to data points yields a conic which is a nonintuitive shape or even of a surprising type (such as a hyperbola when an ellipse was expected). Conic representations and algorithms are explored in Appendix 1.

8.2.6 B-Splines

Interpolative techniques may be used to yield approximate representations. B-splines are a popular choice of piecewise polynomial interpolant. Introduced in computer aided design and computer graphics, these classes of curves provide adequate aesthetic content for much design and also have many useful analytic properties. Usually, the fact that the curves are "interpolating" is not very relevant. What is relevant is that they have predictable properties which make them easy to manipulate in image processing, that they "look good" to human beings, that they closely approximate curves of interest in nature, and so forth. Several schemes exist for constructing complex curves that are useful in geometric modeling, and detailed expositions are to be found in [deBoor 1978; Barnhill and Riesenfeld 1974]. The B-spline formulation is one of the simplest that still has properties useful for interactive modeling and the extraction from raw data.

B-splines are piecewise polynomial curves which are related to a *guiding polygon*. Cubic polynomials are the most frequently used for splines since they are the lowest order in which the curvature can change sign. An example of the relationship between the guiding polygon and its spline curve is shown in Fig. 8.8. Splines are useful in computer vision because they allow accurate, manipulable internal models of complex shapes. The models may be used to guide and monitor segmentation and recognition tasks. Interactive generation of complex shape models is possible with B-splines, and the fact that the complex spline curves have terse representations (as their guiding polygons) allows programs to manipulate them easily.

Spline approximations have good computational properties as well as good representational ones. First, they are *variation diminishing*. This means that the curve is guaranteed to "vary less" than its guiding polygon (many interpolation schemes have a tendency to oscillate between sample points). In fact, the curve is guaranteed to lie between the convex hull of groups of $n + 1$ consecutive points where n is the degree of the interpolating polynomial (Fig. 8.9.) The second advan-

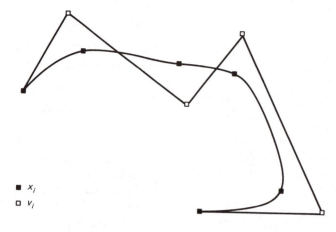

■ x_i
□ v_i

Fig. 8.8 A spline curve and its guiding polygon.

tage is that the interpolation is local; if a point on the guiding polygon is moved, the effects are intuitive and limited to nearby points on the spline. A third advantage is directly related to its use in vision; a technique for matching a spline-represented boundary curve against raw data is to search perpendicular to the spline for edges whose direction is parallel to the spline curve and location perpendicular to the spline curve. Perpendicular and parallel directions are computable directly from the parameters representing the spline.

B-Spline Mathematics

The interpolant through a given set of points x_i, $i = 1, ..., n$ is $x(s)$, a vector valued piecewise polynomial function of the parameter s; s changes uniformly between data points. For convenience, assume that $x(i) = x_i$, that is: s assumes integer values at data points, and $s = 1, ..., n$. Each *piece* of $x(s)$ is a cubic polyno-

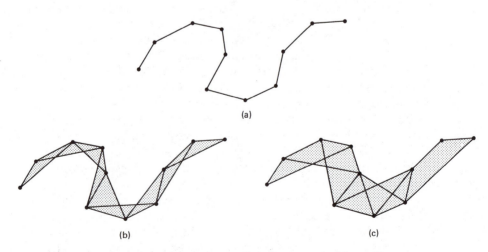

(a)

(b)   (c)

Fig. 8.9 The spline of degree n must lie in the convex hull formed by consecutive groups of $n + 1$ points. (a) $n = 1$ (linear). (b) $n = 2$ (quadratic). (c) $n = 3$ (cubic).

mial. Globally, $\mathbf{x}(s)$ has three orders of continuity across data points (i.e., up to continuity of second derivative: curvature). Formally, $\mathbf{x}(s)$ is defined as

$$\mathbf{x}(s) = \sum_{i=0}^{n+1} \mathbf{v}_i B_i(s) \tag{8.4}$$

The $\mathbf{v}_i$ are *coefficients* representing the curve $\mathbf{x}(s)$. They also turn out to be the vertices of the guiding polygon. They are a dual to the set of points $\mathbf{x}_i$; each can be derived from the other. The n data points $\mathbf{x}$ determine n $\mathbf{v}$'s. There are actually $n + 2$ $\mathbf{v}$'s; the additional two coefficients are determined from *boundary conditions*. For example, if the curvature at the end points is to be 0,

$$\mathbf{v}_1 = \frac{(\mathbf{v}_0 + \mathbf{v}_2)}{2} \tag{8.5}$$

$$\mathbf{v}_n = \frac{(\mathbf{v}_{n-1} + \mathbf{v}_{n+1})}{2}$$

Thus only n of the $n + 2$ coefficients are selectable.

The basis functions $B_i(s)$ are nonnegative and have a *limited support*, that is, each B_i is non-zero only for s between $i - 2$ and $i + 2$, as shown in Fig. 8.10. The limited support means that on a given span $(i, i + 1)$ there are only four basis functions that are nonzero, namely: $B_{i-1}(s)$, $B_i(s)$, $B_{i+1}(s)$, and $B_{i+2}(s)$. Figure 8.11 shows this configuration. Thus, to calculate $\mathbf{x}(s_0)$ for some s_0, simply find in which span it resides, and then use only four terms in the summation (8.4), since there are only four basis functions which are non-zero there.

The basis functions $B_i(s)$ are, themselves, piecewise cubic polynomials and their definition depends on the relative size (in parameter space) of the spans under their support. If the spans are of uniform size (e.g., unity), then all the basis functions have the same *form* and are merely translates of each other. Moreover, each of the basis functions, on its nonzero support, is made of four pieces. So, in Fig. 8.11 in the span $(i, i + 1)$ appear: the fourth piece of $B_{i-1}(s)$, the third piece of $B_i(s)$, the second piece of $B_{i+1}(s)$, and the first piece of $B_{i+2}(s)$. Call these pieces $C_{i,0}(s), ..., C_{i,3}(s)$ respectively; then $\mathbf{x}(s)$ on the interval $(i, i + 1)$ is given by:

$$\mathbf{x}(s) = C_{i-1,3}(s)\mathbf{v}_{i-1} + C_{i,2}(s)\mathbf{v}_i$$
$$+ C_{i+1,1}(s)\mathbf{v}_{i+1} + C_{i+2,0}(s)\mathbf{v}_{i+2}$$

No matter what i is, $C_{i,j}$ will have the same shape; this property allows a simplification in calculations. Define four *primitive basis functions*, and interpolate along the curve by parameter shifting:

$$C_{i,j}(s) = C_j(s - i) \quad i = 0, ..., n + 1; \quad j = 0, 1, 2, 3 \tag{8.6}$$

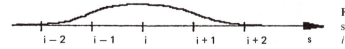

Fig. 8.10 Uniform B-spline: $B_i(s)$. Its support is non-zero only for s between $i - 2$ and $i + 2$.

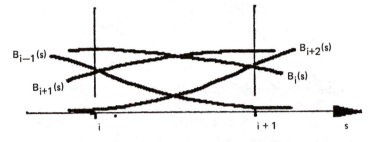

Fig. 8.11 The only four basis functions that are non-zero over the span $(i, i+1)$. Only the overlapping parts on this span are shown.

To find $\mathbf{x}(s)$, if s is in the span $(i, i+1)$, use the formula:

$$\mathbf{x}(s) = \mathbf{v}_{i-1}C_3(s-i) + \mathbf{v}_i C_2(s-i) + \mathbf{v}_{i+1}C_1(s-i) + \mathbf{v}_{i+2}C_0(s-i) \qquad (8.7)$$

where the $C_i(t)$ are given by:

$$C_0(t) = \frac{t^3}{6}$$

$$C_1(t) = \frac{-3t^3 + 3t^2 + 3t + 1}{6}$$

$$C_2(t) = \frac{3t^3 - 6t^2 + 4}{6}$$

$$C_3(t) = \frac{-t^3 + 3t^2 - 3t + 1}{6}$$

Formal derivations may be found in [Barnhill and Riesenfeld 1974; deBoor 1978].

Useful Formulae

The formulae may be simplified still further. $\mathbf{x}(s)$ is calculated in pieces (segments); define the segments $\mathbf{x}_i(t)$ where t ranges from 0 to 1. Then

$$\mathbf{x}_i(0) = \mathbf{x}_i \qquad \text{for } i = 1, \ldots, n-1$$

and

$$\mathbf{x}_{n-1}(1) = \mathbf{x}_n \qquad (8.8)$$

In matrix notation, and explicitly calculating the definition of the cubic polynomials $C_i(t)$,

$$\mathbf{x}_i(t) = [t^3, t^2, t, 1][C][\mathbf{v}_{i-1}, \mathbf{v}_i, \mathbf{v}_{i+1}, \mathbf{v}_{i+2}]^T \qquad (8.9)$$

where $[C]$ is the matrix:

$$\frac{1}{6} \begin{bmatrix} -1 & 3 & -3 & 1 \\ 3 & -6 & 3 & 0 \\ -3 & 0 & 3 & 0 \\ 1 & 4 & 1 & 0 \end{bmatrix}$$

Ch. 8 Representation of Two-Dimensional Geometric Structures

The ith column in the matrix $[C]$ in Eq. (8.9) above is the coefficients of the cubic polynomial $C_i(t)$ ($i = 0, 1, 2, 3$).

There is a distinction between *open* and *closed* curves. For open curves the boundary conditions must be used to solve for the two additional coefficients, as above. For closed curves, simply

$$\mathbf{v}_0 = \mathbf{v}_n \quad \text{and} \quad \mathbf{v}_{n+1} = \mathbf{v}_1 \tag{8.10}$$

The relation between the different $\mathbf{v}_i$ and $\mathbf{x}_i$ is summarized as follows. For open curves with zero curvature at the endpoints:

$$\frac{1}{6}
\begin{bmatrix}
6 & 0 & & & & \\
1 & 4 & 1 & & & \\
& & \cdot & & & \\
& & & \cdot & & \\
& & & & \cdot & \\
& & & 1 & 4 & 1 \\
& & & & 0 & 6
\end{bmatrix}
\begin{bmatrix}
\mathbf{v}_0 \\ \mathbf{v}_1 \\ \cdot \\ \cdot \\ \cdot \\ \mathbf{v}_{n-1} \\ \mathbf{v}_n
\end{bmatrix}
=
\begin{bmatrix}
\mathbf{x}_0 \\ \mathbf{x}_1 \\ \cdot \\ \cdot \\ \cdot \\ \mathbf{x}_{n-1} \\ \mathbf{x}_n
\end{bmatrix}$$

and for closed curves:

$$\frac{1}{6}
\begin{bmatrix}
4 & 1 & & & & 1 \\
1 & 4 & 1 & & & \\
& & \cdot & & & \\
& & & \cdot & & \\
& & & 1 & 4 & 1 \\
1 & & & & 1 & 4
\end{bmatrix}
\begin{bmatrix}
\mathbf{v}_0 \\ \mathbf{v}_1 \\ \cdot \\ \cdot \\ \mathbf{v}_{n-1} \\ \mathbf{v}_n
\end{bmatrix}
=
\begin{bmatrix}
\mathbf{x}_0 \\ \mathbf{x}_1 \\ \cdot \\ \cdot \\ \mathbf{x}_{n-1} \\ \mathbf{x}_n
\end{bmatrix} \tag{8.11}$$

Equation (8.10) gives the relationship between the points on the guiding polygon and the points on the spline. It may be derived from Eq. (8.9) with $t = 0$ (see exercises). To interpolate between these points, use a value of t between the extremes of 0 and 1. Choosing $t = k\,dt$ for $k = 0, ..., n$ where $n\,dt = 1$ and substituting into Eq. (8.9) yields

$$\mathbf{x}_i(k\,dt) = [(k\,dt)^3 (k\,dt)^2 (k\,dt)1]\,[C]\,[\mathbf{v}_{i-1}, \mathbf{v}_i, \mathbf{v}_{i+1}, \mathbf{v}_{i+2}]^t \tag{8.12}$$

This can be decomposed [Wu et al. 1977; Gordon 1969] into the following equation.

$$\mathbf{x}_i(k\,dt) =
\begin{bmatrix} 0 \\ 0 \\ 0 \\ 1 \end{bmatrix}^T
\begin{bmatrix} 1 & & & \\ 1 & 1 & & \\ 1 & 1 & 1 & \\ 1 & 1 & 1 & 1 \end{bmatrix}^k
\begin{bmatrix} 6 & & & \\ -6 & 2 & & \\ 1 & -1 & 1 & \\ 0 & 0 & 0 & 1 \end{bmatrix}
\begin{bmatrix} dt^3 \\ dt^2 \\ dt \\ 1 \end{bmatrix}
[C]
\begin{bmatrix} \mathbf{v}_{i-1} \\ \mathbf{v}_i \\ \mathbf{v}_{i+1} \\ \mathbf{v}_{i+2} \end{bmatrix} \tag{8.13}$$

The tangent at a curve is obtained by differentiation:

$$\mathbf{x}'_i(k\,dt) = \frac{1}{6}
\begin{bmatrix} 0 \\ 0 \\ 1 \end{bmatrix}^T
\begin{bmatrix} 1 & & \\ 1 & 1 & \\ 1 & 1 & 1 \end{bmatrix}^k
\begin{bmatrix} 2 & 0 & 0 \\ -1 & 1 & 0 \\ 0 & 0 & 1 \end{bmatrix}
\begin{bmatrix} 3dt^2 \\ 2dt \\ 1 \end{bmatrix}
\begin{bmatrix} -1 & 3 & -3 & 1 \\ 3 & -6 & 3 & 0 \\ -3 & 0 & 3 & 0 \end{bmatrix}
\begin{bmatrix} \mathbf{v}_{i-1} \\ \mathbf{v}_i \\ \mathbf{v}_{i+1} \\ \mathbf{v}_{i+2} \end{bmatrix} \tag{8.14}$$

8.2.7 Strip Trees

In many computational problems there are space-time trade-offs. A nonredundant explicit representation for a general discrete curve, such as a chain code, is terse but may be difficult to use for certain computations. On the other hand, a representation for curves may take up much space but allow operations on those curves to be very efficient. A representation with the latter property is *strip trees* [Ballard 1981]. Strip tress are closed under intersection and union operations, and these operations may be efficiently implemented.

A strip tree is a binary tree. The datum at each node is a eight-tuple, of which six entries define a strip (rectangle) and two denote addresses of the sons (if any). Thus each strip is defined by a six-tuple $S(\mathbf{x}_b, \mathbf{x}_e, \mathbf{w})$ as shown in Fig. 8.12. (Only five parameters are necessary to define an arbitrary rectangle, but the redundant representation proves useful in union and intersection algorithms to follow.)

The tree can be created from any curve by the following recursive procedure, which is very similar to Algorithm 8.1.

Algorithm 8.3: Making a Strip Tree

Find the smallest rectangle with a side parallel to the line segment $[\mathbf{x}_0, \mathbf{x}_n)$ that just covers all the points. This rectangle is the datum for the root node of a tree. Pick a point $\mathbf{x}_k$ that touches one of the sides of the rectangle. Repeat the above process for the two sublists $[\mathbf{x}_0, ..., \mathbf{x}_k)$ and $[\mathbf{x}_k, ..., \mathbf{x}_n)$. These become sons of the root node. Repeat the process until the approximation is accurate enough.

The half-open interval facilitates the computations to follow. In the example above the point $\mathbf{x}_k$ explicitly appears in both subtrees but implementationally need not be part of the left one. Figure 8.13 shows the strip tree construction process.

Intersecting Two Curves via Strip Trees

Consider what happens when a strip from one tree intersects a strip from another, as shown in Fig. 8.14. If the strips do not intersect, the underlying curves

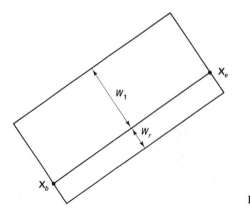

Fig. 8.12 Strip definition.

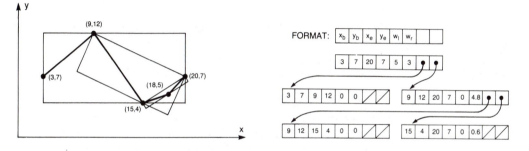

Fig. 8.13 Strip tree construction process.

do not intersect. If the strips do intersect, the underlying curves may or may not. To determine which, the computation may be applied recursively. At the leaf level of the tree defined as the *primitive* level, the problem can always be resolved.

Algorithm 8.4: Intersecting Two Strip Trees Representing Curves

Boolean Procedure TreeInt (*T*1, *T*2, *L*)
Begin
  case intersection type of two strips *T*1 and *T*2 *of*
  begin case
  [primitive] *return* (true)
  [null] *return* (false)
  [possible] *If T*2 is the "fatter" strip
  return (TreeInt(*T*1,*L*Son(*T*2) or TreeInt(*T*1,*R*Son(*T*2))
  Else return (TreeInt(*L*Son(*T*1),*T*2) or TreeInt(*R*Son(*T*1),*T*2));
  end case;
end;

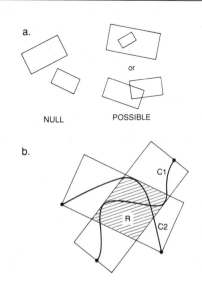

Fig. 8.14 Types of strip intersections. (a) Two kinds of intersections: NULL on the left; various POSSIBLE intersections on the right. (b) Under certain conditions the underlying curves must intersect.

The "Union" of Two Strip Trees

The "union" of two strip trees may be defined as a strip that covers both of the two root strips. The two curves defined by $[\mathbf{x}'_0, ..., \mathbf{x}'_n)$, $[\mathbf{x}''_0, ..., \mathbf{x}''_m)$ are treated as two concatenated lists. That is, the resultant ordering is such that $\mathbf{x}_0 = \mathbf{x}'_0$, $\mathbf{x}_{m+n+1} = \mathbf{x}''_m$. This construction is shown in Fig. 8.15.

Closed Curves Represented by Strip Trees

A region may be represented by its (closed) boundary. The strip-tree construction method described in Algorithm 8.3 works for closed curves and, incidentally, also for self-intersecting curves. Furthermore, if a region is not simply connected (has "holes") it can still be represented as a strip tree which at some level has connected primitives.

Many useful operations on regions can be carried out with strip trees. Examples are intersection between a curve and a region and intersecting two regions. Another example is the determination of whether a point is inside a region. Roughly, if any semi-infinite line terminating at the point intersects the boundary of the region an odd number of times, the point is inside. The implied algorithm is computationally simplified for strip trees in the following manner:

> *Point Membership Property.* To decide whether a point z is a member of a region represented by a strip tree, compute the number of nondegenerate intersections of the strip tree with any semi-infinite strip L which has $\|w\| = 0$ and emanates from z. If this number is odd, the point is inside the region.

This is because for clear intersections the underlying curves may intersect more than once but must intersect an odd number of times. A potential difficulty exists when the strip L is tangent to the curve. To overcome this difficulty in practice, a different L may be used.

Intersecting a Curve with a Region

The strategy behind intersecting a strip tree representing a curve with a strip tree representing a region is to create a new tree for the portion of the curve that overlaps the region. This can be done by trimming the original curve strip tree. Trimming is done efficiently by taking advantage of an obvious property of the intersection process:

> *Pruning Property:* Consider two strips S_C from T_C and S_a from T_a. If the intersection of S_C with T_a is null, then (a) if any point on S_C is inside T_a, the entire tree whose root strip is S_C is inside or on T_a, and (b) if any point on S_C is outside of T_a then the entire tree whose root strip is S_C is outside T_a.

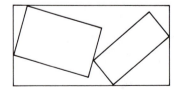

Fig. 8.15 Construction for "union" of strip trees representing two curves.

This leads to the Algorithm 8.5 for curve-region intersection using trees. If the curve strip is "fatter" (i.e., has more area), copy the node and resolve the intersection at lower levels. In the converse case prune the tree sequentially by first intersecting the resultant pruned tree with the right region strip.

Algorithm 8.5: Curve–Region Intersection

comment A Reference Procedure returns a pointer;
reference procedure CurveRegionInt($T1, T2$)
  begin
  $R := T2$;
  comment R is a global used by CRInt;
  return (CRInt($T1, T2$));
  end;

reference procedure CRInt(T1,T2)
  begin
  begin Case StripInt($T1, T2$) of
    [Null or Primitive]
      if intersection ($T1, R$, TRUE) = null *then*
        if Inside($T1, R$) *then* return ($T1$)
        else return (null);
      else return ($T1$);
    [Possible] *if* $T1$ is "fatter" *then*
    begin
      NT := NewRecord;
      $\mathbf{x}_b$(NT) := $\mathbf{x}_b$ (T);
      $\mathbf{x}_e$(NT) := $\mathbf{x}_e$ (T);
      w_l(NT) := w_l(T);
      w_r(NT) := w_r(T);
      $LSon$(NT) := CRInt ($LSon(T1), T2$);
      $RSon$(NT) := CRInt ($RSon(T1), T2$);
      return(NT);
    end
    else comment $T2$ is "fatter"
      Return (CRInt(CRInt($T1, LSon(T2)$), $RSon(T2)$));
    end;
  end Case;
  end;

The problem of intersecting two regions can be decomposed into two curve-region intersection problems (Fig. 8.16). Thus algorithm 8.5 can also be used to solve the region-region intersection problem.

8.3 REGION REPRESENTATIONS

8.3.1 Spatial Occupancy Array

The most obvious and quite a useful representation for a region on a raster is a membership predicate $p(x, y)$ which takes the value 1 when point (x, y) is in the

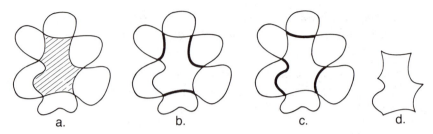

Fig. 8.16 Decomposition of Region-Region Intersection. (a) Desired result. (b) Portion of boundary generated by treating three-lobed region as a curve. (c) Portion of boundary generated by treating five-lobed region as a curve. (d) Result of union operation.

region and the value 0 otherwise. One easy way to implement such a function is with a *membership array*, an array of 1's and 0's with the obvious interpretation. Such arrays are quicky interrogated and also quite easily unioned, merged and intersected by AND and OR operations, applied elementwise on the operand arrays. The disadvantages of this representation are that it requires much space and does not represent the boundary in a useful way.

8.3.2 *y*-Axis

A representation that is more compact and which offers reasonable algorithms for intersection, merging, and union is the *y*-axis representation [Merrill 1973]. This is a run-length encoding of the membership array, and as such it provides no explicit boundary information. It is a list of lists. Each element on the main list corresponds to a row of constant *y* in the image raster. Each row of constant *y* is encoded as a list of *x*-coordinate points; the first *x* point at which the region is entered while moving along that *y* row, then the *x* point at which the region is exited, then the *x* point at which it next is entered, and so forth. The *y*-rows with no region points are omitted from the main list. Thus, in a notation where successive levels of sublist are surrounded by successive levels of parentheses, the *y*-axis encoding of a region is shown in Fig. 8.17; here the first element of each sublist is the *y* coordinate, followed by a list of "into" and "out of" *x* coordinates. Where a *y* coordinate con-

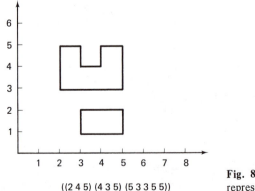

((2 4 5) (4 3 5) (5 3 3 5 5))

Fig. 8.17 *y*-axis region representation.

Ch. 8 Representation of Two-Dimensional Geometric Structures

tains an isolated point in the region, this point is repeated in the x-axis representation, as shown by the example in Fig. 8.17. Thus "lines" (regions of unit width) can be easily (although not efficiently) represented in this system.

Union and intersection are implemented on y-axis representations as merge-like operations which take time linearly proportional to the number of y rows. Two instances of y-axis representations and the representation of their union are shown in Fig. 8.18. Note that the union amounts to a merge of x elements along rows organized within a merge of rows themselves.

The y-axis representation is wasteful of space if the region being represented is long, thin, and parallel to the y axis. In this case one is invited to encode it in x-axis format, in an obvious extension. Working with mixed x-axis and y-axis formats presents no conceptual difficulties, but considerable loss of convenience.

8.3.3 Quad Trees

Quad trees [Samet 1980] are a useful encoding of the spatial occupancy array. The easiest way to understand quad trees is to consider pyramids as an intermediate representation of the binary array. Figure 8.19 shows a pyramid (Section 3.7) made from the base image (on the left). Each pixel in images above the lowest level has one of three values, BLACK, WHITE, or GRAY. A pixel in a level above the base is BLACK or WHITE if all its corresponding pixels in the next lower level are BLACK or WHITE respectively. If some of the lower level pixels are BLACK and others are WHITE, the corresponding pixel in the higher level is GRAY.

Such a pyramid is easy to construct. To convert the pyramid to a quad tree, simply search the pyramid recursively from the top to the base. If an array element in the pyramid is either BLACK or WHITE, form a terminal node of the corresponding type. Otherwise, form a GRAY node with pointers to the results of

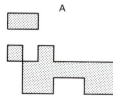
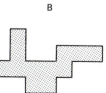

A   B

((1 2 3 6 7) (2 2 7) (3 1 1 3 3) (5 1 2))   ((1 3 4) (2 1 5) (3 2 2 5 7) (4 2 2))

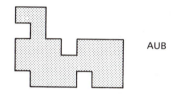

AUB

((1 2 4 6 7) (2 1 7) (3 1 3 5 7) (4 2 2) (5 1 2))

Fig. 8.18 Two point sets *A*, *B*, and *A* ∪ *B*, with their *y*-axis representations.

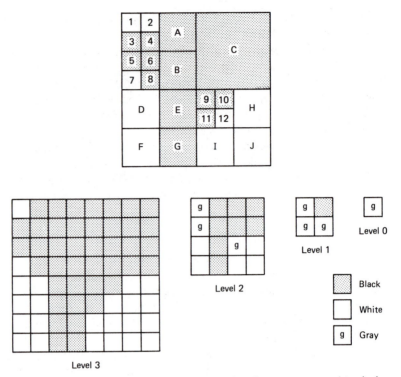

Fig. 8.19 Pyramid used in quad tree construction. Letters correspond to pixels in the pyramid that are either BLACK or WHITE.

the recursive examination of the four elements at the next level in the tree (Algorithm 8.6).

Algorithm 8.6: Quad Tree Generation

Reference Procedure QuadTree (*integer array* pyramid; *integer x, y,* level);
Comment NW, NE, SW, SE are fields denoting the sons of a quadtree node;
Newnode(*P*);
TYPE(*P*) := Pyramid(IND (x,y,Level));
if TYPE(*P*) = BLACK or WHITE *then return* (*P*)
else begin
  SW(*P*):=QuadTree(Pyramid, 2∗*x*, 2∗*y*, Level + 1);

  SE(*P*):=QuadTree(Pyramid, 2∗*x* + 1, 2∗*y*, Level + 1);

NW(*P*):=QuadTree(Pyramid, 2∗*x*, 2∗*y* + 1, Level + 1);

NE(*P*):=QuadTree(Pyramid, 2∗*x* + 1, 2∗*y* + 1, Level + 1);
  return (*P*)
  end;

Here an implementational point is that the entire pyramid fits into a linear array of size $2(2^{2 \times \text{level}})$. IND is an indexing function which extracts the appropriate value given the x, y and level coordinates. The reader can apply this algorithm to the example in Fig. 8.19 to verify that it creates the tree in Fig. 8.20.

The quad tree can be created directly from the base of the pyramid, but the algorithm is more involved. This is because proceeding upward from the base, one must sometimes defer the creation of black and white nodes. This algorithm is left for the exercises [Samet 1980].

Many operations on quad trees are simple and elegant. For example, consider the calculation of area [Schneier 1979]:

Algorithm 8.7:   Area of a Quad Tree

Integer Procedure Area (reference QuadTree; integer height)
 Begin
 Comment NW, NE, SW, SE are fields denoting the sons of
 a quadtree node;
 BlackArea := 0;
 if TYPE(QuadTree) = GRAY *then*
  for I in the set {NW, NE, SW, SE} *do*
  BlackArea = BlackArea + Area(I(QuadTree), height-1)
 else if TYPE(QuadTree) = BLACK *then*
  BlackArea = BlackArea + $2^{2*\text{height}}$;
 return(BlackArea)
 end;

Other examples may be found in the References and are pursued in the Exercises.

The quad tree and the associated pyramid have two related disadvantages as a representation. The first is that the resolution cannot be extended to finer resolution after a grid size has been chosen. The second is that operations between quad

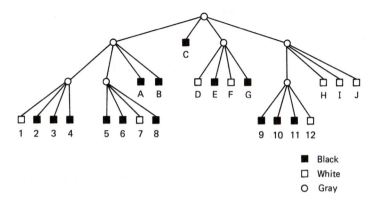

Fig. 8.20   Quad tree for the example in Fig. 8.19.

trees tacitly assume that their pyramids are defined on the same grids. The grids cannot be shifted or scaled without cumbersome conversion routines.

8.3.4 Medial Axis Transform

If the region is made of thin components, it can be well described for many purposes by a "stick-figure" *skeleton*. Skeletons may be derived by thinning algorithms that preserve connectivity of regions; the medial axis transform (MAT), of [Blum 1973; Marr 1977] is a well-known thinning algorithm.

The skeleton is defined in terms of the distance of a point $\mathbf{x}$ to a set A:

$$d_s(\mathbf{x}, A) = \inf\{d(\mathbf{x}, \mathbf{z}) \,|\, \mathbf{z} \text{ in } A\} \qquad (8.15)$$

Popular metrics are the Euclidean, city block, and chessboard metrics described in Chapter 2.

Let B be the set of boundary points. For each point P in a region, find its closest neighbors (by some metric) on the region boundary. If *more than one* boundary point is the minimum distance from $\mathbf{x}$, then $\mathbf{x}$ is on the skeleton of the region. The skeleton is the set of pairs $\{\mathbf{x}, d_s(\mathbf{x}, B)\}$ where $d_s(\mathbf{x}, B)$ is the distance from $\mathbf{x}$ to the boundary, as defined above (this is a definition, not an efficient algorithm.) Since each $\mathbf{x}$ in the skeleton retains the information on the minimum distance to the boundary, the original region may be recovered (conceptually) as the union of "disks" (in the proper metric) centered on the skeleton points.

Some common shapes have simply structured medial axis transform skeletons. In the Euclidean metric, a circle has a skeleton consisting of its central point. A convex polygon has a skeleton consisting of linear segments; if the polygon is nonconvex, the segments may be parabolic or linear. A simply connected polygon has a skeleton that is a tree (a graph with no cycles). Some examples of medial axis transform skeletons appear in Fig. 8.21.

The figure shows that the skeleton is sensitive to noise in the boundary. Reducing this sensitivity may be accomplished by smoothing the boundary, using a polygonal boundary approximation, or including only those points in the skeleton that are greater than some distance from the boundary. The latter scheme can lead to disconnected skeletons.

Algorithm 8.8: *Medial Axis Transformation* [Rosenfeld and Kak 1976]

Let region points have value 1 and exterior points value 0. These points define an image $f^0(\mathbf{x})$. Let $f^k(\mathbf{x})$ be given by

$$f^k(\mathbf{x}) = f^0(\mathbf{x}) + \min_{d(x,z) \leqslant 1} [f^{k-1}(\mathbf{z})], \qquad k > 0$$

The points $f^k(\mathbf{x})$ will converge when k is equal to the maximum thickness of the region. Where $f^k(\mathbf{x})$ has converged, the skeleton is defined as all points $\mathbf{x}$ such that

$$f^k(\mathbf{x}) \geqslant f^k(\mathbf{z}), \qquad d(\mathbf{x}, \mathbf{z}) \leqslant 1.$$

Ch. 8   Representation of Two-Dimensional Geometric Structures

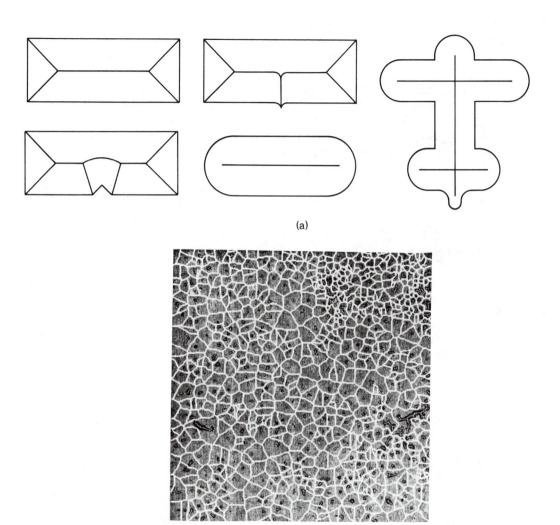

(a)

(b)

Fig. 8.21 Medial Axis Transform skeletons (a), and the technique applied to human cell nuclei (b). Shown in (b) are both the "normal" skeleton obtained by measuring distances interior to the boundaries, and the exo-skeleton, obtained by measuring distances exterior to the boundary.

This algorithm can produce disconnected skeletons for excursions or lobes off the main body of the region. Elegant thinning algorithms to compute skeletons are given in [Pavlidis 1977].

8.3.5 Decomposing Complex Regions

Much work has been done on the decomposition of point sets (usually polygons) into a union of convex polygons. Such convex decompositions provide structural analysis of a complex region that may be useful for matching different point sets.

An example of the desired result in two dimensions is presented here, and the interested reader may refer to [Pavlidis 1977] for the details. Such a decomposition is not unique in general and in three dimensions, such difficulties arise that the problem is often called ill-formed or intractable [Voelcker and Requicha 1977].

The shapes of Fig. 8.22 have three "primary convex subsets" labeled *X, Y,* and *Z*. They form different numbers of "nuclei" (roughly, intersection sets). The shape is described by a graph that has nodes for nuclei and primary convex subsets and an arc between intersecting sets (Fig. 8.22c). Without nodes for the nuclei (i.e., if only primary convex subsets and their intersections are represented), regions with different topological connectedness can produce identical graphs (Fig. 8.22b).

8.4 SIMPLE SHAPE PROPERTIES

8.4.1 Area

The *area* of a region is a basic descriptive property. It is easily computed from curve boundary representations (8.2.1) and thus also for chain codes (8.2.2); their con-

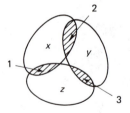
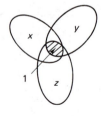

(a)

(b)

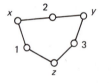

(c)

Fig. 8.22 Decomposition of polygon into primary convex subsets and nuclei (see text).

tinuous analog is also useful. Consider a curve parameterized on arc length s so that points (x, y) are given by functions $(x(s), y(s))$

$$\text{area} = \int_0^P (x \, \frac{dy}{ds} - y \, \frac{dx}{ds}) \, ds \qquad (8.16)$$

where P is the perimeter.

8.4.2 Eccentricity

There are several measures of *eccentricity*, or "elongation". One of them is the ratio of the length of maximum chord A to maximum chord B perpendicular to A (Fig. 8.23).

Another reasonable measure is the ratio of the principal axes of inertia; this measure can be based on boundary points or the entire region [Brown 1979]. An (approximate) formula due to Tenenbaum for an arbitrary set of points starts with the mean vector

$$\mathbf{x}_0 = \frac{1}{n} \sum_{\mathbf{x} \text{ in } R} \mathbf{x} \qquad (8.17)$$

To compute the remaining parameters, first compute the ijth moments M_{ij} defined by

$$M_{ij} = \sum_{\mathbf{x} \text{ in } R} (x_0 - x)^i (y_0 - y)^j \qquad (8.18)$$

The orientation, θ, is given by

$$\theta = \frac{1}{2} \tan^{-1}(\frac{2M_{11}}{M_{20} - M_{02}}) + n(\frac{\pi}{2}) \qquad (8.19)$$

and the approximate eccentricity e is

$$e = \frac{(M_{20} - M_{02})^2 + 4M_{11}}{\text{area}} \qquad (8.20)$$

8.4.3 Euler Number

The Euler number is a topological property defining the set of objects that are equivalent under "rubber-sheet" deformations of the plane. It describes the connectedness of a region, not its shape. A *connected region* is one in which all pairs of

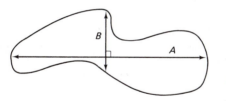

Fig. 8.23 An eccentricity measure: A / B.

points may be connected by a curve lying entirely in the region. If a complex two-dimensional object is considered to be a set of connected regions, where each one can have holes, the *Euler number* for such an object is defined as

(number of connected regions) − (number of holes)

The number of holes is one less than the connected regions in the set complement of the object.

8.4.4 Compactness

One measure of *compactness* (not compactness in the sense of point-set topology) is the ratio (perimeter^2)/area, which is dimensionless and minimized by a disk. This measure is computed easily from the chain-code representation of the boundary where the length of an individual segment of eight-neighbor chain code is given by $(\sqrt{2})$ if the (eight-neighbor) direction is odd and by 1 if the direction is even. The area is computed by a modification of Algorithm 8.2 and the perimeter may be accumulated at the same time.

For small discrete objects, this measure may not be satisfactory; another measure is based on a model of the boundary as a thin springy wire [Young et al. 1974]. The normalized "bending energy" of the wire is given by

$$E = \frac{1}{P} \int_0^P |\kappa(s)|^2 ds \tag{8.21}$$

where κ is *curvature*. This measure is minimized by a circle. E can be computed from the chain code representation by recognizing that $\kappa = d\theta/ds$, and also from the Fourier coefficients mentioned below since

$$|\kappa(s)|^2 = \left[\frac{d^2x}{ds^2}\right]^2 + \left[\frac{d^2y}{ds^2}\right]^2 \tag{8.22}$$

so that E, using Parseval's theorem, is

$$\sum_{k=-\infty}^{\infty} (kw_0)^4 (|X_k|^2 + |Y_k|^2) \tag{8.23}$$

where $\mathbf{X}_k = (X_k, Y_k)$ are the Fourier descriptor coefficients in (8.2).

8.4.5 Slope Density Function

The $\psi-s$ curve can be the basis for the *slope density function* (SDF) [Nahin 1974]. The SDF is the histogram or frequency distribution of ψ collected over the boundary. An example is shown in Fig. 8.24. The SDF is flat for a circle (or in a continuous universe, any shape with a monotonically varying ψ); straight sides stand out sharply, as do sharp corners, which in a continuous universe leave gaps in the histogram. The SDF is the signature of the $\psi-s$ curve along the ψ axis.

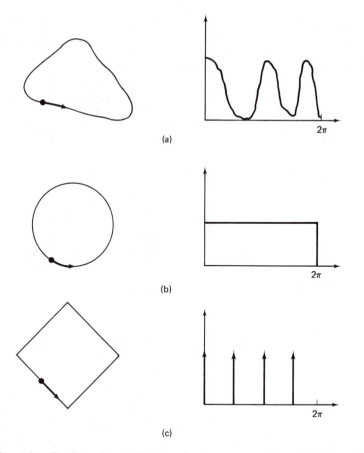

Fig. 8.24 The Slope Density Function for three curves: a triangular blob, a circle, and a square.

8.4.6 Signatures

By definition, a projection is not an information-preserving transformation. But Section 2.3.4 showed that (as with Fourier descriptors,) enough projections allow reconstruction of the region to any desired degree of accuracy. (This observation forms the basis for computer assisted tomography.)

Given a binary image $f(\mathbf{x}) = 0$ or 1, define the horizontal *signature p* (x) as

$$p(x) = \int_y f(x, y) \qquad (8.24)$$

$p(x)$ is simply the projection of f onto the x axis. Similarly, define $p(y)$, the vertical signature, as

$$p(y) = \int_x f(x, y) \qquad (8.25)$$

Maxima and minima of signatures are often useful for establishing preliminary landmarks in an image to reduce subsequent search effort [Kruger et al. 1972] (Fig. 8.25). If the region is not binary, but consists of a density function, Eq. (8.24) may still be used. Polar projections may be useful characterizations if the point of projection is chosen carefully.

Another idea is to provide a number of projections, q_1, ..., q_n, the ith one based on the ith sublist in each row in a y-axis-like region representation. This technique is more sensitive to non-convexities and holes than is a regular projection (Fig. 8.26).

8.4.7 Concavity Tree

Concavity trees [Sklansky 1972] represent information necessary to fill in local indentations of the boundary as far as the convex hull and to study the shape of the resultant concavities.

A region S is *convex* iff for any x_1 and x_2 in S, the straight line segment connecting x_1 and x_2 is also contained in S. The *convex hull* of an object S is the smallest H such that

$$S \subset H$$

and H is convex.

Figure 8.27 shows a region, the steps in the derivation of the concavity tree, and the concavity tree itself.

8.4.8 Shape Numbers

For closed curves and a 3-bit chain code (together with a controlled digitization scheme), many chain-coded boundaries can be given a unique *shape number* [Bri-

Fig. 8.25 The use of signatures to locate a left ventricle cross section in ultrasound data. (Outer curves are smoothed versions of inner signatures.)

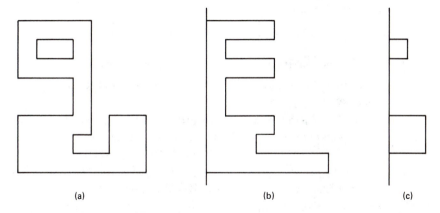

(a)   (b)   (c)

Fig. 8.26   A shape (a) and projections; from the first (b) and second (c) sublists of the *y*-axis representation.

biesca and Guzman 1979]. The shape number is related to the resolution of the digitization scheme. In a multiple resolution pyramid of digitization grids, every possible shape can be represented as a path through a tree. At each grid resolution corresponding to a level in the tree, there are a finite number of possible shapes. Moving up the tree, the coarser grids tend to blur distinctions between different shapes until at some resolution they are identical. This level can be used as a similarity measure between shapes. The basic idea behind shape numbers is the following. Consider all the possible closed boundaries with n chain segments. These form the possible shapes of "order n." The chain encoding for a particular boundary can be made unique by interpreting the chain-code direction sequence as a number and picking the start point that minimizes this number. Notice that the orders of shape numbers must be even on rectangular grids since a curve of odd order cannot close.

Algorithm 8.9 generates a shape number of order n.

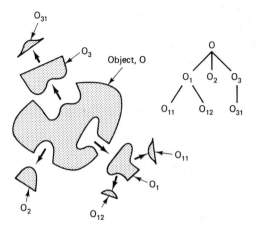

Fig. 8.27   Concavities of an object and the concavity tree.

Algorithm 8.9: Making a Shape Number of Order n

1. Choose the maximal diameter of the shape as one of the coordinate axes.
2. Find the smallest rectangle that has a side parallel to this axis and just covers the shape.
3. From the possible rectangles of order n, find the one that best approximates the rectangle in step 2. Scale this rectangle so that the length of the longest side equals that of the major axis, and center it over the shape.
4. Set all the pixels falling more than 50% inside the region to 1, and the rest to 0. NOTE: The derivative is defined slightly differently from p. 237.
5. Find the derivative of the chain encoded boundary of the region of 1's from step 4.
6. Normalize this number by rotating the digits until the number is minimum. The normalized number is the shape number.

Figure 8.28 shows these steps.

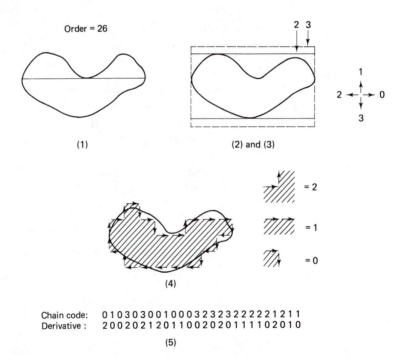

Fig. 8.28   Steps in determining a shape number (see text).

Ch. 8   Representation of Two-Dimensional Geometric Structures

Generating a shape number of a specific order may be tricky, as there is a chance that the resulting shape number may be greater than order n due to deep concavities in the boundary. In this case, the generation procedure can be repeated for smaller values of n until a shape number of n digits is found. Even this strategy may sometimes fail. The shape number may not exist in special cases such as boundaries with narrow indentations. These features may cause step 4 in Algorithm 8.9 to fail in the following way. Even though the rectangle of step 3 was of order n, the resultant boundary may have a different order. Nevertheless, for the vast majority of cases, a shape number can be computed.

The degree of similarity for two shapes is the largest order for which their shape numbers are the same. The "distance" between two shapes is the inverse of their degree of similarity. This distance is an *ultradistance* rather than a norm:

$$
\begin{aligned}
d(S, S) &= 0 \\
d(S_1, S_2) &\geq 0 \qquad \text{for } S_1 \neq S_2 \\
d(S_1, S_3) &\leq \max(d(S_1, S_2), d(S_2, S_3))
\end{aligned}
\qquad (8.26)
$$

Figure 8.29 shows the similarity tree for six shapes as computed from their shape numbers. When the shape number is well defined, it is a useful measure since it is unique (for each order), it is invariant under rotation and scale changes of an object, and it provides a metric by which shapes can be compared.

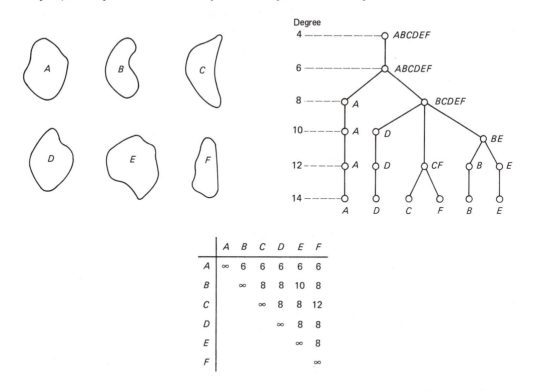

Fig. 8.29 Six shapes, their similarity trees, and the degree of similarity between the shapes.

EXERCISES

8.1 Consider a region segmentation where regions are of two types: (1) filled in and (2) with holes. Relate the number of junctions, boundaries, and filled-in regions to the Euler number.

8.2 Write a procedure for finding where two chain codes intersect.

8.3 Devise algorithms to intersect and union two regions in the y-axis representation.

8.4 Show that the number of intersections of the curves under a clear strip intersection is odd.

8.5 Modify Algorithm 8.4 to work with strip trees with varying numbers of sons.

8.6 Derive Eq. (8.9) from Eq. (8.7).

8.7 Show that Eqs. (8.12) and (8.13) are equivalent.

8.8 Given two points $\mathbf{x}_1$ and $\mathbf{x}_2$ and slopes $\phi(\mathbf{x}_1)$ and $\phi(\mathbf{x}_2)$, find the ellipse with major axis a that fits the points.

8.9 Write a procedure to intersect two regions represented by quad trees, producing the quad tree of the intersection.

8.10 Determine the shape numbers for (a) a circle and (b) an octagon. What is the distance between them?

REFERENCES

AMBLER, A. P., H. G. BARROW, C. M. BROWN, R. M. BURSTALL, and R. J. POPPLESTONE. "A versatile system for computer controlled assembly." *Artificial Intelligence 6*, 2, 1975, 129–156.

BALLARD, D. H. "Strip trees: A hierarchical representation for curves." *Comm. ACM 24*, 5, May 1981, 310–321.

BARNHILL, R. E. and R. F. RIESENFELD. *Computer Aided Geometric Design.* New York: Academic Press, 1974, 160.

BARROW, H. G. and R. J. POPPLESTONE. "Relational descriptions in picture processing." In *MI6*, 1971.

BLUM, H. "Biological shape and visual science (Part I)." *J. Theoretical Biology 38*, 1973, 205–287.

BRIBIESCA, E. and A. GUZMAN. "How to describe pure form and how to measure differences in shapes using shape numbers." *Proc.*, PRIP, August 1979, 427–436.

BRICE, C. R. and C. L. FENNEMA. "Scene analysis using regions." *Artificial Intelligence 1*, 3, Fall 1970, 205–226.

BROWN, C. M. "Two descriptions and a two-sample test for 3-d vector data." TR49, Computer Science Dept., Univ. Rochester, February 1979.

DEBOOR, C. *A Practical Guide to Splines.* New York: Springer-Verlag, 1978.

DUDA, R. O. and P. E. HART. *Pattern Recognition and Scene Analysis.* New York: Wiley, 1973.

FREEMAN, H. "Computer processing of line drawing images." *Computer Surveys 6*, 1, March 1974, 57–98.

GALLUS, G. and P. W. NEURATH. "Improved computer chromosome analysis incorporating preprocessing and boundary analysis." *Physics in Medicine and Biology 15*, 1970, 435.

GORDON, W. J. "Spline-blended surface interpolation through curve networks." *J. Mathematics and Mechanics 18*, 10, 1969, 931–952.

HOROWITZ, S. L. and T. P. PAVLIDIS. "Picture segmentation by a tree traversal algorithm." *J. ACM 23*, 2, April 1976, 368–388.

KRUGER, R. P., J. R. TOWNE, D. L. HALL, S. J. DWYER, and G. S. LUDWICK, "Automatic radiographic diagnosis via feature extraction and classification of cardiac size and shape descriptors." *IEEE Trans. Biomedical Engineering 19*, 3, May 1972.

MARR, D. "Representing visual information." AI Memo 415, AI Lab, MIT, May 1977.

MERRILL, R. D. "Representations of contours and regions for efficient computer search." *Comm. ACM 16*, 2, February 1973, 69–82.

NAHIN, P. J. "The theory and measurement of a silhouette descriptor for image preprocessing and recognition." *Pattern Recognition 6*, 2, October 1974.

PATON, K. A. "Conic sections in automatic chromosome analysis." In *MI5*, 1970.

PAVLIDIS, T. *Structural Pattern Recognition*. New York: Springer-Verlag, 1977.

PERSOON, E. and K. S. FU. "Shape discrimination using Fourier descriptors." *Proc.*, 2nd IJCPR, August 1974, 126–130.

REQUICHA, A. A. G. "Mathematical models of rigid solid objects." TM-28, Production Automation Project, Univ. Rochester, November 1977.

ROBERTS, L. G. "Machine perception of three-dimensional solids." In *Optical and Electro-optical Information Processing*, J.P. Tippett et al. (Eds.). Cambridge, MA: MIT Press, 1965.

ROSENFELD, A. and A. C. KAK. *Digital Picture Processing*. New York: Academic Press, 1976.

SAMET, H. "Region representation: quadtrees from boundary codes." *Comm. ACM 23*, 3, March 1980, 163–170.

SCHNEIER, M. "Linear time calculations of geometric properties using quadtrees." TR-770, Computer Science Center, Univ. Maryland, May 1979.

SHIRAI, Y. "Analyzing intensity arrays using knowledge about scenes." In *PCV*, 1975.

SKLANSKY, J. "Measuring concavity on a rectangular mosaic." *IEEE Trans. Computers 21*, 12, December 1972.

SKLANSKY, J. and D. P. KIBLER. "A theory of non-uniformly digitizing binary pictures." *IEEE Trans. SMC 6*, 9, September 1976, 637–647.

TOMEK, I. "Two algorithms for piecewise linear continuous approximation of functions of one variable." *IEEE Trans. Computers 23*, 4, April 1974, 445–448.

TURNER, K. J. "Computer perception of curved objects using a television camera." Ph.D. dissertation, Univ. Edinburgh, 1974.

VOELCKER, H. B. and A. A. G. REQUICHA. "Geometric modelling of mechanical parts and processes." *Computer 10*, December 1977, 48–57.

WU, S., J. F. ABEL, and D. P. GREENBERG. "An interactive computer graphics approach to surface representations." *Comm. ACM 20*, 10, October 1977, 703–711.

YOUNG, I. T., J. E. WALKER, and J. E. BOWIE. "An analysis technique for biological shape I." *Information and Control 25*, 1974.

Representations of
Three-Dimensional Structures 9

9.1 SOLIDS AND THEIR REPRESENTATION

We consider three general classes of representations for rigid solids·

1. Surface or boundary
2. Sweep (in general, generalized cylinders)
3. Volumetric (in general, constructive solid geometry)

The semantics of solid representations is intuitively clear but sometimes mathematically tricky. The representations have different computational properties, and readers should keep this in mind when assessing a representation for possible use. As a simple example, a surface representation can describe how an object looks; a volumetric version, which expresses the solid as a combination of subparts, may not explicitly contain information about the surface of the object. However, the solid representation may be better for matching, if it can be structured to reflect functional subparts.

Certainly we believe, as do others, that model-based vision will ultimately have to confront the issues of geometric modeling in three dimensions [Nishihara 1979]. Ultimately, nonrigid as well as rigid solids will have to be represented. The characterization of nonrigid solids presents very challenging problems.

Nonrigid solids are often a useful way to model time-varying aspects of objects. Here, again, the kind of model that is best depends heavily on the domain. For example, a useful mammal model may be one with a piecewise rigid linkage (for the skeleton) and some elastic covering (for the flesh). Computer vision in the domain of mammals, either static in various positions or actually moving, might be based on generalized cylinders (Section 9.3). However, another nonrigid domain is that of heart chambers, that change through time as the heart beats. Here the skeleton is a much less intuitive notion, so a different model of nonrigidity may apply. In most cases, nonrigid objects are modeled as parameterized rigid objects. In the example of the human figure, the parameters may be joint angles for linkages representing the skeleton.

The last part of this chapter deals with understanding line drawings, an influential and well-publicized subfield of computer vision. This seemingly simple and accessible domain avoids many of the problems involving early processing and segmentation, yet it is important because it has furnished several important algorithmic and geometric insights. An important breakthrough in this domain was a move from "image understanding" in two dimensions to to an approach based on the three-dimensional world and laws governing three-dimensional solids.

9.2 SURFACE REPRESENTATIONS

The *enclosing surface*, or *boundary*, of a well-behaved three-dimensional object should unambiguously specify the object [Requicha 1980]. Since surfaces are what is seen, these representations are important for computer vision. Section 9.2.1 considers mainly planar polyhedral surface representations. More complex "sculptured surfaces" [Forrest 1972; Barnhill and Riesenfeld 1974; Barnhill 1977] are treated in Section 9.2.2. Some useful surfaces are defined as functions of three-dimensional directions from a central point of origin. Two of these are mentioned in Section 9.2.3.

9.2.1 Surfaces with Faces

Figure 9.1 shows the solid representation scheme most familiar to computer scientists. Solids are represented by their boundaries, or enclosing surfaces, which are represented in terms of such primitive entities as unbounded mathematical surfaces, curves, and points which together may be used to define "faces."

In general, a boundary is made up of a number of faces; faces are represented by mathematical surfaces and by information about their own boundaries (consisting of edges and possibly vertices). A closed surface such as the sphere or a spherical harmonic surface of Section 9.2.3 may be thought of as having only one face.

To specify a boundary representation, one must answer several important questions of representation design. What is a face, and how are faces represented? What is an edge, and how are edges represented? How much extra information (i.e., useful but redundant relationships and geometric data) should be kept?

What is a face? "Face" is an initially appealing but imprecise notion; it is at its clearest in the context of planar polyhedra. A face should probably always be a subset of the boundary of an object; presumably, it should have area but no dangling edges or isolated points, and the union of all the faces should make up the boundary or the object. Beyond this little can be said. For many purposes it makes sense to have faces overlap; it may be elegant to consider the letter on an alphabet block a special kind of face on the block that is a subset of the face making up the side of the block. On the other hand, it is easy to imagine applications in which faces should not overlap in area (then one easily can compute the surface area of a solid from its faces). In some objects, just what the faces are is purely a matter of

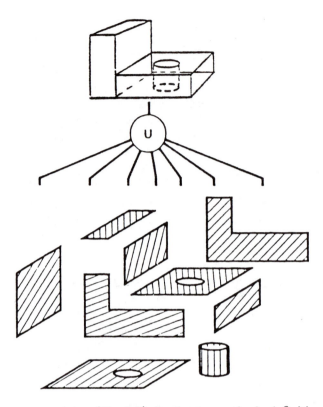

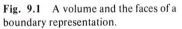

Fig. 9.1 A volume and the faces of a boundary representation.

opinion (Fig. 9.2). In short, any single definition of face is likely to be inadequate for some important application.

The availability of explicit representations of edges, faces, and vertices makes boundary representations quite useful in computer vision and graphics. The computational advantages of polyhedral surfaces are so great that they are often pressed into service as approximate representations of nonpolyhedra (Fig. 9.3).

An influential system for using face-based representations for planar polyhedral objects is the "winged edge" representation [Baumgart 1972]. Included in the system is an editor for creating complex polyhedral objects (such as that of Fig. 9.3) interactively. The system uses rules for construction based on the theorem of Euler that if V is the number of vertices in a polyhedron, E the number of edges, and F the number of faces, then $V - E + F = 2$. In fact, the formula can be extended to deal with non-simply connected bodies. The extended relation is $V - E + F = 2(B - H)$, with B being the number of bodies and H being the

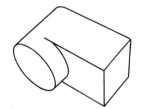

Fig. 9.2 What are the faces?

Ch. 9 Representations of Three-Dimensional Structures

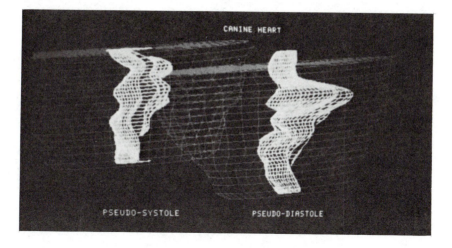

Fig. 9.3 A polyhedral approximation to a portion of a canine heart at systole and diastole. Both exterior (coarse grid) and interior surfaces (fine grid) are shown.

number of holes, or "handles," each resulting from a hole through a body [Lakatos 1976]. Baumgart's system uses these rules to oversee and check certain validity conditions on the constructions made by the editor.

The "winged edge" polyhedron representation achieves many desiderata for boundary representations in an elegant way. This representation is presented below to give a flavor of the features that have been traditionally found useful. Given as primitives the vertices, edges, faces, and polyhedra themselves, and given various relations between these primitives, one is naturally thinks of a record and pointer (relational) structure in which the pointers capture the binary relations and the records represent primitives and contain data about their locations or parameters.

In the winged edge representation, there are data structure records, or nodes, which contain fields holding data or links (pointers) to other nodes. An example using this structure to describe a tetrahedron is shown in Fig. 9.4. There are four kinds of nodes: vertices, edges, faces, and bodies. To allow convenient access to these nodes, they are arranged in a circular doubly linked list. The body nodes are actually the heads of circular structures for the faces, edges, and vertices of the body. Each face points to one of its perimeter edges, and each vertex points to one of the edges impinging on it. Each edge node has links to the faces on each side of it, and the vertices at either end.

Figure 9.4 shows only the last-mentioned links associated with each edge node. The reader may notice the similarity of this data structure with the data structure for region merging in Section 5.4. They are topologically equivalent. Each edge also has associated four links which give the name "winged edge" to the representation. These links specify neighboring edges in order around the two faces which are associated with the edge. The complete link set for an edge is shown in Fig. 9.5, together with the link information for bodies, vertices, and faces. To allow unambiguous traversal around faces, and to preserve the notion of

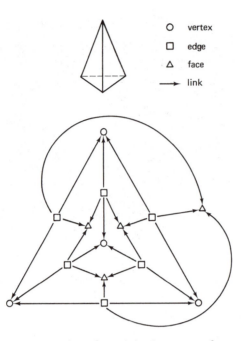

Fig. 9.4 A subset of edge links for a tetrahedron using the "winged edge" representation.

interior and exterior of a polyhedron, a preferential ordering of vertices and lines is picked (counterclockwise, say, as seen from outside the polyhedron).

Data fields in each vertex allow storage of three-dimensional world coordinates, and also of three-dimensional perspective coordinates for display. Each node has fields specifying its node type, hidden line elimination information, and other general information. Faces have fields for surface normal vector information, surface reflectance, and color characteristics. Body nodes carry links to relate them to a tree structure of bodies in a scene, allowing for hierarchical arrangement of subbodies into complex bodies. Thus body node data describe the scene structure; face node data describe surface characteristics; edge node data give the topological information needed to relate faces, edges, and vertices; and vertex node data describe the three-dimensional vertex location.

This rich and redundant structure lends itself to efficient calculation of useful functions involving these bodies. For instance, one can easily follow pointers to extract the list of points around a face, faces around a point, or lines around a face. Winged edges are not a universal boundary representation for polyhedra, but they do give an idea of the components to a representation that are likely to be useful. Such a representation can be made efficient for accessing all faces, edges, or vertices; for accessing vertex or edge perimeters; for polyhedron building; and for splitting edges and faces (useful in construction and hidden-line picture production, for instance).

9.2.2 Surfaces Based on Splines

The natural extension of polyhedral surfaces is to allow the surfaces to be curved. However, with an arbitrary number of edges for the surface, the interpolation of

FIG. 2-7a

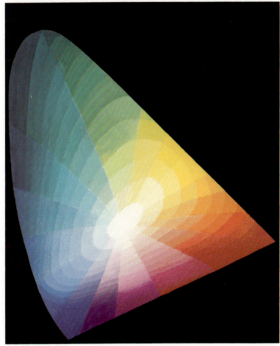

A painting by Louis Condax; courtesy of Eastman Kodak Company and the Optical Society of America.

FIG. 2-8a

Courtesy of D. Greenberg and G. Joblove, Cornell Program of Computer Graphics.

FIG. 2-8b

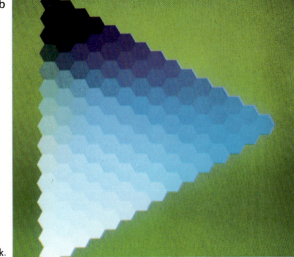

Courtesy of Tom Check.

FIG. 5-4a

Courtesy of Sam Kapilivsky.

FIG. 5-4b

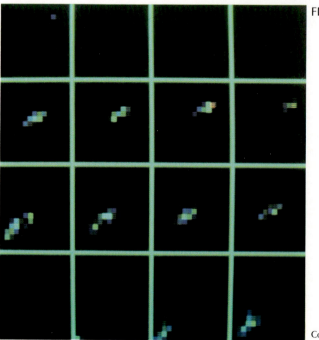

Courtesy of Sam Kapilivsky.

FIG. 5-4c

Courtesy of Sam Kapilivsky.

FIG. 9-10

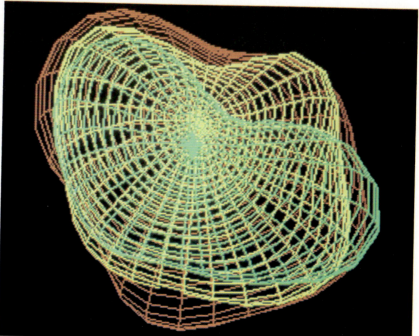

Courtesy of Robert Schudy.

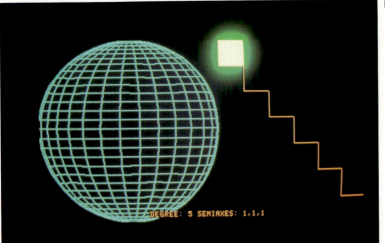

FIG. 11-3a

DEGREE: 5 SEMIAXES: 1,1,1

Courtesy of Robert Schudy.

FIG. 11-3b

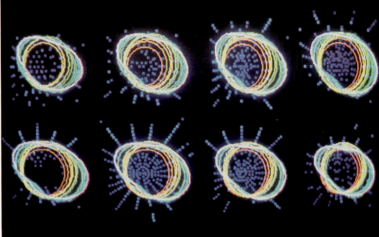

Courtesy of Robert Schudy.

FIG. 11-3c

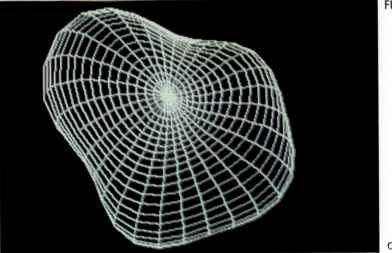

Courtesy of Robert Schudy.

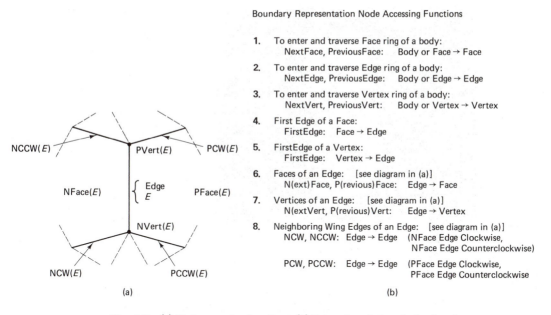

1. To enter and traverse Face ring of a body:
     NextFace, PreviousFace: Body or Face → Face

2. To enter and traverse Edge ring of a body:
     NextEdge, PreviousEdge: Body or Edge → Edge

3. To enter and traverse Vertex ring of a body:
     NextVert, PreviousVert: Body or Vertex → Vertex

4. First Edge of a Face:
     FirstEdge: Face → Edge

5. FirstEdge of a Vertex:
     FirstEdge: Vertex → Edge

6. Faces of an Edge: [see diagram in (a)]
     N(ext)Face, P(revious)Face: Edge → Face

7. Vertices of an Edge: [see diagram in (a)]
     N(ext)Vert, P(revious)Vert: Edge → Vertex

8. Neighboring Wing Edges of an Edge: [see diagram in (a)]
     NCW, NCCW: Edge → Edge   (NFace Edge Clockwise,
                                                    NFace Edge Counterclockwise)

     PCW, PCCW: Edge → Edge   (PFace Edge Clockwise,
                                                    PFace Edge Counterclockwise

(a)   (b)

Fig. 9.5   (a) Node accessing functions. (b) Semantics of winged edge functions.

interior face points becomes impractically complex. For that reason, the number of edges for a curved face is usually restricted to three or four.

A general technique for approximating surfaces with four-sided surface *patches* is that of Coons [Coons 1974]. Coons specifies the four sides of the patch with polynomials. These polynomials are used to interpolate interior points. Although this is appropriate for synthesis, it is not so easy to use for analysis. This is because of the difficulty of registering the patch edges with image data. A given surface will admit to many patch decompositions.

An attractive representation for patches is splines (Fig. 9.6). In general, two-dimensional spline interpolation is complex: For two parameters u and v interpolate with

$$\mathbf{x}(u, v) = \sum_i \sum_j V_{ij} B_{ij}(u, v) \tag{9.1}$$

similar to Eq. (8.4). However, for certain applications a further simplification can be made. In a manner analogous to (8.9) define a grid of knot points $\mathbf{v}_{ij}$ corresponding to $\mathbf{x}_{ij}$ and related by

$$\mathbf{x}_{ij} = M\mathbf{v}_{ij} \tag{9.2}$$

Now rather than interpolating in two dimensions simultaneously, interpolate in one direction, say t, to obtain

$$\mathbf{x}_{ij}(t) = [t^3 \quad t^2 \quad t \quad 1][C][\mathbf{v}_{i-1,j_0}, \mathbf{v}_{i,j_0}, \mathbf{v}_{i+1,j_0}, \mathbf{v}_{i+2,j_0}]^T \tag{9.3}$$

for each value of j. Now compute $\mathbf{v}_{ij}(t)$ by solving

$$\mathbf{x}_{ij}(t) = M\mathbf{v}_{ij}(t) \tag{9.4}$$

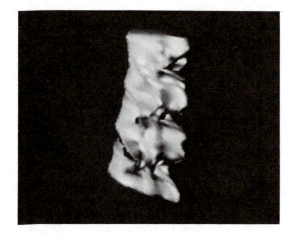

Fig. 9.6 Using spline curves to model the surface of an object: a portion of a human spinal column taken from CAT data.

for each value of t. Finally, interpolate in the other direction and solve:

$$\mathbf{x}_{ij}(s, t) = [s^3 \quad s^2 \quad s \quad 1][C][\mathbf{v}_{i-1,j}(t), \mathbf{v}_{i,j}(t), \mathbf{v}_{i+1,j}(t), v_{i+2,j}(t)] \quad (9.5)$$

This is the basis for the spline filtering algorithm discussed in Section 3.2.3.

Some advantages of spline surfaces for vision are the following.

1. The spline representation is economical: the space curves are represented as a sparse set of knot points from which the underlying curves can be interpolated.

2. It is easy to define splines interactively by giving the knot points; reference representations may be built up easily.

3. It is often useful to search the image in a direction perpendicular to the model reference surface. This direction is a simple function of the local knot points.

9.2.3 Surfaces That Are Functions on the Sphere

Some surfaces can be expressed as functions on the "Gaussian sphere." (the distance from the origin to a point on the surface is a function of the direction of the point, or of its longitude and latitude if it were radially projected on a sphere with the center at the origin.) This class of surfaces, although restricted, is useful in some application areas [Schudy and Ballard 1978, 1979]. This section explores briefly two schemes for representation of these surfaces. The first specifies explicitly the distance of the surface from the origin for a set of vector directions from the origin. The second is akin to Fourier descriptors; an economically specified set of coefficients characterizes the surface with greater accuracy as the number of coefficients increases.

Direction–Magnitude Sets

One approximation to a spherical function is to specify a number of three-dimensional direction vectors from the origin and for each a magnitude. This is equivalent to specifying a set of (θ, ϕ, ρ) points in a spherical coordinate system (Appendix 1). These points are on the surface to be represented; connecting them yields an approximation.

It is often convenient to represent directions as points on the unit (Gaussian) sphere centered on the origin. The points may be connected by straight lines to form a polyhedron with triangular, hexagonal or rhomboidal faces. Moving the points on the sphere out (or in) by their associated magnitude distorts this polyhedron, moving its vertices radically out or in.

The spherical function determines the distance of face vertices from the origin. Resolution at the surface increases with the number of faces. An approximately isotropic distribution of directions over the surface may be obtained by placing the face vertices (directions) in accordance with "geodesic dome"–like calculations which make the faces approximately equilateral triangles [Clinton 1971].

Although the geodesic tesselation of the sphere's surface is more complex than a straightforward (latitude and longitude, say) division, its pleasant properties of isotropy and display [Brown 1979a; 1979b; Schudy and Ballard 1978] sometimes recommend it. Some example shapes indicating the range of representable surfaces are given in Fig. 9.7. Methods for tesselating the sphere are given in Appendix 1.

Spherical Harmonic Surfaces

In two dimensions, Fourier coefficients can give approximations to certain curved boundaries (Section 8.3.4). Analogously in three dimensions, a set of orthogonal functions may be used to express a closed boundary as a set of coefficients when the boundary is a function on the sphere. One such decomposition is *spherical harmonics*. Low order coefficients capture gross shape characteristics; higher order coefficients represent surface shape variations of higher spatial frequency. The function with $m = 0$ is a sphere, the three with $m = 1$ represent translation about the origin, the five with $m = 2$ are similar to prolate and oblate spheroids, and so forth, the lobedness of the surfaces increasing with m. A sample three dimensional shape and its "description" is shown in Fig. 9.8.

Spherical harmonics are analogs on the sphere of Fourier functions on the plane; like Fourier functions, they are smooth and continuous to every order. They may be parameterized by two numbers, m and n; thus they are a doubly infinite set of functions which are continuous, orthogonal, single-valued, and complete on the

Fig. 9.7 Sample surfaces described by some 320 triangular facets in a geodesic tesselation.

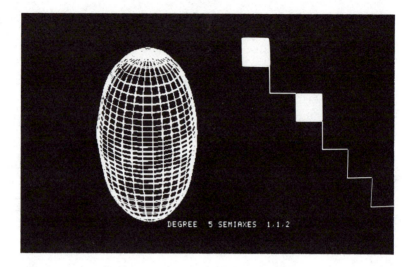

Fig. 9.8 A spherical harmonic function description of an ellipsoid. Coefficients are displayed on the right as grey levels in the matrix format

$$u_{00}$$

$$
\begin{array}{ccccc}
u_{01} & v_{11} & & & \\
u_{11} & u_{02} & v_{12} & v_{22} \\
& u_{12} & & & \\
& u_{21} & & &
\end{array}
$$

sphere. In combination, the harmonics can thus produce all "well-behaved" spherical functions.

The spherical harmonic functions $U_{mn}(\theta,\phi)$ and $V_{mn}(\theta,\phi)$ are defined in polar coordinates by:

$$U_{mn}(\theta,\phi) = \cos(n\theta)\sin^n(\phi)P(m,\,n,\,\cos(\phi)) \tag{9.6}$$

$$V_{mn}(\theta,\phi) = \sin(n\theta)\sin^n(\phi)P(m,\,n,\,\cos(\phi)) \tag{9.7}$$

with $m = 0, 1, 2, ..., M$; $n = 0, 1, ..., m$. Here $P(m,\,n,\,x)$ is the nth derivative of the mth Legendre polynomial as a function of x. To represent an arbitrary shape, let the radius R in polar coordinates be a linear sum of these spherical harmonics:

$$R(\theta,\phi) = \sum_{m=0}^{M} \sum_{n=0}^{m} A_{mn} U_{mn}(\theta,\phi) + B_{mn} V_{mn}(\theta,\phi) \tag{9.8}$$

Any continuous surface on the sphere may be represented by a set of these real constants; reasonable approximations to heart volumes are obtained with $m \leqslant 5$ [Schudy and Ballard 1979].

Figure 9.9 shows a few simple combinations of functions of low values of $(m,\,n)$. The sphere, or $(0,0)$ surface, is added to the more complex ones to ensure positive volumes and drawable surfaces.

Spherical harmonics have the following attractive properties.

1. They are orthogonal on the sphere under the inner product;

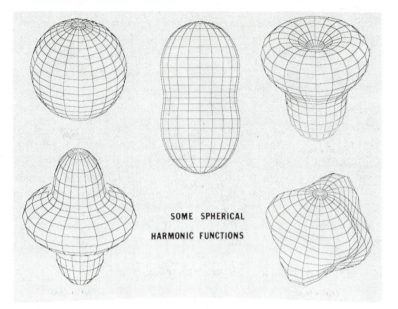

SOME SPHERICAL
HARMONIC FUNCTIONS

Fig. 9.9 Simple combinations of functions.

$$(u, v) = \int uv \, \sin\phi \, d\theta \, d\phi$$

2. The functions are arranged in increasing order of spatial complexity.

3. The whole set is complete; any twice-differentiable function on the sphere can be approximated arbitrarily closely.

Spherical harmonics can provide compact, nonredundant descriptions of surfaces that are useful for analysis of shape, but are less useful for synthesis. The principal disadvantages are that the primitive functions are not necessarily related to the desired final shape in an intuitive way, and changing a single coefficient affects the entire resulting surface.

An example of the use of spherical harmonics as a volume representation is the representation of heart volume [Schudy and Ballard 1978, 1979]. In extracting a volume associated with the heart from ultrasound data, a large mass of data is involved. The data is originally in the form of echo measurements taken in a set of two-dimensional planes through the heart. The task is to choose a surface surrounding the heart volume of interest by optimization techniques that will fit three dimensional time-varying data. The optimization involved is to find the best coefficients for the spherical harmonics that define the surface. The goodness of fit of a surface is measured by how well it matches the edge of the volume as it appears in the data slices. To extend spherical harmonics to time-varying periodic data, let the radius R in polar coordinates be a linear sum of these spherical harmonics:

$$R(\theta, \phi, t) = \sum_{m=0}^{M} \sum_{n=0}^{m} A_{mn}(t) U_{mn}(\theta, \phi) + B_{mn}(t) V_{mn}(\theta, \phi) \qquad (9.9)$$

The functions $A(t)$ and $B(t)$ are given by Fourier time series:

$$A_{mn}(t) = a_{mno} + \sum_{i=1}^{I} a_{mni} \cos(2\pi t/\tau) + b_{mni} \sin(2\pi t/\tau) \qquad (9.10)$$

$$B_{mn}(t) = b_{mno} + \sum_{i=1}^{I} c_{mni} \cos(2\pi t/\tau) + d_{mni} \sin(2\pi t/\tau) \qquad (9.11)$$

where t is time, the a_{mni}, b_{mni}, c_{mni}, and d_{mni} are arbitrary real constants, and τ the period. Any continuous periodically moving surface on the sphere may be represented by some selection of these real constants; in the cardiac application, reasonable approximations to the temporal behavior are obtained with $t \leqslant 3$. Figure 9.10 shows three stages from a moving-harmonic-surface representation of the heart in early systole. The atria, at the top, contract and pump blood into the ventricles below, after which there is a ventricular contraction.

9.3 GENERALIZED CYLINDER REPRESENTATIONS

The volume of many biological and manufactured objects is naturally described as the "swept volume" of a two-dimensional set moved along some three-space curve. Figure 9.11 shows a "translational sweep" wherein a solid is represented as the volume swept by a two-dimensional set when it is translated along a line. A "rotational sweep" is similarly defined by rotating the two-dimensional set around an axis. In "three-dimensional sweeps," volumes are swept. In a "general" sweep scheme, the two-dimensional set or volume is swept along an arbitrary space curve, and the set may vary parametrically along the curve [Binford 1971; Soroka and Bajcsy 1976; Soroka 1979a; 1979b; Shani 1980]. General sweeps are quite a popular representation in computer vision, where they go by the name *generalized cylinders* (sometimes "generalized cones").

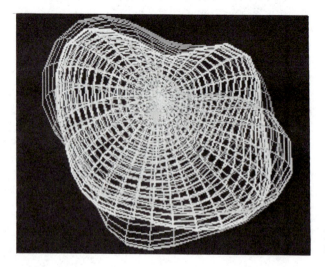

Fig. 9.10 Three stages from a moving harmonic surface (*see text and color insert*).

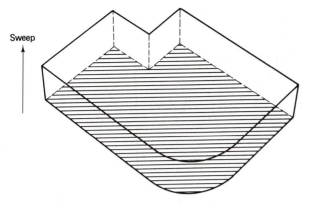

Sweep

Fig. 9.11 A translational sweep.

A generalized cylinder (GC) is a solid whose axis is a 3-D space curve (Fig. 9.12a). At any point on the axis a closed cross section is defined. A usual restriction is that the axis be normal to the cross section. Usually it is easiest to think of an axis space curve and a cross section point set function, both parameterized by arc length along the axis curve. For any solid, there are infinitely many pairs of axis and cross section functions that can define it.

Generalized cylinders present certain technical subtleties in their definition. For instance, can it be determined whether any two cross sections intersect, as they would if the axis of a circular cylinder were sharply bent (Fig. 9.12b)? If the solid is defined as the volume swept by the cross section, there is no conceptual or computational problem. A problem might occur when computing the surface of such an object. If the surface is expressed in terms of the axis and cross-section functions (as below), the domain of objects must be limited so that the boundary formula indeed gives only points on the boundary.

Generalized cylinders are intuitive and appealing. Let us grant that "pathological" cases are barred, so that relatively simple mathematics is adequate for representing them. There are still technical decisions to make about the representation. The axis curve presents no difficulties, but a usable representation for the cross-section set is often not so straightforward. The main problem is to choose a usable coordinate system in which to express the cross section.

9.3.1 Generalized Cylinder Coordinate Systems and Properties

Two mathematical functions defining axis and cross section for each point define a unique solid with the "sweeping" semantics described above. In a fixed Cartesian coordinate system x, y, z, the axis may be represented parametrically as a function of arc length s:

$$\mathbf{a}(s) = (x(s), y(s), z(s)) \qquad (9.12)$$

It is convenient to have a local coordinate system defined with origin at each point of $\mathbf{a}(s)$. It is in this coordinate system that the cross section is defined. This system may change in orientation as the axis winds through space, or it may be most natural for it not to be tied to the local behavior of the axis. For instance, imagine tying a knot in a solid rubber bar of square cross section. The cross section

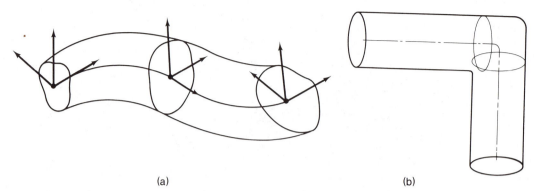

<table>
<tr><td>(a)</td><td>(b)</td></tr>
</table>

Fig. 9.12 (a) A generalized cylinder and some cross-sectional coordinate systems. (b) A possibly "pathological" situation. Cross sections may be simply described as circles centered on the axis, but then their intersection makes volume calculations (for instance) less straightforward.

will stay approximately a square, and (this is the point) will remain approximately fixed in a coordinate system that twists and turns through space with the axis of the bar. On the other hand, imagine bolt threads. They can be described by a single cross section that stays fixed in a coordinate system that rotates as it moves along the straight axis of the bolt. There is no a priori reason to suppose that such a useful local coordinate system should twist along the GC axis.

A coordinate system that mirrors the local behavior of the GC axis space curve is the "Frenet frame," defined at each point on the GC axis. This frame provides much information about the GC-axis behavior. The GC axis point forms the origin, and the three orthogonal directions are given by the vectors $(\boldsymbol{\xi}, \boldsymbol{\nu}, \boldsymbol{\zeta})$, where

$\boldsymbol{\xi}$ = unit vector tangent to axis

$\boldsymbol{\nu}$ = unit vector direction of center of curvature of axis

   normal curve

$\boldsymbol{\zeta}$ = unit vector direction of center of torsion of axis

Consider the curve to be produced by a point moving at constant speed through space; the distance the point travels is the parameter of the space curve [O'Neill 1966]. Since $\boldsymbol{\xi}$ is of constant length, its derivative measures the way the GC axis turns in space. Its derivative $\boldsymbol{\xi}'$ is orthogonal to $\boldsymbol{\xi}$ and the length of $\boldsymbol{\xi}'$ measures the curvature κ of the axis at that point. The unit vector in the direction of $\boldsymbol{\xi}'$ is $\boldsymbol{\nu}$. Where the curvature is not zero, a binormal vector $\boldsymbol{\zeta}$ orthogonal to $\boldsymbol{\xi}$ and $\boldsymbol{\nu}$ is defined. This binormal $\boldsymbol{\zeta}$ is used to define the torsion τ of the curve. The vectors $\boldsymbol{\xi}$, $\boldsymbol{\nu}$, $\boldsymbol{\zeta}$ obey Frenet's formulae:

$$\boldsymbol{\xi}' = \kappa\boldsymbol{\nu}$$
$$\boldsymbol{\nu}' = -\kappa\boldsymbol{\xi} + \tau\boldsymbol{\zeta} \qquad (9.13)$$
$$\boldsymbol{\zeta}' = -\tau\boldsymbol{\nu}$$

where

$$\kappa = \text{curvature} = -\boldsymbol{\nu}' \cdot \boldsymbol{\xi} = \boldsymbol{\nu} \cdot \boldsymbol{\xi}' \qquad (9.14)$$

$$\tau = \text{torsion} = \boldsymbol{\nu}' \cdot \boldsymbol{\zeta} = -\boldsymbol{\nu} \cdot \boldsymbol{\zeta}' \qquad (9.15)$$

The Frenet frame gives good information about the axis of the GC, but it has certain problems. First, it is not well defined when the curvature of the GC axis is zero. Second, it may not reflect known underlying physical principles that generate the cross sections (as in the bolt thread example). A solution, adopted in [Agin 1972, Shani 1980], is to introduce an additional parameter that allows the cross section to rotate about the local axis by an arbitrary amount. With this additional degree of freedom comes an additional problem: How are successive cross sections registered? Figure 9.13 shows two solutions in addition to the Frenet frame solution.

The cross sectional curve is usually defined to be in the $\boldsymbol{\nu}$–$\boldsymbol{\zeta}$ plane, normal to $\boldsymbol{\xi}$, the local GC axis direction. The cross section may be described as a point set in this plane, using inequalities expressed in the $\boldsymbol{\nu}$–$\boldsymbol{\zeta}$ coordinate system. The cross section boundary (outline curve) may be used instead, parameterized by another parameter r. Let this curve be given by

$$\text{cross section boundary} = (x(r, s), \, y(r, s))$$

The dependence on s reflects the fact that the cross section shape may vary along the GC axis. The expression above is in world coordinates, but should be moved to

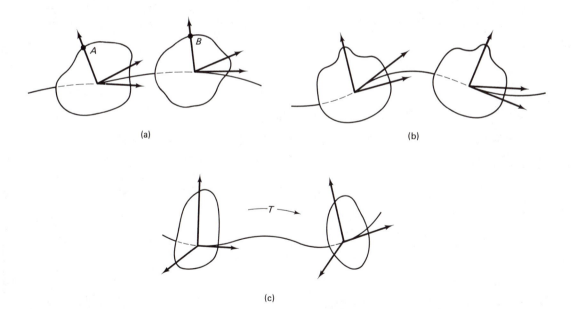

(a)

(b)

(c)

Fig. 9.13 (a) Local coordinates are the Frenet frame. Points A and B must correspond. (b) Local coordinates are determined by the cross sectional shape. (c) Local coordinates are determined by a heuristic transformation from world coordinates.

the local coordinates on the GC axis. A transformation of coordinates allows the GC boundary to be expressed (if the GC is well behaved) as

$$B(r, s) = \mathbf{a}(s) + x(r, s)\, \boldsymbol{\nu}(s) + y(r, s)\, \boldsymbol{\zeta}(s) \tag{9.16}$$

One of the advantages of the generalized cylinder representation is that it allows many parameters of the solid to be easily calculated.

- In matching the GC to image data it is often necessary to search perpendicular to a cross section. This direction is given from $x(r, s)$, $y(r, s)$ by $((dy/ds)\boldsymbol{\nu}, -(dx/ds)\boldsymbol{\zeta})$.

- The area of a cross section may be calculated from Eq. (8.16).

- The volume of a GC is given by the integral of: the area as a function of the axis parameter multipled by the incremental path length of the GC axis, i.e.,

$$\text{volume} = \int_0^L \text{area}(s)\, ds$$

9.3.2 Extracting Generalized Cylinders

Early work in biological form analysis provides an example of the process of fitting a GC to real data and producing a description [Agin 1972]. One of the goals of this work was to infer the stick figure skeleton of biological forms for use in matching models also represented as skeletons. In Fig. 9.14 the process of inferring the axis from the original stripe three-dimensional data is shown; the process iterates toward a satisfactory fit, using only circular cross sections (a common constraint with "generalized" cylinders). Figure 9.15 shows the data and the analysis of a complex

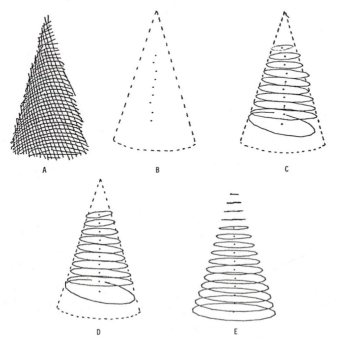

Fig. 9.14 Stages in extracting a generalized cylinder description for a circular cone. (a) Front view. (b) Initial axis estimate. (c) Preliminary center and axis estimate. (d) Cone with smoothed radius function. (e) Completed analysis.

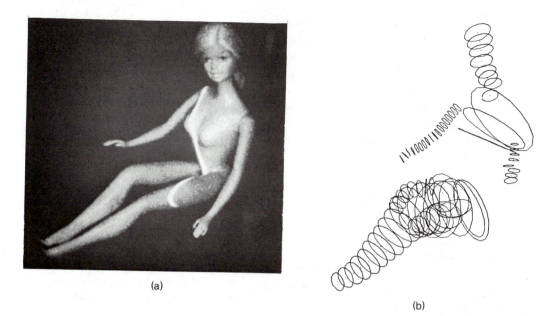

(a)

(b)

Fig. 9.15 (a) TV image of a doll. (b) Completed analysis of doll.

biological form. In real data, complexly interrelated GCs are hard to decompose into satisfactory subparts. Without that, the ability to form a satisfactory articulated skeleton is severely restricted.

In later work, GCs with spline-based axes and cross sections were used to model organs of the human abdomen [Shani 1980]. Figure 9.16 shows a rendition of a GC fit to a human kidney.

9.3.3 A Discrete Volumetric Version of the Skeleton

An approximate volume representation that can be quite useful is based on an articulated wire frame skeleton along which spheres (not cross sections) are placed.

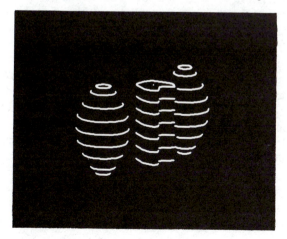

Fig. 9.16 Generalized cylinder representation of two kidneys and a spinal column. This coarse, nominal model is refined during examination of CAT data (see Fig. 9.6).

This representation has some of the flavor of an approximate sweep representation. An example of the use of such a representation and a figure are given in Section 7.3.4. This representation was originally conceived for graphics applications (the spheres look the same from any viewpoint) [Badler and Bajcsy 1978]. Collision detection is easy, and three-dimensional objects can be decomposed into spheres automatically [O'Rourke and Badler 1979]. From the spheres, the skeleton may be derived, and so may the surface of the solid. This representation is especially apt for many computer vision applications involving nonrigid bodies if strict surface and volumetric accuracy is not necessary [Badler and O'Rourke 1979].

9.4 VOLUMETRIC REPRESENTATIONS

Most world objects are solids, although usually only their surfaces are visible. A representation of the objects in terms of more primitive solids is often useful and can have pleasant properties of terseness, validity, and sometimes ease of computation. The representations given here are presented in order of increasing generality; constructive solid geometry includes cell decomposition, which in turn includes spatial occupancy arrays.

Algorithms for processing volume-based representations are often of a different flavor than surface-based algorithms. We give some examples in Section 9.4.4. Objects represented volumetrically can be depicted on raster graphics devices by a "ray-casting" approach in which a line of sight is constructed through the viewing plane for a set of raster points. The surface of the solid at its intersection with the line of sight determines the value of the display at the raster point. Ray casting can produce hidden-line and shaded displays; graphics is only one of its applications (Section 9.4.4).

9.4.1 Spatial Occupancy

Figure 9.17 shows that three-dimensional spatial occupancy representations are the three-dimensional equivalent of the two-dimensional spatial occupancy representations of Chapter 8. Volumes are represented as a three-dimensional array of cells which may be marked as filled with matter or not. Spatial occupancy arrays can require much storage if resolution is high, since space requirements increase as the cube of linear resolution. In low-resolution work with irregular objects, such as arise in computer-aided tomography, spatial occupancy arrays are very common. It is sometimes useful to convert an exact representation into an approximate spatial occupancy representation. Slices or sections through objects may be easily produced. The spatial occupancy array may be run-length encoded (in one dimension), or coded as blocks of different sizes; such schemes are actually cell-decomposition schemes (Section 9.4.2).

With the declining cost of computer memory, explicit spatial occupancy arrays may become increasingly common. The improvement of hardware facilities for parallel computation will encourage the development of parallel algorithms to compute properties of solids from these representations.

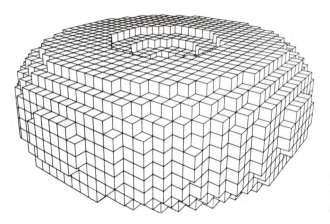

Fig. 9.17 A solid (the shape of a human red blood cell) approximated by a volume occupancy array.

9.4.2 Cell Decomposition

In cell decomposition, cells are more complex in shape but still "quasi-disjoint" (do not share volumes), so the only combining operation is "glue" (Fig. 9.18). Cells are usually restricted to have no holes (they are "simply connected"). Cell decompositions are not particularly concise; their construction (especially for curved cells) is best left to programs. It seems difficult to convert other representations exactly into cell decompositions. Two useful cell decompositions are the "oct-tree" [Jackins and Tanimoto 1980] and the kd-tree [Bentley 1975]. They both can be produced by recursive subdivision of volume; these schemes are the three-dimensional analogs of pyramid data structures for two dimensional binary images.

The quasi-disjointness of cell-decomposition and spatial-occupancy primitives may be helpful in some algorithms. Mass properties (Section 9.4.4) may be computed on the components and summed. It is possible to tell whether a solid is connected and whether it has voids. Inhomogeneous objects (such as human anatomy inside the thorax) can be represented easily with cell decomposition and spa-

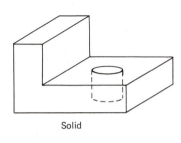

Solid

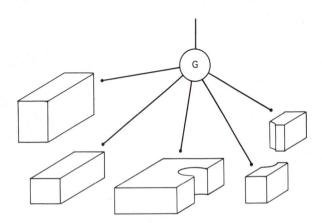

Fig. 9.18 A volume and its cell decomposition.

tial occupancy. The CT number (transparency to x-rays) or a material code can be kept in a cell instead of a single bit indication of "solid or space."

9.4.3 Constructive Solid Geometry

Figure 9.19 shows one constructive solid geometry (CSG) scheme [Voelcker and Requicha 1977; Boyse 1979]. Solids are represented as compositions, via set operations, of other solids which may have undergone rigid motions. At the lowest level are primitive solids, which are bounded intersections of closed half-spaces defined by some $F(x, y, z) \geqslant 0$, where F is well-behaved (e.g., analytic). Usually, primitives are entities such as arbitrarily scaled rectangular blocks, arbitrarily scaled cylinders and cones, and spheres of arbitrary radius. They may be positioned arbitrarily in space.

Figure 9.20 shows a parameterized representation [Marr and Nishihara 1978; Nishihara 1979] based on shapes (here cylinders) that might be extracted from an image.

A CSG representation is an expression involving primitive solid and set operators for combination and motion.

$<CSG\,Rep>$ ::= $<$primitive solid$>$ |

MOVE $<$CSG Rep$>$ BY $<$Motion Params$>$ |

$<$CSG Rep$>$ $<$Combine Op$>$ $<$CSG Rep$>$

The combining operators are best taken to be *regularized* versions of set union, intersection, and difference (the complement is a possible operator, but it allows unbounded solids from bounded primitives).

Regularity is a fundamental property of any set of points that models a solid. In a given space, a set X is regular if $X = kiX$, where k and i denote the *closure* and *interior* operators. Intuitively, a regular set has no isolated or dangling boundary points. The regularization r of a set X is defined by $rX = kiX$. Regularization informally amounts to taking what is inside a set and covering that with a tight skin. Regular sets are not closed under conventional set operations, but *regularized*

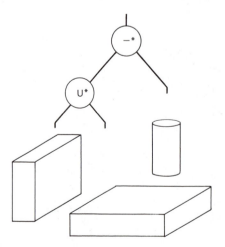

Fig. 9.19 Constructive solid geometry for the volume of Fig. 9.18.

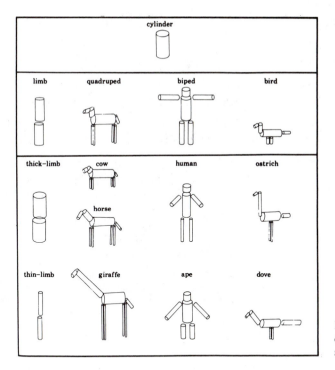

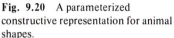

Fig. 9.20 A parameterized constructive representation for animal shapes.

operators do preserve regularity. Regularized operators are defined by

$$X <OP> * Y = r(X <OP> Y)$$

Regularity and regularized set operators provide a natural formalization of the dimension-preserving property exhibited by many geometric algorithms, thus obviating the need to enumerate many annoying "special cases." Figure 9.21 illustrates conventional versus regularized intersection of two sets that are regular in the plane.

If the primitives are unbounded, checking for boundedness of an object can be difficult. If they are bounded, any CSG representation is a valid volume representation. CSG can be inefficient for some geometric applications, such as a line drawing display. (Converting the CSG representation to a boundary representation is the one way to proceed; see Section 9.4.4.)

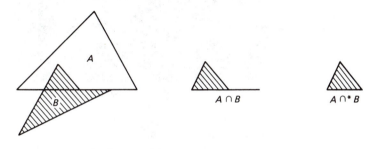

Fig. 9.21 Conventional ($\cap$) and regularized ($\cap$ *) polygon intersection.

9.4.4 Algorithms for Solid Representations

Set Membership Classification

The set membership classification (SMC) function M takes a candidate point set C and a reference set S, and returns the points of C that are in S, out of S, and on the boundary of S.

$$(CinS, \ CoutS, \ ConS) := M(C, \ S)$$

Figure 9.22a shows line–polygon classification.

SMC is a generalization of set intersection [Tilove 1980]. It is a useful geometric utility; polygon–polygon classification is generalized clipping, and volume–volume classification detects solid interference. Line–solid classification

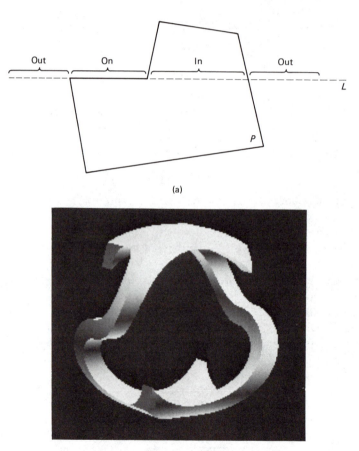

(a)

(b)

Fig. 9.22 (a) The set membership classification (SMC) function M(L, P) finds the portions of the candidate set L (here a line) that are in, on, and out of a reference set (here a polygon) P. (b) Image produced by ray casting, a special case of SMC.

may be used for ray casting visualization techniques to generate images of a known three-dimensional representation (Fig. 9.22b).

An algorithm for SMC illustrates a "divide and conquer" approach to computing on CSG. Recall that CSG is like a tree of set operations, whose leaves are primitive sets which usually are simple solids such as cylinders, spheres, and blocks. Presumably classification can be more easily computed with these simple sets as reference than with complex unions, intersections, and differences as reference.

The idea is that the classification of a set C with respect to a complex object S defined in CSG may be determined recursively. Any internal node S in the CSG tree is an operation node. It has left and right arguments and an operation OpofS. Each subtree is itself a CSG subtree or a primitive.

$$M(X, S) = \text{IF } S \text{ is a primitive THEN prim--}M(X, S)$$

$$\text{ELSE Combine}(M(X, \text{ left--subtree}(S),$$

$$M(X, \text{ right--subtree}(S),$$

$$\text{OPof}S);$$

Prim-M is the easily computed classification with respect to a simple primitive solid. The Combine operation is a nontrivial calculation that combines the subresults to produce a more complex classification. It is illustrated in two dimensions for line classification in Fig. 9.23. Having classified the line L against the polygon $P1$ and $P2$, the classifications can be combined to produce the classification for $P1 \cap P2$. Precise rules for combine may be written for (regularized) union, intersection, and set difference. An important point is that when a point is in the "on" set of S_1 and in the "on" set of S_2, the result of the combination depends on extra information. In Fig. 9.23, segments X and Y both result from this ON–ON case of combine, but segment X is OUT of the boundary of the intersection and Y is IN the intersection. The ambiguity must be resolved by keeping "neighborhood information" (local geometry) attached to point sets, and combining the neighborhoods along with the classifications. The technical problems surrounding combine can be solved, and SMC is basic in several solid geometric modeling systems [Boyse 1979; Voelcker et al. 1978; Brown et al. 1978].

Mass Properties

The analog of many two-dimensional geometric properties is to be found in "mass properties," which are defined by volume integrals over a solid. The four types of mass properties commonly of interest are:

Volume: $V = \int_s du$

Centroid: e.g. $GC_x = \dfrac{\int_s x\, du}{V}$

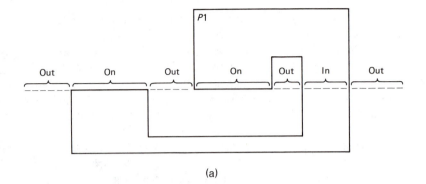

(a)

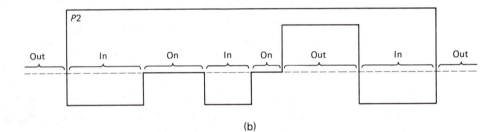

(b)

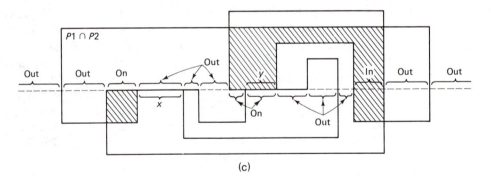

(c)

Fig. 9.23 Combining line-polygon classifications (a) and (b) must produce the classification (c).

Moment of   (9.17)
Inertia: e.g. $I_{xx} = m \int_s (y^2 + z^2) \, du$

Product of
Inertia: e.g. $P_{xy} = m \int_s xy \, du$

where m is a density measure, du the volume differential, and integrals are taken over the volume.

Measures such as these are not necessarily easy to compute from a given representation. The calculation of mass properties of solids from various representations is discussed in [Lee and Requicha 1980]. The approaches suggested by the representations are shown in Fig. 9.24.

One method is based on decomposing the solid into quasi-disjoint cells. An integral property of the cell decomposition is just the sum of the property for each of the cells. Hence if computing the property for the cells is easy, the calculation is easy for the whole volume. One is invited to decompose the body into simple cells, such as columns or cubes, as shown in Fig. 9.25. The resulting calculations, performed to reasonable error bounds on fairly complex volumes, take unacceptably long for the pure spatial occupancy enumeration, but are acceptable for the column and block decompositions. (The column decomposition corresponds to a ray casting approach.) The block decomposition method can be programmed using oct-trees or kd-trees in a manner reminiscent of the Warnock hidden-line algorithm [Warnock 1969], in which the blocks are found automatically, and their size diminishes as increased resolution is needed in the solid. In calculating from a constructive solid geometry representation, the same divide-and-conquer strategy that is useful for SMC may be applied. Again, it recursively solves subproblems induced by the set operators (Fig. 9.26). The strategy is less appealing here since the number of subproblems can grow exponentially in the worst case.

In boundary representations, one can perhaps directly integrate over the boundary in a three-dimensional version of the polygon area calculation given in Chapter 8. This method is often impossible for curved surfaces, which, however, may be approximated by planar faces. An alternative is to use the divergence

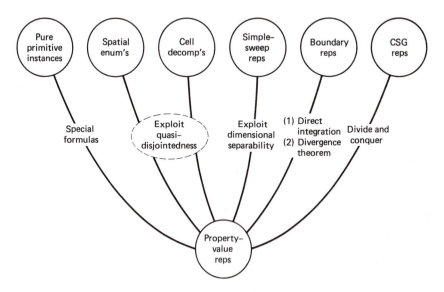

Fig. 9.24 "Natural" approaches to computing mass properties from several representations.

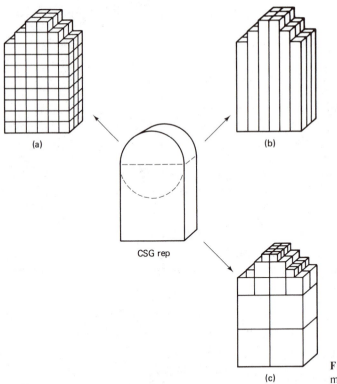

(a)

(b)

CSG rep

(c)

Fig. 9.25 Cell decompositions for mass properties.

theorem (Gauss's theorem). The *divergence* is a scalar quantity defined at any point in a vector field by writing the vector function as

$$\mathbf{G}(x, y, z) = P(x, y, z)\mathbf{i} + Q(x, y, z)\mathbf{j} + R(x, y, z)\mathbf{k}. \qquad (9.18)$$

The divergence is

$$\text{div } \mathbf{G} = \frac{P}{x} + \frac{Q}{y} + \frac{R}{z} \qquad (9.19)$$

There is always a function $\mathbf{G}$ such that div $\mathbf{G} = f(x, y, z)$ for any continuous function f (f computes the integral property of interest.) Thus

$$\int_S f \, dv = \int_S \text{div } \mathbf{G} \, dv \qquad (9.20)$$

But the divergence theorem states that

$$\int_S \text{div } \mathbf{G} \, dv = \Sigma_i \int_{F_i} \mathbf{G}\mathbf{n}_i \, dF_i \qquad (9.21)$$

where F_i is a face of the solid S, $\mathbf{n}_i$ is the unit normal to F_i, and dF_i the surface differential. Again this formula works well for planar faces, but may require approximation techniques for curved faces with complex boundaries.

Boundary Evaluation

The calculation of a face-based surface (boundary) representation from a

• Divide and conquer

Reduction formula

$$\int_{A \cup B} = \int_A + \int_B - \int_{A \cap B}$$

$$\int_{A - B} = \int_A - \int_{A \cap B}$$

Example

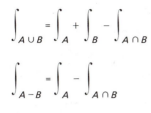

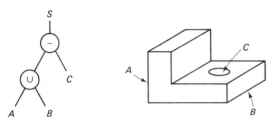

$I_S = I_A + I_B - \underbrace{I_{A \cap B} - I_{A \cap C}}_{\emptyset} - \underbrace{I_{B \cap C} + I_{A \cap B \cap C}}_{\emptyset}$

Fig. 9.26 Recursive problem decomposition for mass property calculation.

CSG representation is called *boundary evaluation*. It is an example of *representation conversion*. Both the CSG and boundary are usually unambiguous representations of a volume; a CSG expression (a solid) has just one boundary, but a boundary (representing a solid) usually has many CSG expressions. Since a solid may be put together from primitives in many ways, the mapping back from boundary to CSG is not usually attempted (but see [Markovsky and Wesley 1980, Wesley and Markovsky 1981]).

One style of boundary evaluation is based on the following observations [Voelcker and Requicha 1980; Boyse 1979].

• Boundaries of composite objects may be computed from certain set-theoretic formulae. For (regularized) intersection of two objects *S* and *T*, the formula is

$$b(S \bigcap {}^* T) = (bS \bigcap {}^* iT) \bigcup {}^* (iS \bigcap {}^* bT)$$

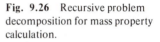

(9.22)

where $\bigcap^*$ and $\bigcup^*$ are regularized intersection and union: *b*, *i*, and *k* are the boundary, interior, and closure operators. (Recall that *ki* is *r*, the regularization operator).

• Faces of composite objects can arise only from faces of primitives.

• Faces are either bounded by edges or are self-closing (as is the sphere).

These observations and the existence of the classification operation motivate the grand strategy that follows (ignoring several important details and concentrating on the core of the algorithm.)

1. Find all possible ("tentative") edges for each face of each primitive in the composite.

2. Classify each tentative edge with respect to the composite solid.

3. The ON portions of those edges must be enough to define the boundary.

Given the grand strategy, several algorithms of varying sophistication are possible, depending on what edges should be classified (how to generate tentative edges), in what order they should be classified, and how classification is done. The following algorithm is very simple (but very inefficient); useful algorithms are rather more complex.

Algorithm 9.1: CSG to Boundary Conversion (top-level control loop)

Input: Solid defined by CSG expression of regularized set operations applied to primitive solids.

Output: "Bfaces" in the object boundary. Bfaces are represented by their bounding edges. They may have little relation to the "intuitive faces" of the boundary; they may overlap each other, and a Bface may be disconnected (specify more than one region). Edges may appear many times. The Bface-oriented boundary may be processed to remove repetition and merge Bfaces into more intuitively appealing boundary faces.

BEGIN

Form a list PFaces of all ("intuitive") faces of primitive solids involved in the CSG expression, and an initially empty list BFaces to hold the output faces.

For every PFace $F1$ in PFaces:

Create a B-Face called ThisBFace, initially with no edges in it.

For every PFace $F2$ after $F1$ in the PFaces list (this generates all distinct pairs of PFaces just once):

Intersect $F1$ and $F2$ to get TEdges, a set of edges tentatively on the boundary of the solid. If $F1$ and $F2$ do not intersect or intersect only in a point, TEdges is empty. If they intersect in a line, TEdges is the single resulting edge. If they intersect in a two-dimensional region, TEdges contains the bounding edges of the intersection region.

Classify every TEdge in TEdges with respect to the whole solid (the CSG expression). Put TEdges that are ON the solid boundary into ThisBFace.

If ThisBFace is not empty, put it into BFaces.

End Inner Loop

End Outer Loop

END

Algorithms such as this involve many technical issues, such as merging coplanar faces, stitching edges together into faces, regularization of faces, removing multiple versions of edges. Boundary evaluation is inherently rather complex, and depends on such things as the definition and representation of faces as well as the geometric utilities taken as basic [Voelcker and Requicha 1981]. Boundary evaluation is an example of exact conversion between significantly different representations. Such conversions are useful, since no single representation seems convenient for all geometric calculations.

9.5 UNDERSTANDING LINE DRAWINGS

"Engineering" line drawings have been (and to a great extent are still) the main medium of communication between human beings about quantitative aspects of three-dimensional objects. The line drawings of this section are only those which are meant to represent a simple domain of polyhedral or simply curved objects. Interpretation of "naturalistic" drawings (such as a sketchmap [Mackworth 1977]) is another matter altogether.

Line drawings (even in a restricted domain) are often ambiguous; interpreting them sometimes takes knowledge of everyday physics, and can require training. Such informed interpretation means that even drawings that are strictly nonsense can be understood and interpreted as they were meant. Missing lines in drawings of polyhedra are often so easy to supply as to pass unnoticed, or be "automatically supplied" by our model-driven perception.

Generalizing the line drawing to three dimensions as a list of lines or points is not enough to make an unambiguous representation, as is shown by Fig. 9.27,

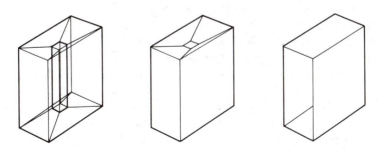

Fig. 9.27 An ambiguous (wireframe) representations of a solid with two of three possible interpretations.

which illustrates that a set of vertices or edges can define many different solids. (It is possible, however, to determine algorithmically all possible polyhedral boundaries described by a three-dimensional wireframe [Markowsky and Wesley 1980].). A line drawing nevertheless does convey three-dimensional information. For any set of N projection specifications (e.g., viewpoint and camera transform), a wire-frame object may be constructed that is ambiguous given the N projections. However, for a given object, there is a maximum number of projections that can determine the object unambiguously. The number depends on the number of edges in the object [Shapira 1974]. Reconstruction of all solids represented by projections is possible [Wesley and Markowsky 1981].

Line drawings were a natural early target for computer vision for the following reasons:

1. They are related closely to surface features of polyhedral scenes.

2. They may be represented exactly; the noise and incomplete visual processing that may have affected the "line drawing extraction" can be modelled at will or completely eliminated.

3. They present an interpretation problem that is significant but seems approachable.

The understanding of simple engineering (3-view) drawings was the first stage in a versatile robot assembly system [Ejiri et al. 1971]. This application underlined the fact that heuristics and conventions are indispensible in engineering drawing understanding. This section deals with the problem of "understanding" a single-view line drawing representation of scenes containing polyhedral and simple curved objects like those in Fig. 9.28.

Our exposition follows a historical path, to show how early heuristic programs in the middle 1960s evolved into more theoretical insights in the early 1970s.

The first real computer vision program with representations of a three-dimensional domain appeared around 1963 [Roberts 1965]. This system, ambitious even by today's standards, was to accept a digitized image of a polyhedral scene and produce a line drawing of the scene as it would appear when viewed from any requested viewpoint. This work addressed basic issues of imaging geometry, feature finding, object representation, matching, and computer graphics.

Since then, several systems have appeared for accomplishing either the same or similar results [Falk 1972; Shirai 1975; Turner 1974]. The line drawings of this section can appear as intermediate representations in a working polyhedral vision system, but they have also been studied in isolation. This topic took on a life of its own and provides a very pretty example of the general idea of going to the three-dimensional world of physics and geometry to understand the appearance of a two-dimensional image. The later results can be used to understand more clearly the successes and failures of early polyhedral vision systems. One form of understanding (line labelling) provided one of the first and most convincing demonstrations of parallel constraint propagation as a control structure for a computer vision process.

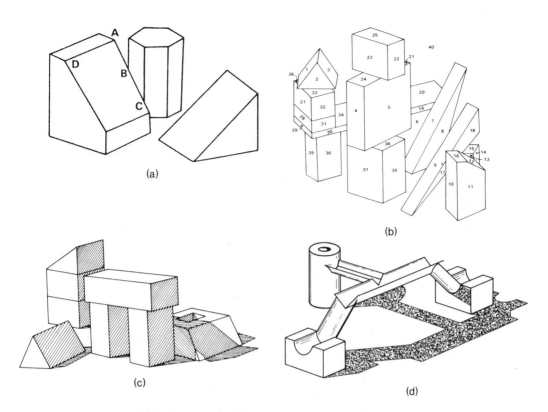

Fig. 9.28 Several typical line drawing scenes for computer understanding.

9.5.1 Matching Line Drawings to Three-dimensional Primitives

Roberts desires to interpret a line drawing such as Fig. 9.28a in terms of a small set of three polyhedral primitives, shown in Fig. 9.29. A simple polyhedron in a scene is regarded as an instance of a transformed primitive, where a transform may involve scaling along the three coordinate axes, translation, and rotation. Compound polyhedra, such as Fig. 9.28a, are regarded as simple polyhedra "glued together." (A cell-decomposition representation is thus used for compound polyhedra.) The program is first to derive from the scene the identity of the primitive objects used to construct it (including details of the construction of compound polyhedra). Next, it is to discover the transformations applied to the primitives to obtain the particular incarnations making up the scene. Finally, to demonstrate its understanding, it should be able to construct a line drawing of the scene from any viewpoint, using its derived description.

To understand a part of the scene, the program first decides which primitive it comes from, and then derives the transformation the primitive underwent to appear as it does in the scene. Identifying primitives is done by matching "topological" features of the line drawing (configurations of faces, lines, and vertices) with those of the model primitives; matching features induce a match between scene and model points. At least four noncoplanar matching points are needed to derive

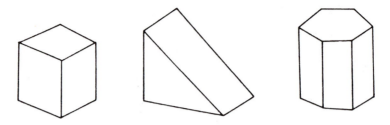

Fig. 9.29 Primitive objects for scene construction.

a transformation. Tentative topological matches are checked by a metrical process which determines whether a primitive can allowably be transformed into the required shape, and if so whether it lies completely inside the observed polyhedra. Since the same image can result from a close small scene or a distant large one, assumptions about the location of the supporting surface and how objects are placed on it are used to fix the distance.

The three primitives all have convex polygons as faces, which project onto the line drawing as convex polygons. The faces all have three, four, or six sides, so faces that have not suffered occlusion or merging with another face while forming a compound polyhedron appear convex, have three, four, or six sides, and have no sides that are the uprights of "T vertices" (which result from occlusion). Polygons that pass these three tests are "approved" and are remembered on a list of possible primitive faces (Fig. 9.30).

In searching for points to identify between the scene and the primitives, the program looks for topological structures (Fig. 9.31) in decreasing order of efficacy, extracts the highest-quality information, reinterprets the scene, and searches again.

When a transformed primitive is identified in the scene, it is notionally unglued and removed, the resulting new visible lines are filled in, and the new scene is analyzed. Roberts's algorithm is not infallible, but it was pioneering work and is a sound starting point for the study of polyhedral scene analysis.

9.5.2 Grouping Regions into Bodies

A program by Guzman [Guzman 1969] takes as input a drawing of a polyhedral scene which may be quite complicated (Fig. 9.28b). The lines divide the drawing into a number of polygonal regions, and the goal of the program is just to group these regions into sets, each set corresponding to one polyhedral "block." Any reasonable description of an ambiguous scene is satisfactory. One could say that Guzman was addressing a polyhedral version of the general question of how human beings segment the world into objects.

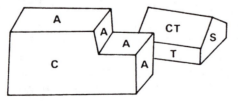

Fig. 9.30 Approved and nonapproved polygons in a line drawing: A: approved; C: Concave; T: T-joint; S: Wrong number of sides.

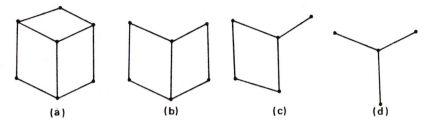

Fig. 9.31 Topological match structures of Roberts.

The idea once again is to accumulate local evidence from the scene, and then to group polygons on the basis of this evidence. The evidence takes the form of "links" which link two regions if they may belong to the same body; links are planted around vertices, which are classified into types, each type always planting the same links (Fig. 9.32). No links are made with the background region.

Scenes are interpreted by grouping according to regions/links, using fairly complex rules, including "inhibitory links" that preclude two neighboring regions from being in the same body.

The final form of the program performs reasonably well on scenes without accidents of visual alignment, but it is a maze of special cases and exceptions, and seems to shed little light on what is going on in known polyhedral line-drawing perception. One might well ask where the links come from; no justification of why they are correct is given. Further ([Mackworth 1973]), Guzman can accept as one body the two regions in Fig. 9.33a. Finally, one feels a little dissatisfied with a scheme that just answers "one body" to a scene like Fig. 9.33b, instead of answering "pyramid on cube" or "two wedges," for example.

Guzman's method is correct for a world of convex isolated trihedral polyhedra: it is extended by ad hoc adjustments based on various potentially conflicting items of evidence from the line drawing. Ultimately it performs adequately with a much increased range of scenes, albeit not very elegantly. Further progress in the line drawing domain came about when attention was directed at the three-dimensional causes of the different vertex types.

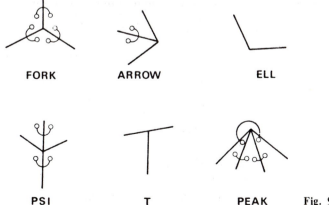

FORK   ARROW   ELL

PSI   T   PEAK   Fig. 9.32 Links around vertices.

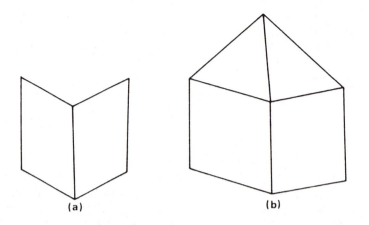

Fig. 9.33 (a) Non-polyhedral scene. (b) Two wedges or a pyramid on cube.

9.5.3 Labeling Lines

Huffman and Clowes independently concerned themselves with scenes similar to Guzman's, not excluding non-simply connected polyhedra, but excluding accidents of alignment [Huffman 1971; Clowes 1971]. They desired to say more about the scene than just which regions arose from single bodies; they wanted to ascribe interpretations to the lines. Figure 9.34 shows a cube resting on the floor; lines labeled with a + are caused by a convex edge, those labeled with a − are caused by a concave edge, and those labeled with a > are caused by matter occluding a surface behind it. The occluding matter is to the right of the line looking in the direction of the >, the occluded surface is to the left. If the cube were floating, one would label the lowest lines with < instead of with −. The shadow line labels (arrows) were not used by Huffman.

A systematic investigation can find the types of lines possibly seen around a trihedral corner; such corners can be classified by how many octants of space are filled by matter around them (one for the corner of a cube, seven for the inside corner of a room, etc.). By considering all possible trihedral corners as seen from

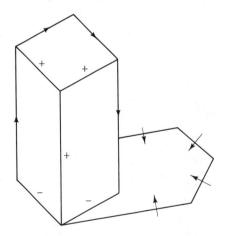

Fig. 9.34 A block resting on its bottom surface.

Ch. 9 Representations of Three-Dimensional Structures

all possible viewpoints, Huffman and Clowes found that without occlusion, just four vertex types and only a few of the possible labelings of lines meeting at a vertex can occur. Figure 9.35 shows views of one- and three-octant corners which give rise to all possible vertices for these corner types. The vertices appear in the first two rows of Table 9.1, which is a catalogue of all possible vertices, including those arising from occlusion, in this restricted world of trihedral polyhedra. It is easy to imagine extending the catalog to include vertices for other corner types.

It is important to note that there are four possible labels for each line $(+ - >$ $<)$, and thus $4^3 = 64$ possible labels for the fork, arrow, and T and 16 possible labels for the ell. In the catalog, however, only 3/64, 3/64, 4/64, and 6/16, respectively, of the possible labels actually occur. Thus only a small fraction of possible labels can occur in a scene.

The main observation that lets line-labeling analysis work is the coherence rule: In a real polyhedral scene, *no line may change its interpretation (label) between vertices*. For example, what is wrong with scenes like Fig. 9.36 is that they cannot be coherently labeled; lines change their interpretation within the impossible object. Perhaps the lines in drawings of real scenes can be interpreted quickly because the small percentage of meaningful labelings interacts with the coherence rule to reduce drastically the number of explanations for the scene.

How does line labeling relate to Guzman? A labeled-line description clearly indicates the grouping of regions into bodies, and also rejects scenes like Fig. 9.33a, which cannot be coherently labeled with labels from the catalog. The origin of Guzman's links can be explained this way: consider again the world of convex polyhedra; the only labels from the catalog that are possible are shown in Fig. 9.37a. Further, it is clear that a convex edge has two faces of the same body on either side of it, and an occluding edge has faces from two different bodies on either side of it. A convex label means the regions on either side of it should be linked; this is Guzman's link-planting rule (Fig. 9.37b). The inhibition rules are a further corollary of the labels; they are to suppress links across an edge if evidence that it

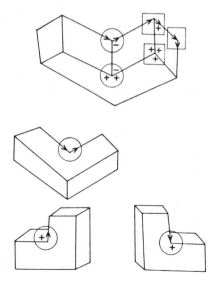

Fig. 9.35 Different views of various corner types.

Table 9.1
VERTEX CATALOGUE

Octants filed \ Visible surfaces	3	2	1	0
1	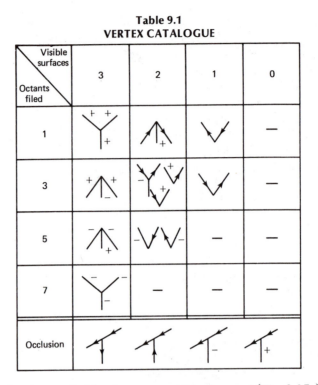			—
3				—
5			—	—
7		—	—	—
Occlusion				

must be occluding is supplied by the vertex at its other end (Fig. 9.37c). When vertices at both ends of a line agree that the line is convex, Guzman would have planted two links; this is in fact the strongest evidence that the regions are part of the same body. If just one vertex gives evidence that the edge has a link, a decision based on heuristics is made; the coherence rule is being used implicitly by Guzman. The same physical and geometric reality is driving both his scheme and that of Huffman.

The labeling scheme explained here still has problems: syntactically nonsensical scenes are coherently labeled (Fig. 9.38a); scenes are given geometrically impossible labels (Fig. 9.38b); and scenes that cannot arise from polyhedra are easily labelled (Fig. 9.38c). It is very hard to see how a labeling scheme can detect the illegality of scenes like (Fig. 9.38c); the problem is not that the edges are incorrectly labeled, but that the faces cannot be planar.

Concern with this last-mentioned problem led to a program (see the next section) that can obtain information about a polyhedral scene equivalent to labeling it,

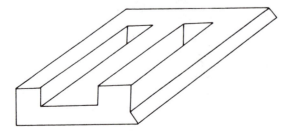

Fig. 9.36 An impossible object.

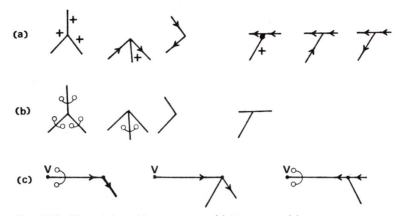

Fig. 9.37 The relation of links to labels. (a) Line labels. (b) Link planting vertices. (c) Inhibitory links.

and also can reject non-polyhedra as impossible. There has also been an exciting denoument to the line-labeling idea [Waltz 1975; Turner 1974].

Waltz extends the line labels to include shadows, three illumination codes for each face on the side of an edge, and the separability of bodies in the scene at cracks and concave edges; this brings the number of line labels possible up to just below 100. He also extends the possible vertex types, so that many vertices of four lines occur. He can deal with scenes such as the one shown in Fig. 9.28c.

The combinatorial consequence of these extensions is clear; the possible vertex labelings multiply enormously. The first interesting thing Waltz discovered was that despite the combinatorics, as more information is coded into the lines, the smaller becomes the percentage of geometrically meaningful labels for a vertex. In his final version, only approximately 0.03 percent of the possible arrow labels can occur, and for some vertices the percentage is approximately 0.000001.

The second interesting thing Waltz did was to use a constraint-propagating labeling algorithm which very quickly eliminates labels for a vertex that is impossible given the neighboring vertices and the coherence rule, which places *constraints* on labelings. The small number of meaningful labels for a vertex imposes severe constraints on the labeling of neighboring vertices. By the coherence rule, the constraints may be passed around the scene from each vertex to its neighbors; eliminating a label for a vertex may render neighboring labels illegal as well, and so on recursively.

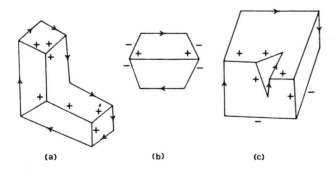

(a) (b) (c)

Fig. 9.38 Nonsense labelings and nonpolyhedra.

Waltz found that for scenes of moderate complexity, eliminating all impossible labelings left only one, the correct one. The labeling process, which might have been expected to involve much search, usually involved none. This constraint propagation is an example of parallel constraint satisfaction, and is discussed in Chapter 12 in a broader context. In the event that a vertex is left with several labels after all junction coherence constraints have been applied, they all participate in *some* legal labeling. At this point one can resort to tree search to find the explicit labelings, or one can apply more constraints. Many such constraints, heuristic and geometric, may be imagined. For instance, a constraint could involve color edge profiles. If two aligned edges are separated by some (possibly occluding) structure, but still divide faces of the same color, they should have the same label. Another important constraint concerns how face planarity constrains line orientations.

Scenes with missing lines may be labeled; one merely adds to the legal vertex catalog the vertices that result if lines are missing from legal vertices. This idea has the drawbacks of increasing the vertex catalog and widening the notion of consistency, but can be useful.

Another extension to line labeling is that of [Kanade 1978]. This extension considers not only solid polyhedra but objects (including nonclosed "shells") made up of planar faces. This extension has been called *origami world* after the art of making objects from folded (mostly planar) paper. An example from origami world is the box in Fig. 9.39a. A quick check shows that this cannot be labeled with the Huffman-Clowes label set. It can be labeled using the origami world label set (Table 9.2) and its interpretation is shown in Fig. 9.39b.

Table 9.2

EXPANDED JUNCTION TABLE

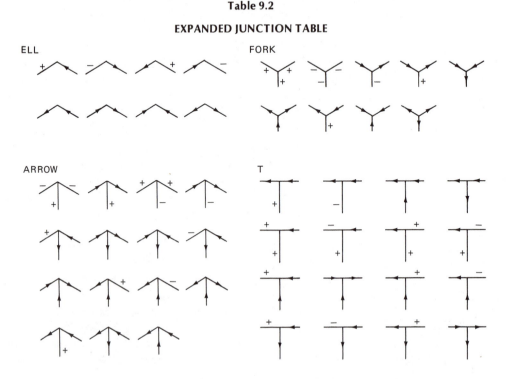

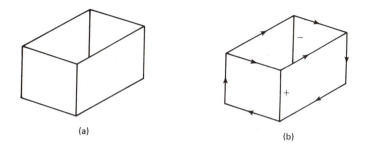

(a)   (b)

Fig. 9.39 (a) Box. (b) Labeled edges according to origami world label set.

The vertex labels may be extended to include scenes with cylinders, cones, spheres, tori, and other simple curves. In expanded domains the notion of "legal line drawing" becomes very imprecise. In any event the number of vertex types and labels grow explosively, and the coherence rule must be modified to cope with the fact that lines can change their interpretation between vertices and can tail off into nothing, and that one region can attain all three of Waltz's illumination types [Turner 1974, Chakravarty 1979]. The domain is of scenes such as appear in Fig. 9.28d.

9.5.4 Reasoning About Planes

The deficiencies in the scene line-labeling algorithms prompted a consideration of the geometrical foundations of the junction labels [Mackworth 1973, Sugihara 1981]. This work seeks to answer the same sorts of questions as do labeling programs, but also to take account of objects that cannot possibly be planar polyhedra, such as those of Fig. 9.40. Neither approach uses a catalog of junction labels, but relies instead on ideas of geometric coherence. The basis is a plane-oriented formulation rather than a line-oriented one.

Gradient Space

Mackworth's program relies heavily on the relation of polyhedral surface gradients to the lines in the image (recall section 3.5.2). Image information from orthographic projections of planar polyhedral scenes may be related to gradient information in a useful way. An image line L is the projection of a three-space line M arising from the intersection of two faces lying in distinct planes Π_1 and Π_2 of gradients (p_1, q_1) and (p_2, q_2). With the (p, q) coordinate system superimposed on the image (x, y) coordinate system, there is the following constraint. The orientation of L constrains the gradients of Π_1 and Π_2; specifically, the line L is perpendicular to the line G between (p_1, q_1) and (p_2, q_2) (Fig. 9.41).

Fig. 9.40 Labelable but not planar polyhedra.

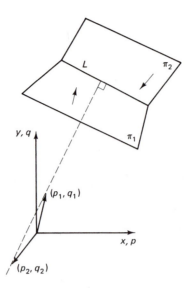

Fig. 9.41 Gradient space constraint.

The result is easily shown. With orthographic projection, the origin may be moved of the image plane to be in L without loss of generality. Then L is defined by its direction vector $(\lambda, \mu) = (\cos\theta, \sin\theta)$. The three-space point on Π_1 corresponding to $(0, 0)$ may be expressed as $(0, 0, k_1)$, and at (λ, μ) the corresponding point is $(\lambda, \mu, \lambda p_1 + \mu q_1 + k_1)$. Thus moving along M (which is in Π_1) from $(x, y) = (0, 0)$ to $(x, y) - (\lambda, \mu)$ moves along $-z$ by $\lambda p_1 + \mu q_1$. The coordinates of a unit vector on L can then be expressed as $(\lambda, \mu, \lambda p_1 + \mu q_1)$. But L is also in Π_2, and this argument may be repeated for Π_2, using p_2 and q_2. Thus

$$\lambda p_1 + \mu q_1 = \lambda p_2 + \mu q_2 \tag{9.23}$$

or

$$(\lambda, \mu) \cdot (p_2 - p_1, \ q_2 - q_1) = 0 \tag{9.24}$$

Equation (9.24) is a dot product set equal to zero, showing that its two vector operands are orthogonal, which was to be shown.

Every picture line results from the intersection of two planes, and so it has a line associated with it in gradient space which is perpendicular to it. Furthermore, if the gradients of the surfaces are on the same side of the picture line as their surfaces, the edge was convex; if the gradients are on opposite sides of the line from their causing surfaces, the edge was concave (Fig. 9.42). For every junction in the image there are just two ways the gradients can be arranged to satisfy the perpendicularity requirement (Fig. 9.43). In the first, all edges are convex, in the second, concave. Switching interpretations from one to the other by negating gradients is the psychological "Necker reversal."

Notice that if an image junction is a three-space polyhedral vertex, each edge of the vertex is the intersection of two face planes. If the corresponding gradients are connected, a "dual" (p, q) space representation of the (x, y) space junction is formed. The connected (p, q) gradient points form a polygon whose edges are perpendicular to the junction lines in (x, y) space. The polygon is larger if the three-

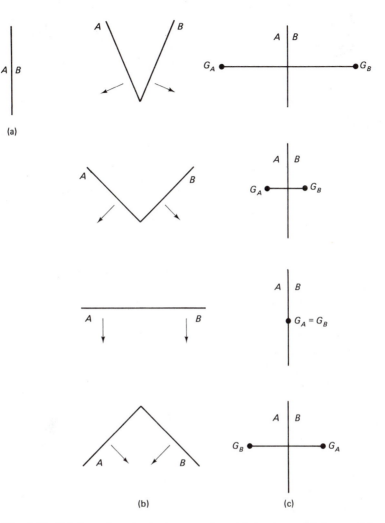

Fig. 9.42 Relation of gradients, image and world structures. (a) Image. (b) World. (c) Gradients.

dimensional corner is sharper, and shrinks toward the junction point as the corner gets blunter.

Interpreting Drawings

It is possible to use these geometric results to interpret the lines in orthogonally projected polyhedral scenes as being "connect" (i.e., as being between two connected faces) or occluding. It can also be determined if connect edges are convex or concave, and for occluding edges which surface is in front. Hidden parts of the scene may sometimes be reconstructed. The orientation of each surface and edge in the scene may be found. Thus a program can determine that input such as Fig. 9.40 is not a planar-faced polyhedron [Mackworth 1973]. Sugihara's work generalizes Mackworth's; it does not use gradient space and does not rely on orthographic projection.

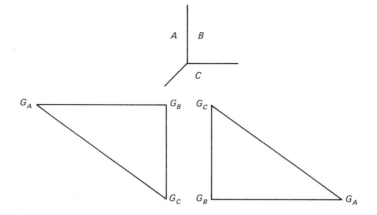

Fig. 9.43 A scene junction and two resulting triangles in gradient space.

Mackworth's procedure to establish connect edges produces the most connected interpretation first (a nonconnected interpretation is just a collection of floating faces which line up by accident to give the line drawing). The background region is the first to be interpreted; that is, means to have its gradient fixed in gradient space. After a region is interpreted, the region having the most lines in common with regions so far interpreted is interpreted next.

The image of a scene is given in Fig. 9.44a; it is interpreted as follows. No coherent interpretation is possible with five or four connect edges. Trying for three connect edges, the program interprets A by arbitrarily picking a gradient for the surface A represents (the background). It picks the origin of gradient space. In order to be able to reason about lines in the image, it needs to have an interpreted region on either side of the line, so it must interpret another region. It picks B (C would be as good).

The lines bounding B are examined to see if they are connect. Line 1 is considered. If it is connect, the gradient space dual of it will be perpendicular to it through the gradient space point representing surface A (i.e., the origin). Now another arbitrary choice: The gradient corresponding to surface B is placed at unit distance from the origin, thus "imagining" the second gradient in a row. From now on, the gradients are more strongly located. The arbitrary scaling and point of origin imposed by these first two choices can be changed later if that is important.

In gradient space, the situation is now shown in Fig. 9.44b. Now consider line 2; to establish it as a connect edge, $G_B = (p_B, 1_B)$ (the gradient space point corresponding to the surface B) must lie on a line perpendicular to 2 through G_A (Fig. 9.44c). This cannot happen; the situation with 1 and 2 both connect is incoherent. Thus, with a line 1 connect edge, 2 must be occluding. This sort of incoherency result was what kept the program from finding four or five edges connect. Further interpretation involves assigning gradients and vertices into the developing diagram in a noncontradictory, maximally connected manner (Fig. 9.44d).

The next part of the program determines convexity or concavity of the lines. The final part of the program looks at occlusion. It also suggests hidden surfaces

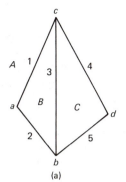

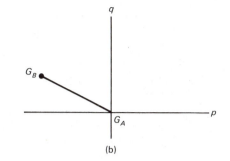

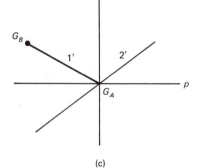

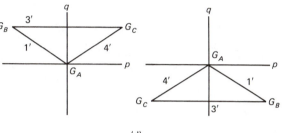

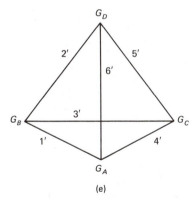

Fig. 9.44 (a) Polyhedral scene considered by Mackworth. (b) Partial interpretation. (c) Continued interpretation. (d) Occluding and connect interpretations. (e) Final interpretation.

and thus hidden lines that are consistent with the interpretation (Fig. 9.44e). This figure in gradient space resembles a tetrahedron, as well it might; it is formed in the same way as the graph-theoretic dual (point per face, edge per edge, face per point) which defines dual graphs and dual polyhedra; the tetrahedron is self-dual. The arbitrary choices of gradient reflect degrees of freedom in the drawing that are also identified by Sugihara.

Skewed Symmetry

Many planar objects are symmetrical about an axis. This axis and another, which is perpendicular to the first and in the plane of the object, form a natural orthogonal coordinate system for the object. If the plane of the object is perpendicular to the line of sight from the viewpoint, the coordinate axes appear to be at right angles. If the object is tilted from this position, the axes appear skewed. Some examples are shown in Fig. 9.45.

A skewed symmetry may or may not reflect a real symmetry; the object may itself be skewed. However, if the skewed symmetry results from a tilted real symmetry, a constraint in gradient space may be developed for the object's orientation [Kanade 1979].

An imaged unit vector inclined at α inscribed on a plane at orientation (p, q) must have three-dimensional coordinates given by

$$(\cos \alpha, \ \sin \alpha, \ p \cos \alpha + q \sin \alpha)$$

Thus if the two axes of skewed symmetry make angles of α and β with the image x axis, the two vectors in three-space a and b must have coordinates

$$\mathbf{a} = (\cos \alpha, \ \sin \alpha, \ p \cos \alpha + q \sin \alpha)$$

and

$$\mathbf{b} = (\cos \beta, \ \sin \beta, \ p \cos \beta + q \sin \beta)$$

Since these vectors reflect a real symmetry, they must be perpendicular (i.e., $\mathbf{a} \cdot \mathbf{b} = 0$), or

$$\cos (\alpha - \beta) + (p \cos \alpha + q \sin \alpha) \ (p \cos \beta + q \sin \beta) = 0 \qquad (9.25)$$

By rotating the p and q axes by $\lambda = (\alpha + \beta)/2$, that is

$$p' = p \cos \lambda + q \sin \lambda$$

$$q' = -p \sin \lambda + q \cos \lambda$$

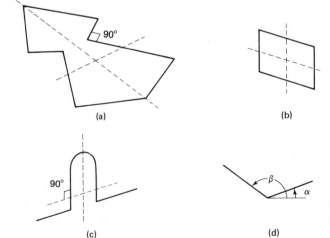

(a)

(b)

(c)

(d)

Fig. 9.45 Skewed symmetries. (a,b,c) are examples. (d) Each skewed symmetry defines two axes.

Equation (9.25) can be put into the form

$$p'^2 \cos^2\left[\frac{\gamma}{2}\right] - q'^2 \sin^2\left[\frac{\gamma}{2}\right] = -\cos(\gamma)$$

where $\gamma = \alpha - \beta$. Thus the gradient of the object must lie on a hyperbola with axis tilted λ from the x axis, and with asymptotes perpendicular to the directions of α and β. This constraint is shown in Fig. 9.46.

To show how skewed symmetry can be exploited to interpret objects with planar faces, reconsider the example of Fig. 9.43. In that example the three convex edges constrained the gradients of the corresponding faces to be at the vertices of a triangle, but the size or position of the triangle in gradient space was unknown. However, skewed symmetry applied to each face introduces three hyperbola upon which the gradients must lie. The only way that both the skewed symmetry constraint and triangle constraint can be satisfied simultaneously is shown in Fig. 9.47 — the combined constraints have uniquely determined the face orientations.

EXERCISES

9.1 Derive an expression for the volume of an object represented by spherical harmonics of order $M = 1$.

9.2 Derive an expression for the perpendicular to the surface of an object represented by spherical harmonics in terms of the appropriate derivatives.

9.3 Derive an expression for the angle centroid of each of the spherical harmonic functions for $M \leqslant 2$.

9.4 Label the lines in the objects of Fig. 9.48.

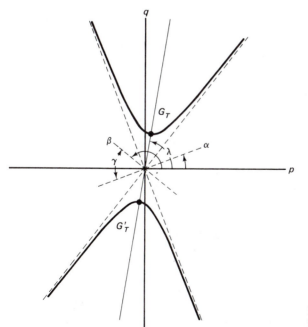

Fig. 9.46 Skewed symmetry constraint in gradient space.

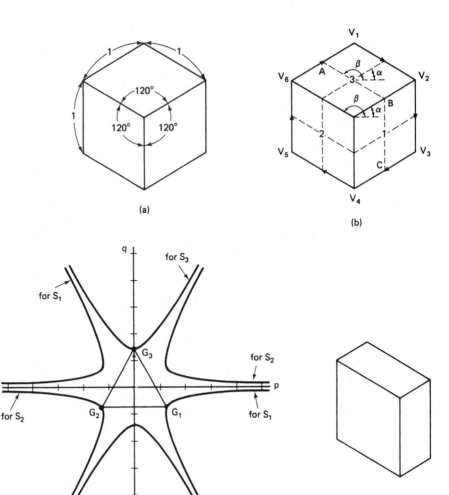

Fig. 9.47 Using skewed symmetry to orient the faces of a cube. (a) The cube. (b) Skewed symmetries. (c) skewed symmetries and junction constraint plotted in gradient space. (d) another possible object obeying the constraints.

9.5 Give two sets of CSG primitives with the same domain.

9.6 Show that the dual of the plane of interpretation for a line and the duals of the two planes that meet in the edge causing the line are all on the dual of the edge.

9.7 Prove (Section 9.3.1) that in the Frenet frame ξ' is perpendicular to ξ.

9.8 Write the precise rules for combining classification results for $\cup^*$, $\cap^*$, and $-^*$ operations.

9.9 Find two interpretations of the tetrahedron of Fig. 9.44a that differ in convexity or concavity of lines. (Hint: The concave interpretation has an accident of alignment.)

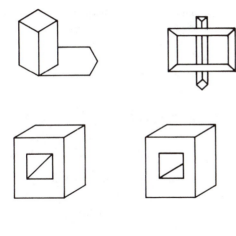

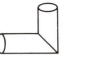

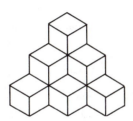

Fig. 9.48 Objects for labeling.

REFERENCES

AGIN, G. J. "Representation and description of curved objects" (Ph.D. dissertation). AIM-173, Stanford AI Lab, October 1972.

BADLER, N. I. and R. K. BAJCSY. "Three-dimensional representations for computer graphics and computer vision." *Computer Graphics 12*, August 1978, 153–160.

BADLER, N. I. and J. O'ROURKE. "Representation of articulable, quasi-rigid, three-dimensional objects." NSF Workshop on the Representation of Three-Dimensional Objects, Univ. Pennsylvania, May 1979.

BARNHILL, R. E. "Representation and approximation of surfaces." In *Mathematical Software III*, J. R. Rice (Ed.). New York: Academic Press, 1977.

BARNHILL, R. E. and R. F. RIESENFELD. *Computer Aided Geometric Design.* New York: Academic Press, 1974.

BAUMGART, B. G. "Winged edge polyhedron representation." STAN-CS-320, AIM-179, Stanford AI Lab, October 1972.

BENTLEY, J. L. Multidimensional search trees used for associative searching, *Comm. ACM 18*, 9, Sept. 1975, 509–517.

BINFORD, T. O. "Visual perception by computer." IEEE Conf. on Systems and Control, Miami, December 1971.

BOYSE, J. W. "Data structure for a solid modeller," NSF Workshop on the Representation of Three-Dimensional Objects, Univ. Pennsylvania, May 1979.

BROWN, C. M. "Two descriptions and a two-sample test for 3-d vector data." TR49, Computer Science Dept., Univ. Rochester, February 1979a.

BROWN, C. M. "Fast display of well-tesselated surfaces." *Computers and Graphics 4*, 2, September 1979b, 77–85.

BROWN C. M., A. A. G. REQUICHA, and H. B. VOELCKER. "Geometric modelling systems for mechanical design and manufacturing." *Proc.*, 1978 Annual Conference of the ACM, Washington, DC, December 1978, 770–778.

CHAKRAVARTY, I. "A generalized line and junction labelling scheme with applications to scene analysis," IEEE Trans. PAMI, April 1979, 202–205.

CLINTON, J. D. "Advanced structural geometry studies, Part I: Polyhedral subdivision concepts for structural applications." NASA CR-1734/35, September 1971.

CLOWES, M. B. "On seeing things." *Artificial Intelligence 2*, 1, Spring 1971, 79–116.

COONS, S. A. "Surface patches and B-spline curves." In *Computer Aided Geometric Design*, R. E. Barnhill and R. F. Riesenfeld (Eds.). (*Proc.*, Conference on Computer Aided Geometric Design, Univ. Utah, March 1974.) New York: Academic Press, 1974.

EJIRI, M., T. UNO, H. YODA, T. GOTO, and K. TAKEYASU. "An intelligent robot with cognition and decision-making ability." *Proc.*, 2nd IJCAI, September 1971, 350–358.

FALK, G. "Interpretation of important line data as a three-dimensional scene." *Artificial Intelligence 3*, 1, Spring 1972, 77–100.

FORREST, A. R. "On cones and other methods for the representation of curved surfaces." *CGIP 1*, 4, December 1972, 341–359.

GUZMAN, A. "Decomposition of a visual scene into three-dimensional bodies" (Ph.D. dissertation). In *Automatic Interpretation and Classification of Images*, A. Grasseli (Ed.). New York: Academic Press, 1969.

HUFFMAN, D. A. "Impossible objects as nonsense sentences." In *MI6*, 1971.

JACKINS, C. L., and S. L. TANIMOTO. Oct-trees and their use in representing three-dimensional objects, *CGIP 14*, 3, Nov. 1980, 249–270.

KANADE, T. "A theory of Origami world." CMU-CS-78-144, Computer Science Dept., Carnegie-Mellon Univ., 1978.

KANADE, T. "Recovery of the three-dimensional shape of an object from a single view." CMU-CS-79-153, Computer Science Dept., Carnegie-Mellon Univ., October 1979.

LAKATOS, I. *Proofs and Refutations.* Cambridge, MA: Cambridge University Press, 1976.

LEE, Y. T. and A. A. G. REQUICHA. "Algorithms for computing the volume and other integral properties of solid objects." Tech. Memo 35, Production Automation Project, Univ. Rochester, Rochester NY, Feb. 1980.

MACKWORTH, A. K. "Interpreting pictures of polyhedral scenes." *Artificial Intelligence 4*, 2, June 1973, 121–137.

MACKWORTH, A. K. "On reading sketch maps." *Proc.*, 5th IJCAI, August 1977, 598–606.

MARKOWSKY, G. and M. A. WESLEY. "Fleshing out wire frames." *IBM J. Res. Devel. 24*, 1 (Jan. 1980) 64–74.

MARR, D. and H. K. NISHIHARA. "Representation and recognition of the spatial organization of three-dimensional shapes." *Proc., Royal Society of London B 200*, 1978, 269–294.

NISHIHARA, H. K. "Intensity, visible surface and volumetric representations." NSF Workshop on the Representation of Three-Dimensional Objects, U. Pennsylvania, May 1979.

O'NEILL, B. *Elementary Differential Geometry.* New York: Academic Press, 1966.

O'ROURKE, J. and N. I. BADLER. "Decomposition of three-dimensional objects into spheres." *IEEE Trans. PAMI 1*, July 1979.

REQUICHA, A. A. G. "Representations of rigid solid objects." *Computer Surveys 12*, 4, December 1980.

ROBERTS, L. G. "Machine perception of three-dimensional solids." In *Optical and Electro-optical Information Processing*, J.P. Tippett et al. (Eds.). Cambridge, MA: MIT Press, 1965.

SCHUDY, R. B. and D. H. BALLARD. "Model-detection of cardiac chambers in ultrasound images." TR12, Computer Science Dept., Univ. Rochester, November 1978.

SCHUDY, R. B. and D. H. BALLARD. "Towards an anatomical model of heart motion as seen in 4-d cardiac ultrasound data." *Proc.*, 6th Conf. on Computer Applications in Radiology and Computer-Aided Analysis of Radiological Images, June 1979.

SHANI, U. "A 3-d model-driven system for the recognition of abdominal anatomy from CT scans." TR77, Computer Science Dept., U. Rochester, May 1980; also in Proc. 5th IJCPR, Miami, December 1980, 585–591.

SHAPIRA, R. "A technique for the reconstruction of a straight-edge, wire-frame object from two or more central projections." *CGIP 3*, 4, December 1974, 318–326.

SHIRAI, Y. "Analyzing intensity arrays using knowledge about scenes." In *PCV*, 1975.

SOROKA, B. I. "Generalised cylinders from parallel slices." *Proc.*, PRIP, 1979a, 421–426.

SOROKA, B. I. "Understanding objects from slices." Ph.D. dissertation, Dept. of Computer and Information Science, Univ. Pennsylvania, 1979b.

SOROKA, B. I. and R. K. BAJCSY. "Generalized cylinders from serial sections." *Proc.*, 3rd IJCPR, November 1976, 734–735.

SUGIHARA, K. "Mathematical structures of line drawings of polyhedra," RNS 81-02, Dept. of Info. Science, Nagoya Univ., May 1981.

TILOVE, R. B. "Set membership classification: a unified approach to geometric intersection problems." *IEEE Trans. Computers 29*, 10, October 1980.

TURNER, K. J. "Computer perception of curved objects using a television camera." Ph.D. dissertation, Univ. Edinburgh, 1974.

VOELCKER, H. B. and A. A. G. REQUICHA, Boundary evaluation procedures for objects defined via constructive solid geometry, Tech. Memo 26, Production Automation Project, Univ. Rochester, 1981.

VOELCKER, H. B. and A. A. G. REQUICHA. "Geometric modeling of mechanical parts and processes." *Computer 10*, December 1977, 48–57.

VOELCKER, H. B. and Staff of Production Automation Project, "The PADL-1.0/2 system for defining and displaying solid objects." *Computer Graphics 12*, 3, August 1978, 257–263.

WALTZ, D. I. "Generating semantic descriptions from drawings of scenes with shadows." Ph.D. dissertation, AI Lab, MIT, 1972; also in *PCV*, 1975.

WARNOCK, J. G. "A hidden-surface algorithm for computer-generated halftone pictures." TR 4-15, Computer Science Dept., Univ. Utah, June 1969.

WESLEY, M. A. and S. MARKOWSKY. "Fleshing out projections." *IBM J. Res. Devel. 25*, 6 (Nov. 1981), 934–954.

RELATIONAL STRUCTURES

IV

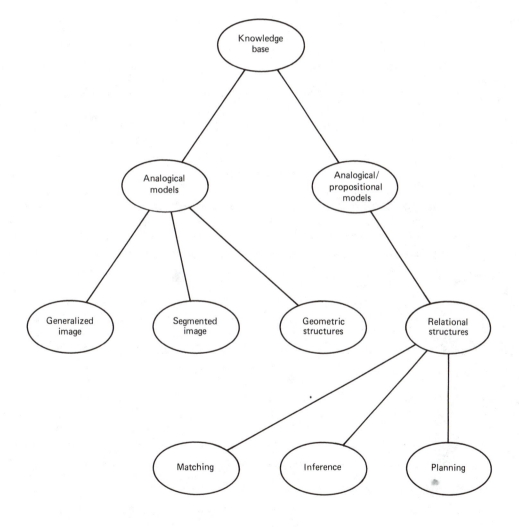

Visual understanding relates input and its implicit structure to explicit structure that already exists in our internal representations of the world. More specifically, vision operations must maintain and update *beliefs* about the world, and achieve specific *goals.*

To consider how higher processes can influence and use vision, one must confront the nonvisual world and powers of reasoning that have more general applicability. The world models that are capable of supporting advanced application-dependent calculations about objects in the visual domain are quite complex. General techniques of *knowledge representation* developed in other fields of artificial intelligence can be brought to bear on them. Similarly, much research has been invested in the basic processes of *inference* and *planning.* These techniques may be used in the visual domain to manipulate beliefs and achieve goals, as well as reasoning for other purposes.

The organization of a complex visual system (Fig. 1.5 or Fig. 10.1), is a loose hierarchy of models of world phenomena. The *relational models* that concern us in this chapter are removed from direct perceptual experience — they are used mainly for the last, highest-level stages of perception. Also, they are used for knowledge attained prior to the visual experience currently being processed. The representations involved may be *analogical* or *propositional.* Analogical representations allow *simulations* of important·physical and geometric properties of objects. Propositions are assertions that are either true or false with respect to the world (or a world model). Each form is useful for different purposes, and one is not necessarily "higher" than the other. The techniques and representations of Part IV are mainly propositional in flavor. Sometimes the reasoning they implement (say about geometrical entities) would seem better suited to analogical calculations; however, technical difficulties can render that impossible.

Part IV is concerned with techniques for making the "motivation" and "world view" of a vision system explicit and available. Such explicit models would be interesting from a scientific standpoint even if they were not directly useful. But explicitly available models are decidedly useful. They are useful to the system designer who desires to reconfigure or extend a system. They are useful to the system itself, which can use them to reason about its own actions, flexibly control its own resources in accordance with higher goals, dynamically change its goals, recover from mistakes, and so forth.

We organize the major topics of Part IV as follows.

1. Knowledge representation (Chapter 10). *Semantic nets* are an important technique for structuring complex knowledge, and can be used as a knowledge representation formalism in their own right.

2. Matching (Chapter 11). *Matching* puts a derived representation of an image into correspondence with an existing representation. This style of processing representations is more pronounced as domain-dependent knowledge, idiosyncratic goals, and experience begin to dominate the ultimate use (or understanding) of the visual input.

3. Inference (Chapter 12). Classical *logical inference* (a technique for manipulating purely propositional knowledge representations) is a well-understood and elegant reasoning technique. It has good formal properties, but occasionally seems restricted in its power to duplicate the range of human processing. *Extended inference* techniques such as *production systems* are those in which the inference process as well as the propositions may contribute materially to the derived knowledge. *Labeling* techniques can "infer" consistent or likely interpretations for an input from given rules about the domain. Inference can be used for both problem solving and belief-maintenance activity.

4. Planning (Chapter 13). *Planning* techniques are useful for problem solving, and are especially tailored to integrating vision with real-world *action*. Planning can be used for resource allocation and attentional mechanisms.

5. Control (Chapter 10; Appendix 2). Control *strategies* and *mechanisms* are of vital concern in any complex artificial intelligence system, and are particularly important when the computation is as expensive as that of vision processing.

Learning is missing from the list above. Disappointing as it is, at this writing the problem of learning is so difficult that we can say very little about it in the domain of vision.

Knowledge
Representation
and Use

10

10.1 REPRESENTATIONS

An internal representation of the world can help an intelligent system plan its actions and foresee their consequences, anticipate dangers, and use knowledge acquired in the past. In Part IV we investigate the creation, maintenance, and use of a *knowledge base*, an abstract representation of the world useful for computer vision. Chapter 1 introduced a layered organization for the knowledge base and divided its contents into "analogical" and "propositional" models. In this section we consider this high-level division more deeply.

The outside world is accessible to a computer vision program through the imaging process. Otherwise, the program is manipulating its internal representations, which should correspond to the world in understood ways. In this sense, the knowledge base of generalized images, segmented images, and geometric entities contains "models" of the phenomena in the world. Another more abstract sense of "model" is high-level, prior expectations about how the world fits together. Such a high-level model is often much more complex than the lower-level representations, often has a large "propositional" component, and is often manipulated by "inference-like" procedures. Explicit knowledge and belief structures are a relatively new phenomenon in computer vision, but are playing an increasingly important role.

The goals of this chapter are three.

1. To develop in more depth some issues of high-level models (Section 10.1).

2. To describe *semantic nets*—an important and general tool for both organizing and representing models (Sections 10.2 and 10.3).

3. To address issues of *control*, at both abstract and implementational levels (Section 10.4 augmented by Appendix 2).

10.1.1 The Knowledge Base—Models and Processes

Figure 10.1 shows the representational layers in the knowledge base as we have developed it through the book, and shows the place of important processes. This organization might be compared with that in [Barrow and Tenenbaum 1981].

The knowledge base organization is mirrored in the organization of the book. Parts I to III dealt with analogical models and their construction; Part IV is concerned with propositional and complex analogical models. In Chapters 11 to 13, the emphasis moves from the structure of models to the processes (matching, inference, and planning) needed to manipulate and use them.

The knowledge base should have the following properties.

- Represent analogical, propositional, and procedural structures
- Allow quick access to information
- Be easily and gracefully extensible
- Support inquiries to the analogical structures
- Associate and convert between structures
- Support belief maintenance, inference, and planning

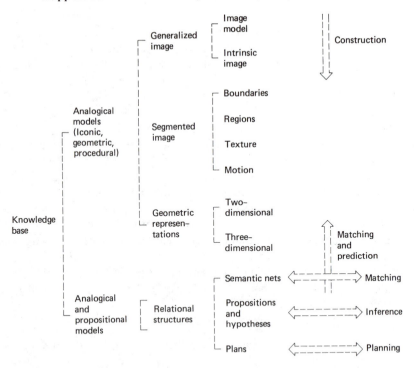

Fig. 10.1 The knowledge base and associated processes in a computer vision system.

The highest levels of the knowledge base contain both *analogical* and *propositional* models. Analogical tools do not exist for many important activities, and when they do exist they are often computationally intensive. A three-dimensional geometric modeling system for automatic manufacturing has very complex data structures and algorithms compared to their elegant and terse counterparts in a propositional model that may be used to plan the highest-level actions. In general it makes sense to do some computation at the analogical level and some at the propositional. This multiple-representation strategy seems more efficient than translating all problems into one representation or the other.

The computations in a vision system should be organized so that information can flow efficiently and unnecessary computation is kept to a minimum. This is the function of the *control* disciplines that allocate effort to different processes. Even the simplest biological vision systems exhibit sophisticated control of processing.

Constructive processes dominate the activity in building lower-level models, and *matching* processes become more important as prior expectations and models are brought into play. Chapter 11 is devoted to the process of matching.

We postulate that an advanced vision system is engaged in two sorts of high-level activity: *belief maintenance* and *goal achievement.* The former is a more or less passive, data-driven, background activity that keeps beliefs consistent and updated. The latter is an active, knowledge-driven, foreground activity that consists of planning future activities. Planning is a problem-solving and simulation activity that anticipates future world states; in computer vision it can determine how the visual environment is expected to change if certain actions are performed. Planning can occur with symbolic, propositional representations (Chapter 13) or in a more analogical vein with such simulations as trajectory planning [Lozano-Perez and Wesley 1979]. Planning is useful as an implementational mechanism even in contexts that are not analogous to human "conscious" problem solving [Garvey 1976]. Helmholtz likened the results of perception to "unconscious conclusions" [Helmholtz 1925]. Similarly even "primitive" vision processes (computer or biological) may use planning techniques to accomplish their ends.

Inference and planning are both classical subfields of artificial intelligence. Neither has seen much application in computer vision. Inference seems useful for belief maintenance. Extended inference can deal with inconsistent beliefs and with beliefs that are maintained with various strengths. We treat inference in Chapter 12. Applications of planning to vision [Garvey 1976; Bolles 1977] show good promise. Planning is treated in Chapter 13.

10.1.2 Analogical and Propositional Representations

Our division of the internal knowledge base into "analogical" and "propositional" reflects a similar division in theories of how human beings represent the world [Johnson-Laird 1980]. Psychological data are not compelling toward either pure theory; there are indications that human beings use both forms of representation. We introduce the division in this book because we find it conceptually useful in the following way. Low-level representations and processes tend to be purely analogical; high-level representations and processes tend to be both analogical and propositional.

Analogical representations have the following characteristics [Kosslyn and Pomerantz 1977; Shepard 1978; Sloman 1971; Kosslyn and Schwartz 1977, 1978; Waltz and Boggess 1979].

1. *Coherence.* Each element of a represented situation appears once, with all its relations to other elements accessible.

2. *Continuity.* Analogous with continuity of motion and time in the physical world; these representations permit continuous change.

3. *Analogy.* The structure of the representation mirrors (and may be isomorphic to) the relational structure of the represented situation. The representation is a description of the situation.

4. *Simulation.* Analogical models are interrogated and manipulated by arbitrarily complex computational procedures that often have the flavor of (physical or geometric) simulation.

Propositional representations have the following characteristics [Anderson and Bower 1973; Palmer 1975; Pylyshyn 1973].

1. *Dispersion.* An element of a represented situation can appear in several propositions. However, the propositions can be represented in a coherent manner by using semantic nets.

2. *Discreteness.* Propositions are not usually used to represent continuous change. However, they may be made to approximate continuous values arbitrarily closely. Small changes in the representation can thus be made to correspond to small changes in the represented situation.

3. *Abstraction.* Propositions are true or false. They do not have a geometric resemblance to the situation; their structure is not analogous to that of the situation.

4. *Inference.* Propositional models are manipulated by more or less uniform computations that implement "rules of inference" allowing new propositions to be developed from old ones.

Each sort of model derives its "meaning" differently; the distinctions are interesting, because they can point out weaknesses in each theory [Johnson-Laird 1980; Schank 1975; Fodor, et al. 1975]. Especially in computer implementations, the two representations only differ essentially in the last two points. It is often possible to transform one representation to another without loss of information.

Some examples are in order. A generalized image (Part I) is an analogical model: to find an object above a given object, a procedure can "search upward" in the image. An unambiguous three-dimensional model of a solid (Chapter 9) is analogical. It may be used to calculate many geometric properties of the solid, even those unimagined by the designer of the representation. A set of predicate calculus clauses (Chapter 12) is a propositional model. Closely related models can be used to solve problems and make plans [Nilsson 1971, 1980; Chapter 13].

A short digression: It is interesting that people do not seem to perform syllogistic inference (formal propositional deduction) in a "mechanical" way. Given two clauses such as "Some appliances are telephones" and "All telephones are black," we are much more likely to conclude "Some appliances are black" than the equally valid "Some black things are appliances." There is not a satisfying theory of the mental processes underlying syllogistic inference. An interesting speculation [Johnson-Laird 1980] is that inference is primarily done through analogical mental models (in which, for example, a population of individuals is conjured up and manipulated). Then syllogistic inference techniques may have arisen as a bookkeeping mechanism to assure that analogical reasoning does not "miss any cases."

10.1.3 Procedural Knowledge

Procedures as explicit elements in a model pose problems because they are not readily "understood" by other knowledge base components. It is very hard to tell what a procedure does by looking at its code.

In our taxonomy we think of "procedural" knowledge as being analogical. The sequential nature of a program's steps is analogous to an ordering of actions in time that can only be clumsily expressed in current propositional representations. Knowledge about "how-to" perform a complex activity is most propitiously represented in the form of explicit process descriptions. Descriptions not involving the element of time may be naturally represented as passive (analogical or propositional) structures.

There have been several attempts to organize chunks of procedural knowledge by associating with the procedure a description of what it is to accomplish. For example, procedural knowledge can be stored in the internal model structure (knowledge base) indexed under *patterns* that correspond to the arguments of the procedure. *Pattern-directed invocation* involves going to the knowledge base for a procedure that matches the given pattern, matching pattern elements to bind arguments, and invoking the procedure. Several advantages accrue in pattern-directed invocation, such as not having to know the "proper names" of procedures, only their descriptions (what they claim to do). Also, when several procedures match a pattern, one either gets nondeterminism or a chance to choose the best. Often system facilities include a procedure to run to choose the best procedure dynamically. Similar pattern matching is involved in resolution theorem provers and production systems (Chapter 12).

As an example, in a program to locate ribs in a chest radiograph [Ballard 1978], procedures to find ribs under different circumstances are attached to nodes in a mixed analogic and propositional model of the ribcage as shown in Fig. 10.2. Each procedure has an associated description which determines whether it can be run. For example, some programs require instances of neighboring ribs to be located before they can run, whereas others can run given only rudimentary scaling information. When invoked, each procedure tries to find a geometric structure corresponding to the associated rib in a radiograph. Instead of searching for ribs in a mechanical order, descriptors allow a choice of order and procedures and hence a

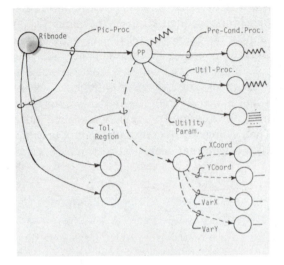

Fig. 10.2 A portion of a ribcage model (see text). Procedural attachment to a model is denoted by jagged lines.

more flexible, efficient and robust program (Appendix 2).

The representation and use of procedural knowledge is an important topic [Schank and Abelson 1977; Winograd 1975; Freuder 1975]. We expect it to be increasingly important for computer vision.

10.1.4 Computer Implementations

A computer implementation can (and often does) obscure the sharp divisions imposed by pure philosophical differences between analogical and propositional models. A propositional representation need not be an unordered set of clauses, but may have a coherent structure; the coherent versus dispersed distinction is thus blurred. A geometry theorem prover or a block-stacking program may manipulate diagrams or simulate physical phenomena such as gravitational stability and wobble in the manipulator [Gelernter 1963; Fahlman 1974; Funt 1977]. "Nonstandard inference" is an important tool that extends classical inference techniques. Although techniques such as production systems and relaxation labeling algorithms (Chapter 11) bear little superficial resemblance to predicate logic, both may be naturally used to manipulate propositional models.

Propositions may be implemented as procedures. If a proposition "evaluates" to true or false, it is perhaps most naturally considered a function from a world (or world model) to a truth value. This is not to say that all such functions exist or are evaluated when the proposition is "brought to mind"; perhaps "understanding a proposition" is like compiling a function and "verifying a proposition" is like evaluating it. The function may be implicit in an evaluation (inference) mechanism or more explicit, as in a "procedural" semantics such as that of the programming languages PLANNER and CONNIVER [Hewitt 1972; Sussman and McDermott 1972; Winograd 1978]. A proposition may thus be encoded as an (analogical!) procedural recipe for establishing the proposition. An example might be this representation of the fact "In California, Grass and Trees produce green regions."

    (To-Establish (GreenRegion x)
      Establish (AND (InCalifornia())
      (OR (Establish (Grass x))
      (Establish (Trees x)))))

This might mean: To infer that x is a green region, establish that you are in California and then try to establish that x arose from grass. Should the grass inference fail, try to establish that x arose from trees. Since the full power of the programming language is available to an Establish statement, it can perform general computations to establish the inference.

The important point here: Rather than a set of clauses whose application must be organized by an interpreter, propositions may be represented by an explicit control sequence, including procedure calls to other programs. In the example, (Grass x) and (Trees x) may be procedures which have their own complicated control structures.

To say that in a computer "everything is propositions" is a truism; any program can be reduced to a Turing machine described by a finite set of "propositions" with a very simple rule of "inference." The issue is at what level the program should be described. A program may be doing propositional resolution theorem proving or analogical trajectory planning with three-dimensional models; it is not helpful to blur this basic functional distinction by appealing to the lowest implementational level.

10.2 SEMANTIC NETS

10.2.1 Semantic Net Basics

Semantic nets were first introduced under that name as a means of modeling human associative memory [Quillian 1968]. Since then they have received much attention [Nilsson 1980; Woods 1975; Brachman 1976; Findler 1979]. We are concerned with three aspects of semantic nets.

1. Semantic nets can be used as a data structure for conveniently accessing both analogical and propositional representations. For the latter their construction is straightforward and based solely on propositional syntax (Chapter 12).

2. Semantic nets can be used as an analogical structure that mirrors the relevant relations between world entities.

3. Semantic nets can be used as a propositional representation with special rules of inference. Both classical and extended inference can be supported, but it is a challenging enterprise to design net structure that provides the properties of formal logic [Schubert 1976; Hendrix 1979].

A semantic network represents objects and relationships between objects as a graph structure of *nodes* and (labeled) *arcs*. The arcs usually represent relations between nodes and may be "followed" to proceed from node to node. A directed arc with label L between nodes X and Y can signify that the predicate $L(X, Y)$ is true. If, in addition, it has a value V, the arc can signify that some function or relation holds: $L(X, Y) = V$.

The *indexing property* of a network is one of its useful aspects. The network can be constructed so that objects that are often associated in computations, or are especially relevant or conceptually close to each other, may be represented by nodes in the network that are near each other in the network (as measured by number of arcs separating them). Figure 10.3 shows these ideas: (a) nodes can be associated by searching outward along arcs and (b) nodes near a specified node are readily available by following arcs. Semantic networks are especially attractive as analogical representations of spatial states of affairs. If we restrict ourselves to binary spatial relations ("above," and "west of," for example), physical objects or parts of objects may be represented by nodes, and their positions with respect to each other by arcs.

Let us look at a semantic net and make some basic observations. Figure 10.4 is meant to be an analogical representation of an arrangement of chairs around a table. The LEFT-OF and RIGHT-OF relations are directed arcs, the ADJACENT relation is undirected; there can be several such undirected arcs between nodes. Note here that the LEFT-OF and RIGHT-OF relations do not behave in their normal way. If they are transitive, as is normal, then every chair is both LEFT-OF and

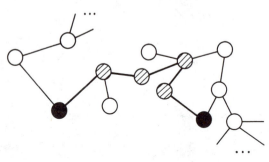

(a)

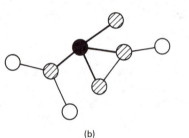

(b)

Fig. 10.3 Semantic networks as structures for associative search. (a) Associating two nodes. (b) Retrieving nearby nodes.

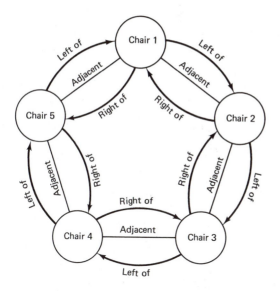

Fig. 10.4 A representation of chairs at a table.

RIGHT-OF every other chair. Flexible treatment of this sort of phenomenon is sometimes difficult in propositional representations.

A simple but basic point: The net of Fig. 10.4 seems to say interesting things about furniture in a scene. But notice that merely by rewriting labels the same net could be "about" modular arithmetic, a string of pearls, or any number of things. There are two morals here. First, a sparsely connected representation (analogical or propositional) may have several equally good interpretations. Second, a net without any interpretation procedures essentially represents nothing [McDermott 1976].

Now consider three neighboring chairs described by the following relations.

1. CHAIR(Armchair), CHAIR(Highchair), CHAIR(Stool)
2. WIDE(Armchair)
3. HIGH(Highchair)
4. LOW(Stool)
5. LEFT-OF(Armchair, Highchair)
6. LEFT-OF(Highchair, Stool)
7. BETWEEN(Highchair, Armchair, Stool)

The relations include four properties (relations with "one argument"), a two-argument and a three-argument relation. One way to encode this information in a net is shown in Fig. 10.5a. Nodes represent individuals, and properties are kept as node contents. The directed arcs represent only binary relations, and "betweenness" is left implicit. Properties can equally well be represented as labeled arcs (Fig. 10.5b).

Relations are encoded as nodes in Fig. 10.6. Here the BETWEEN relation is encoded asymmetrically: it is not possible to tell by arcs emanating from the stool that it is in a "between" relationship.

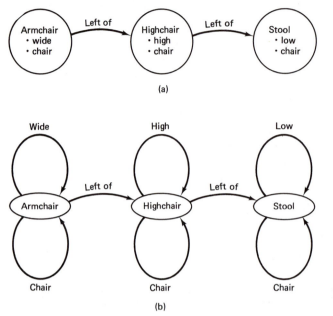

(a)

(b)

Fig. 10.5 (a) A simple semantic net.
(b) An equivalent net.

The three-place relation is treated more symmetrically in Fig. 10.7. In general, *n*-place relations may be "binarized" this way; create a node for the "relation instance" and new (relation) nodes for each distinct argument role in the *n*-ary relation.

An important point: Arcs and nodes had a uniform semantics in Figs. 10.4 and 10.5. This property was lost in the succeeding nets; nodes are either "things" or relations, and arcs leading into relations are not the same as those leading out. For such nets to be useful, the net interpreter (a program that manipulates the net) must keep these things straight. It is possible but not easy to devise a rich and uniform network semantics [Brachman 1979].

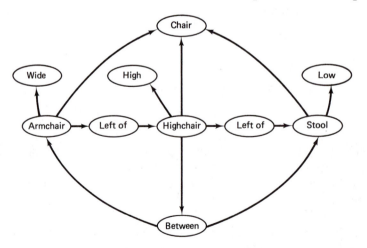

Fig. 10.6 A net with more explicit information.

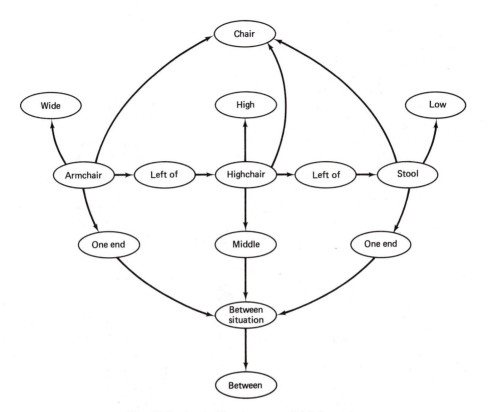

Fig. 10.7 A net with yet more explicit information.

10.2.2 Semantic Nets for Inference

This section explores some further important issues in the semantics of semantic nets. In Chapter 12 semantic nets are used as an indexing mechanism in predicate calculus theorem proving. In some applications an inference system with provably good formal properties may be too restrictive. Some formal properties (such as maintaining consistency by not deducing contradictions) may be considered vital, however. How can "good behavior" be obtained from a representation that may contain "inconsistent" information?

One example of an "inconsistent" representation is the net of Fig. 10.4, with its LEFT-OF and RIGHT-OF problem. Another example is a net version of the propositions "All birds fly," "Penguins are birds," "Penguins do not fly." The generalization is useful "commonsense" knowledge, but the rare exceptions may be important, too. Network interpreters can cope with these sorts of problems by a number of methods, such as only accessing a consistent subnetwork, making deductions from the particular toward the general (this takes care of penguins), and so forth. All these techniques depend on the structure imposed by the net.

Some more subtle aspects of net representations appear below.

Nodes

The basic notation of Fig. 10.4 may tempt us to produce a net such as that shown in Fig. 10.8. Consider the object node sky in Fig. 10.8. Does it stand for the generic sky concept or for a particular sky at a particular time and location? Clearly both meanings cannot be embodied in the same node because they are used in such different ways in reasoning. The standard solution is to use nodes to differentiate between a *type*, or generic concept, and a *token*, or instance of it. Figure 10.9 shows this modification using the *e* (element of) relation to relate the individual to the generic concept. In this simple case, the node sky stands for the *type*, and the empty node stands for a *token*, or instance of the sky concept.

The distinction between type and token is related to the distinction between *intensional* and *extensional* concepts. In analyzing an aerial image there is a difference between

"All bridges span roads or rivers." (10.1)

and

"All bridges (found so far) span roads or rivers." (10.2)

If "bridges" in (10.1) means *any* bridge that might be found, "bridges" is used in an *intensional* sense. If "bridges" means a particular set, it is used it in an *extensional* sense. Normally relations between *type* nodes are used in an intensional sense and relationships between *token* nodes have the extensional sense.

Virtual nodes are objects that are not explicitly represented as object nodes. The need for them arises in expressing complex relations. For example, consider

"The bridge that is at the intersection of road 57
and river 3 is near building 30." (10.3)

which may be represented as shown in Fig. 10.10. The node labeled *x* is the result of intersecting a particular road with a particular river. It is not represented explicitly as an instance of any generic concept; it is a *virtual node*. Virtual nodes can be eliminated by introducing very complex relations, but this would sacrifice an important property of networks, the ability to build up a very large number of com-

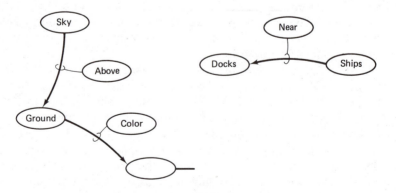

Fig. 10.8 Type or token nodes?

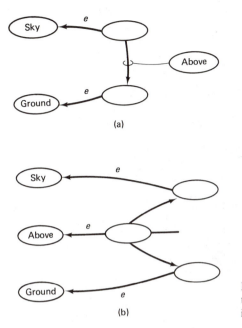

(a)

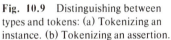

(b)

Fig. 10.9 Distinguishing between types and tokens: (a) Tokenizing an instance. (b) Tokenizing an assertion.

plex relations from a small set of primitives. Virtual nodes enhance this ability by referring to portions of complicated relations.

Nodes in the network can also be used as *variables*. These variables can match other nodes which represent constants. In Fig. 10.11, *x* and *y* are variables and the rest of the nodes are constants. If node *x* is matched to the "telephone" node, then *x* can be regarded as a "telephone" node.

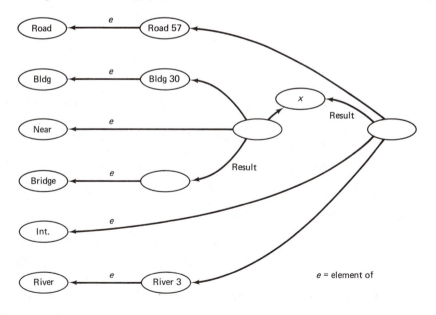

Fig. 10.10 Virtual nodes.

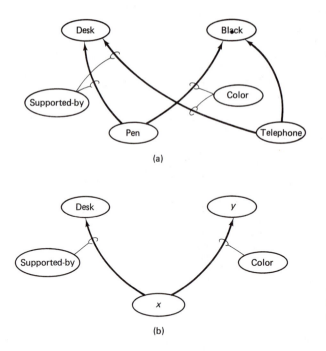

(a)

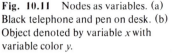

(b)

Fig. 10.11 Nodes as variables. (a) Black telephone and pen on desk. (b) Object denoted by variable x with variable color y.

Often, it is useful to have numerical values as node properties. This can extend the discrete representation of nodes and arcs to a continuous one. For example, in addition to "color of x is red37" we may also associate the particular value of red that we mean with node red37. A special kind of value is a *default value*. If a value can be found for the node in the course of matching other nodes with values or by examining image data, then that value is used for the node value. Otherwise, the default is used.

Relations

Complex relations of many arguments are not uncommon in the world, but for the bulk of practical work, relations of only a few arguments seem to suffice. Semantic nets can clearly represent two-argument relations through their nodes and arcs. More complex relations may be dealt with by various devices. The links to multiple arguments may be ordered within a relation node, or new nodes may be introduced to label the roles of multiple arguments (Fig. 10.7).

If inference mechanisms are to manipulate semantic nets, certain important relations deserve special treatment. One such relation is the "IS-A" relation. The basic issue addressed by this relation is *property inheritance* [Moore 1979]. That is, if Fred IS-A Camel and a Camel IS-A Mammal, then presumably Fred has the properties associated with mammals. It often seems necessary to differentiate between various senses of "IS-A." One basic sense of "X IS-A Y" is "X is an element of the set Y "; others are "X denotes Y," "X is a subset of Y," and "Y is an abstraction of X." Notice that each sense depends on differently "typed" arguments; in the first three cases X is, respectively, an individual, a name, and a set. Deeper treatments of these issues are readily available [Brachman 1979; Hayes* 1977; Nilsson 1980].

It is particularly helpful to have a denotion link to keep perceptual structures separate from model structures. Then if mistakes are made by the vision automaton, a correction mechanism can either sever the denotation link completely or create a new denotation link between the correct model and image structures.

When dealing with many spatial relations, it is economical to recognize that many relations are "inverses" of each other. That is, LEFT-OF(x,y) is the "inverse" of RIGHT-OF(x,y);

$$\text{LEFT--OF}(x,y) \iff \text{RIGHT--OF}(y,x)$$

and also

$$\text{ADJACENT}(x,y) \iff \text{ADJACENT}(y,x)$$

Rather than double the number of these kinds of links, one can *normalize* them. That is, only one half of the inverse pair is used, and the interpreter infers the inverse relation when necessary.

Properties have a different semantics depending on the type of object that has the property. An "abstract" node can have a property that gives one aspect or refinement of the represented concept. A property of a "concrete" node presumably means an established and quantified property of the individual.

Partitions

Partitions are a powerful notion in networks. "Partition" is not used in the sense of a mathematical partition, but in the sense of a barrier. Since the network is a graph, it contains no intrinsic method of delimiting subgraphs of nodes and arcs. Such subgraphs are useful for two reasons:

1. *Syntactic.* It is useful to delimit that part of the network which represents the results of specific inferences.

2. *Semantic.* It is useful to delimit that part of the network which represents knowledge about specific objects. Partitions may then be used to impose a hierarchy upon an otherwise "flat" structure of nodes.

The simple way of representing partitions in a net is to create an additional node to represent the partition and introduce additional arcs from that node to every node or arc in the partition. Partitions allow the nodes and relations in them to be manipulated as a unit.

Notationally, it is cleaner to draw a labeled boundary enclosing the relevant nodes (or arcs). An example is shown by Fig. 10.12 where we consider two objects each made up of several parts with one object entirely left of the other. Rather than use a separate LEFT-OF relation for each of the parts, a single relation can be used between the two partitions. Any pair of parts (one from each object) should inherit the LEFT-OF relation. Partitions may be used to implement quantification in semantic net representations of predicate calculus [Hendrix 1975, 1979]. They may be used to implement frames (Section 10.3.1).

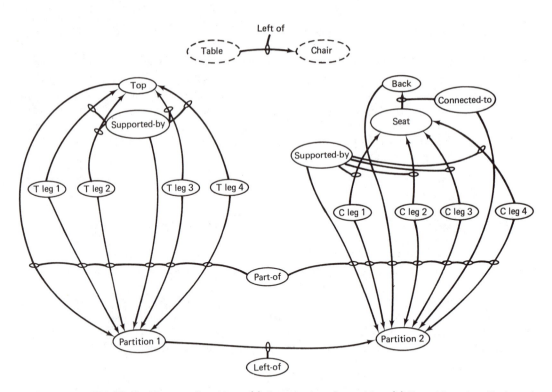

Fig. 10.12 The use of partitions. (a) Construction of a partition. (b) Two objects described by partitions.

Conversions

It is important to be able to transform from geometric (and logical) representations to propositional abstract representations and vice versa. For example, in Fig. 10.13 the problem is to find the exact location of a telephone on a previously located desk. In this case, propositional knowledge that telephones are usually on desktops, together with the desk top location and knowledge about the size of telephones, define a search area in the image.

Converting image data about a particular group of objects into relational form involves the inverse problem. The problem is to perform a level of abstraction to remove the specificity of the geometric knowledge and derive a relation that is appropriate in a larger context. For example, the following program fragment creates the relations ABOVE(A,B), where A and B are world objects.

Comment: assume a world coordinate system where Z is the positive vertical.

Find ZA_{min} for Z in A and ZB_{max} for Z in B.
If $ZA_{min} > ZB_{max}$, then make ABOVE (A,B) true.

Many other definitions of ABOVE, one of which compares centers of gravity, are possible. In most cases, the conversion from continuous geometric relations to discrete propositional relations involves more or less arbitrary conventions. To appreciate this further, consult Fig. 10.14 and try to determine in which of the cases

Fig. 10.13 Search area defined by relational bindings.

block *A* is LEFT-OF block *B*. Figure 10.14d shows a case where different answers are obtained depending on whether a two-dimensional or three-dimensional interpretation is used. Also, when relations are used to encode what is *usually* true of the world, it is often easy to construct a counterexample. Winston [Winston 1975] used

$$\text{SUPPORTS (B,A)} \implies \text{ABOVE (A,B)}$$

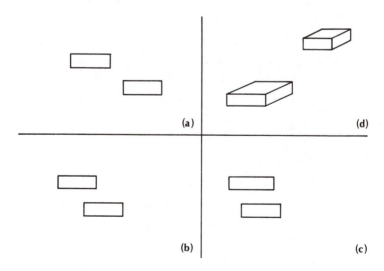

Fig. 10.14 Examples to demonstrate difficulties in encoding spatial relation LEFT-OF (see text).

which is contradicted by Fig. 10.15, given the previous definition of ABOVE.

One common way around these problems is to associate quantitative, "continuous" information with relations (section 10.3.2 and later examples).

10.3 SEMANTIC NET EXAMPLES

Examples of semantic nets abound throughout Part IV. Two more examples illustrate the power of the notions. The first example is described very generally, the second in detail.

10.3.1 Frame Implementations

Frame system theory [Minsky 1975] is a way of explaining our quick access to important aspects of a (perhaps perceptual) situation. It is a provocative and controversial idea, and the reader should consult the References for a full treatment. Implementationally, a frame may be realized by a partition; a frame is a "chunk" of related structure.

Associating related "chunks" of knowledge into manipulable units is a powerful and widespread idea in artificial intelligence [Hayes 1980; Hendrix 1979] as well as psychology. These chunks go by several names: units, frames, partitions, schemata, depictions, scripts, and so forth [Schank and Abelson 1977; Moore and Newell 1973; Roberts and Goldstein 1977; Hayes* 1977; Bobrow and Winograd 1977, 1979; Stefik 1979; Lehnert and Wilks 1979; Rumelhart et al. 1972].

Frames systems incorporate a theory of associative recall in which one selects frames from memory that are relevant to the situation in which one finds oneself. These frames include several kinds of information. Most important, frames have *slots* which contain details of the viewing situation. Frame theory dictates a strictly specific and prototypical structure for frames. That is, the number and type of slots for a particular type of frame are immutable and specified in advance. Further, frames represent specific prototype situations; many slots have default values; this is where expectations and prior knowledge come from. These default values may be disconfirmed by perceptual evidence; if they are, the frame can contain information about what actions to take to fill the slot. Some slots are to be filled in by investigation. Thus a frame is a set of expectations to be confirmed or disconfirmed

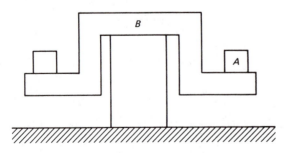

Fig. 10.15 A counterexample to SUPPORTS$(B, A) =>$ ABOVE(A, B).

and actions to pursue in various contingencies. One common action is to "bring in another frame."

The theory is that based on a partial match of a frame's defining slots, a frame can be "brought to mind." The retrieval is much like jumping to a conclusion based on partial evidence. Once the frame is proposed, its slots must be matched up with reality; thus we have the initial major hypothesis that the frame represents, which itself consists of a number of minor subhypotheses to be verified. A frame may have other frames in its slots, and so frames may be linked into "frame systems" that are themselves associatively related. (Consider, for example, the linked perceptual frames for being just outside a theater and for being just inside.) Transformations between frames correspond to the effects of relevant actions. Thus the hypotheses can suggest one another. "Thinking always begins with suggestive but imperfect plans and images; these are progressively replaced by better—but usually still imperfect—ideas" [Minsky 1975].

Frame theory is controversial and has its share of technical problems [Hinton 1977]. The most important of these are the following.

1. Multiple instances of concepts seem to call for copying frames (since the instances may have different slotfillers). Hence, one loses the economy of a preexisting structure.

2. Often, objects have variable numbers of components (wheels on a truck, runways in an airport). The natural representation seems to be a rule for constructing examples, not some specific example.

3. Default values seem inadequate to express legal ranges of slot-filling values or dependencies between their properties.

4. Property inheritance is an important capability that semantic nets can implement with "is a" or "element-of" hierarchies. However, such hierarchies raise the question of which frame to copy when a particular individual is being perceived. Should one copy the generic Mammal frame or the more specific Camel frame, for instance. Surely, it is redundant for the Camel frame to duplicate all the slots in the Mammal frame. Yet our perceptual task may call for a particular slot to be filled, and it is painful not to be able to tell where any particular slot resides.

Nevertheless, where these disadvantages can be circumvented or are irrelevant, frames are seeing increasing use. They are a natural organizing tool for complex data.

10.3.2 Location Networks

This section describes a system for associating geometric analogical data with a semantic net structure which is sometimes like a frame with special "evaluation" rules. The system is a geometrical inference mechanism that computes (or infers) two-dimensional search areas in an image [Russell 1979]. Such networks have found use in both aerial image applications [Brooks and Binford 1980; Nevatia and Price 1978] and medical image applications [Ballard et al. 1979].

The Network

A *location network* is a network representation of geometric point sets related by set-theoretic and geometric operations such as set intersection and union, distance calculation, and so forth. The operations correspond to restrictions on the location of objects in the world. These restrictions, or rules, are dictated by cultural or physical facts.

Each internal node of the location network contains a geometric *operation*, a list of *arguments* for the operation, and a *result* of the operation. For instance, a node might represent the set-theoretic union of two argument point sets, and the result would be a point set. Inference is performed by *evaluating* the net; evaluating all its operations to derive a point set for the top (root) operation.

The network thus has a hierarchy of ancestors and descendents imposed on it through the argument links. At the bottom of this hierarchy are *data nodes* which contain no operation or arguments, only geometric data. Each node is in one of three states: A node is *up-to-date* if the data attached to it are currently considered to be accurate. It is *out-of-date* if the data in it are known to be incomplete, inaccurate, or missing. It is *hypothesized* if its contents have been created by net evaluation but not verified in the image.

In a common application, the expected relative locations of features in a scene are encoded in a network, which thus models the expected structure of the image. The primitive set of geometric relations between objects is made up of four different types of operations.

1. *Directional* operations (left, reflect, north, up, down, and so on) specify a point set with the obvious locations and orientations to another.

2. *Area* operations (close-to, in-quadrilateral, in-circle and so on) create a point set with a non-directional relation to another.

3. *Set* operations (union, difference and intersection) perform the obvious set operations.

4. *Predicates* on areas allow point sets to be filtered out of consideration by measuring some characteristic of the data. For example, a predicate testing width, length, or area against some value restricts the size of sets to be those within a permissible range.

The location of the aeration tank in a sewage treatment plant provides a specific example. The aeration tank is often a rectangular tank surrounded on either end by circular sludge and sedimentation tanks (Fig. 10.16). As a general rule, sewage flows from the sedimentation tanks to aeration tanks and finally through to the sludge tanks. This design permits the use of the following types of restrictions on the location of the aeration tanks.

> *Rule 1*: "Aeration tanks are located somewhere close to both the sludge tanks and the sedimentation tanks."

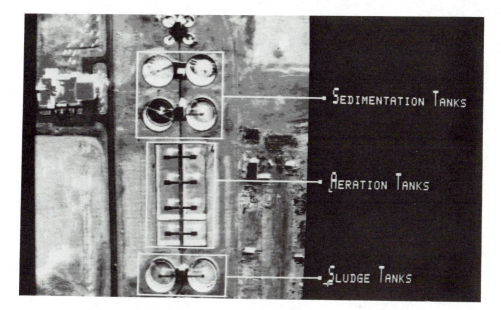

Fig. 10.16 Aerial image of a sewage plant.

The various tanks cannot occupy the same space, so:

> *Rule 2*: "Aeration tanks must not be too close to either the sludge or sedimentation tanks."

Rule 1 is translated to the following network relations.

CLOSE-TO(Union (LocSludgeTanks, LocSedTanks), Distance X)

Rule 2 is translated to

NOT-IN(Union (LocSludgeTanks,LocSedTanks), Distance Y)

The network describing the probable location of the aeration tanks embodies both of these rules. Rule 1 determines an area that is close to both groupings of tanks and Rule 2 eliminates a portion of that area. Thinking of the image as a point set, a set difference operation can remove the area given by Rule 2 from that specified by Rule 1. Figure 10.17 shows the final network that incorporates both rules.

Of course, there could be places where the aeration tanks might be located very far away or perhaps violate some other rule. It is important to note that, like the frames of Section 10.3.1, location networks give prototypical, likely locations for an object. They can work very well for stereotyped scenes, and might fail to perform in novel situations.

The Evaluation Mechanism

The network is interpreted (evaluated) by a program that works top-down in a recursive fashion, storing the partial results of each rule at the topmost node as-

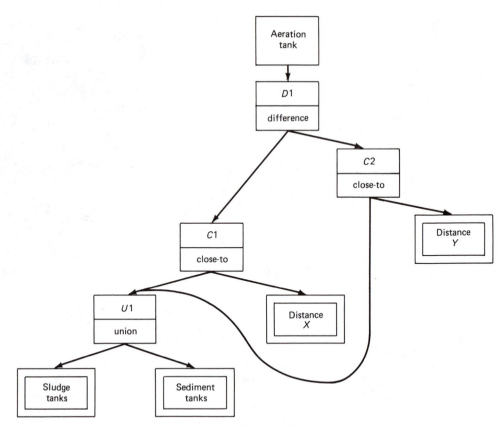

Fig. 10.17 Constraint network for aeration tank.

sociated with that rule (with a few exceptions). Evaluation starts with the root node. In most networks, this node is an operation node. An operation node is evaluated by first evaluating all its arguments, and then applying its operation to those results. Its own result is then available to the node of the network that called for its evaluation.

Data nodes may already contain results which might come from a map or from the previous application of vision operators. At some point in the course of the evaluation, the evaluator may reach a node that has already been evaluated and is marked up-to-date or hypothesized (such a node contains the results of evaluation below that point). The results of this node are returned and used exactly as if it were a data node. Out-of-date nodes cause the evaluation mechanism to execute a low-level procedure to establish the location of the feature. If the procedure is unable to establish the status of the object firmly within its resource limits, the status will remain out-of-date. At any time, out-of-date nodes may be processed without having to recompute any up-to-date nodes. A node marked hypothesized has a value, usually supplied by an inference process, and not verified by low-level image analysis. Hypothesized data may be used in inferences: the results of all inferences based on hypothesized data are marked hypothesized as well.

If a data node ever has its value changed (say, by an independent process that adds new information), all its ancestors are marked out-of-date. Thus the root node will indicate an out-of-date status, but only those nodes on the out-of-date path must be reevaluated to bring the network up to date. Figure 10.18 shows the operation of the aeration tank network of Fig. 10.17 on the input of Fig. 10.16. In this case the initial feature data were a single sludge tank and a single sedimentation tank. Suppose additional work is done to find the location of the remaining sludge and sediment tanks in the image. This causes a reevaluation of the network, and the new result more accurately reflects the actual location of the aeration tanks.

Properties of Location Networks

The location network provides a very general example of use of semantic nets in computer vision.

1. It serves as a data base of point sets and geometric information. The truth status of items in the network is explicitly maintained and depends on incoming information and operations performed on the net.

2. It is an expansion of a geometric expression into a tree, which makes the order of evaluation explicit and in which the partial results are kept for each geometric calculation. Thus it provides efficient updating when some but not all the partial results change in a reevaluation.

3. It provides a way to make geometrical inferences without losing track of the hypothetical nature of assumptions. The tree structure records dependencies among hypotheses and geometrical results, and so upon invalidation of a geometric hypothesis the consequences (here, what other nodes have their values affected) are explicit. The record of dependencies solves a major problem in automated inference systems.

4. It reflects implicit universal quantification. The network claims to represent true relations whose explicit arguments must be filled in as the network is "instantiated" with real data.

5. It has a "flat" semantics. There are no element-of hierarchies or partitions.

6. The concept of "individual" is flexible. A point set can contain multiple disconnected components corresponding to different world objects. In set operations, such an assemblage acts like an explicit set union of the components. An "individual" in the network may thus correspond to multiple individual point (sub)sets in the world.

7. The network allows use of partial knowledge. A set-theoretic semantics of existence and location allows modeling of an unknown location by the set-theoretic universe (the possible location is totally unconstrained). If something is known not to exist in a particular image, its "location" is the null set. Generally, a location is a point set.

8. The set-theoretic semantics allows useful punning on set union and the OR operation, and set intersection and the AND operation. If a dock is on the

shoreline AND near a town, the search for docks need only be carried out in the intersection of the locations.

10.4 CONTROL ISSUES IN COMPLEX VISION SYSTEMS

Computer vision involves the control of large, complex information-processing tasks. Intelligent biological systems solve this control problem. They seem to have complicated control strategies, allowing dynamic allocation of computational resources, parallelism, interrupt-driven shifts of attention, and incremental behavior modification. This section explores different strategies for controlling the complex information processing involved in vision. Appendix 2 contains specific

(a)

(b)

(c)

Fig. 10.18 (a) Initial data to be refined by location network inference. (b) Results of evaluating network of (a). (c) Results of evaluating network after additional information is added.

Ch. 10 *Knowledge Representation and Use*

techniques and programming language constructs that have proven to be useful tools in implementing control strategies for artificial intelligence and computer vision.

10.4.1 Parallel and Serial Computation

In *parallel computation*, several computations are done at the same time. For example, different parts of an image may be processed simultaneously. One issue in parallel processing is synchronization: Is the computation such that the different parts can be done at different rates, or must they be kept in step with each other? Usually, the answer is that synchronization is important. Another issue in parallel processing is its implementation. Animal vision systems have the architecture to do parallel processing, whereas most computer systems are serial (although developing computer technologies may allow the practical realization of some parallel processing). On a serial computer parallelism must be simulated—this is not always straightforward.

In *serial computation*, operations are performed sequentially in time whether or not they depend on one another. The implied sequential control mechanism is more closely matched to a (traditional) serial computer than is a parallel mechanism. Sequential algorithms must be stingy with their resources. This fact has had many effects in computer vision. It has led to mechanisms for efficient data access, such as multiple-resolution representations. It has also led some to emphasize cognitive alternatives for low-level visual processing, in the hope that the massive parallel computations performed in biological vision systems could be circumvented. However, this trend is reversing; cheaper computation and more pervasive parallel hardware should increase the commitment of resources to low-level computations. Parallel and serial control mechanisms have both appeared in algorithms in earlier chapters. It seems clear that many low-level operations (correlation, intrinsic image computations) can be implemented with parallel algorithms. High-level operations, such as "planning" (Chapter 13) have inherently serial components. In general, in the low levels of visual processing control is predominately parallel, whereas at the more abstract levels some useful computations are necessarily serial in nature.

10.4.2 Hierarchical and Heterarchical Control

Visual control strategies dictate the flow of information and activity through the representational layers. What triggers processing: a low level input like a color patch on the retina, or a high level expectation (say, expecting to see a red car)? Different emphasis on these extremes is a basic control issue. The two extremes may be characterized as follows.

1. Image data driven. Here the control proceeds from the construction of the generalized image to segmented structures and finally to descriptions. This is also called *bottom-up* control.

2. Internal model driven. Here high-level models in the knowledge base generate expectations or predictions of geometric, segment, or generalized image structure in the input. Image understanding is the verification of these predictions. This is also called *top-down* control.

Top-down and bottom-up control are distinguished not by what they do but rather by the order in which they do it and how much of it they do. Both approaches can utilize all the basic representations—intrinsic images, features, geometric structures, and propositional representations—but the processing within these representations is done in different orders.

The division of control strategies into top-down and bottom-up is a rather simplistic one. There is evidence that attentional mechanisms may be some of the most complicated brain functions that human beings have [Geschwind 1980]. The different representational subsystems in a complex vision system influence each other in sophisticated and intricate ways; whether control flows "up" or "down" is only a broad characterization of local influence in the (loosely ordered) layers of the system.

The term "bottom-up" was originally applied to parsing algorithms for formal languages that worked their way up the parse tree, assembling the input into structures as they did so. "Top-down" parsers, on the other hand, notionally started at the top of the parse tree and worked downward, effectively generating expectations or predictions about the input based on the possibilities allowed by the grammar; the verification of these predictions confirmed a particular parsing.

These two paradigms are still basic in artificial intelligence, and provide powerful analogies and methods for reasoning about and performing many information-processing tasks. The bottom-up paradigm is comparable in spirit with "forward chaining," which derives further consequences from established results. The top-down paradigm is reflected in "backward chaining," which breaks problems up into subproblems to be solved.

These control organizations can be used not only "tactically" to accomplish specific tasks, but they can dictate the whole "strategy" of the vision campaign. We shall discover that in their pure forms the extreme strategies (top-down and bottom-up) appear inadequate to explain or implement vision. More flexible organizations which incorporate both top-down and bottom-up components seem more suited to a broad spectrum of ambitious vision tasks.

Bottom-Up Control

The general outline for bottom-up vision processing is:

1. *PREPROCESS.* Convert raw data into more usable intrinsic forms, to be interpreted by next level. This processing is automatic and domain-independent.

2. *SEGMENT.* Find visually meaningful image objects perhaps corresponding to world objects or their parts. This process is often but not always broken up into (a) the extraction of meaningful visual primitives, such as lines or regions of homogeneous composition (based on their local characteristics); and (b) the agglomeration of local image features into larger segments.

3. *UNDERSTAND.* Relate the image objects to the domain from which the image arose. For instance, identify or classify the objects. As a step in this process, or indeed as the final step in the computer vision program, the image objects and the relations between them may be described.

In pure bottom-up organization each stage yields data for the next. The progression from raw data to interpreted scene may actually proceed in many steps; the different representations at each step allow us to separate the process into the main steps mentioned above.

Bottom-up control is practical if potentially useful "domain-independent" processing is cheap. It is also practical if the input data are accurate and yield reliable and unambiguous information for the higher-level visual processes. For example, the binary images that result from careful illumination engineering and input thresholding can often be processed quite reliably and quickly in a bottom-up mode. If the data are less reliable, bottom-up styles may still work if they make only tolerably few errors on each pass.

Top-Down Control

A bottom-up, hierarchical model of perception is at first glance appealing on neurological and computational grounds, and has influenced much classical philosophical thought and psychological theory. The "classical" explanation of perception has relatively recently been augmented by a more cognition-based one involving (for instance) interaction of knowledge and expectations with the perceptual process in a more top-down manner [Neisser 1967; Bartlett 1932]. A similar evolution of the control of computer vision processing has accounted for the augmentation of the pure "pattern recognition" paradigm with more "cognitive" paradigms. The evidence seems overwhelming that there are vision processes which do not "run bottom-up," and it is one of the major themes of this book that internal models, goals, and cognitive processes must play major roles in computer vision [Gregory 1970; Buckhout 1974; Gombrich 1972]. Of course, there must be a substantial component of biological vision systems which can perform in a noncognitive mode.

There are probably no versions of top-down organization for computer vision that are as pure as the bottom-up ones. The model to keep in mind in top-down perception is that of goal-directed processing. A high-level goal spawns subgoals which are attacked, again perhaps yielding sub-subgoals, and so on, until the goals are simple enough to solve directly. A common top-down technique is "hypothesize-and-verify"; here an internal modeling process makes predictions about the way objects will act and appear. Perception becomes the verifying of predictions or hypotheses that flow from the model, and the updating of the model based on such probes into the perceptual environment [Bolles 1977]. Of course, our goal-driven processes may be interrupted and resources diverted to respond to the interrupt (as when movement in the visual periphery causes us to look toward the moving object). Normally, however, the hypothesis verification paradigm requires relatively little information from the lower levels and in principle it can control the low-level computations.

The desire to circumvent unnecessary low-level processing in computer vision is understandable. Our low-level vision system performs prodigious amounts of information processing in several cascaded parallel layers. With serial computation technology, it is very expensive to duplicate the power of our low-level visual system. Current technological developments are pointing toward making parallel, low-level processing feasible and thus lowering this price. In the past, however, the price has been so heavy that much research has been devoted to avoiding it, often by using domain knowledge to drive a more or less top-down perception paradigm. Thus there are two reasons to use a top-down control mechanism. First, it seems to be something that human beings do and to be of interest in its own right. Second, it seems to offer a chance to accomplish visual tasks without impractical expenditure of resources.

Mixed Top-Down and Bottom-Up Control

In actual computer vision practice, a judicious mixture of data-driven analysis and model-driven prediction often seems to perform better than either style in isolation. This meld of control styles can sometimes be implemented in a complex hierarchy with a simple pass-oriented control structure. An example of mixed organization is provided by a tumor-detection program which locates small nodular tumors in chest radiographs [Ballard 1976]. The data-driven component is needed because it is not known precisely where nodular tumors may be expected in the input radiograph; there is no effective model-driven location-hypothesizing scheme. On the other hand, a distinctly top-down flavor arises from the exploitation of what little is known about lung tumor location (they are found in lungs) and tumor size. The variable-resolution method using pyramids, in which data are examined in increasingly fine detail, also seems top-down. In the example, work done at 1/16 resolution in a consolidated array guides further processing at 1/4 resolution. Only when small windows of the input array are isolated for attention are they considered at full resolution.

The process proceeds in three passes which move from less to greater detail (Fig. 10.19), zooming in on interesting areas of image, and ultimately finding objects of interest (nodules). Two later passes (not shown) "understand" the nodules by classifying them as "ghosts," tumors or nontumors. Within pass II, there is a distinct data-driven (bottom-up) organization, but passes I and III have a model-directed (top-down) philosophy.

This example shows that a relatively simple, pass-oriented control structure may implement a mixture of top-down and bottom-up components which focus attention efficiently and make the computation practical. It also shows a few places where the ordering of steps is not inherently sequential, but could logically proceed in parallel. Two examples are the overlapping of high-pass filtering of pass II with pass I, and parallel exploration of candidate nodule sites in pass III.

Heterarchical Control

The word "heterarchy" seems to be due to McCulloch, who used it to describe the nonhierarchical (i.e., not partially ordered in rank) nature of neural responses implied by their connectivity in the brain. It was used in the early 1970s to characterize a particular style of nonhierarchical, non-pass-structured control

	PREPROCESS	SEGMENT	CONTROL
Pass 0 (Digitize radiograph)	The digitizer has a hardware attechment which produces the optical density.		
Pass I (Find lung boundaries)	In 64 × 56 consolidated array, apply gradient at proper resolution	In 64 × 56 array, find rough lung outline; in 256 × 224 array, refine lung outline	TOP–DOWN
Pass II (Find candidate nodule sites and large tumors)	In 256 × 224 array, apply high-pass filter to enhance edges, then inside lung boundaries; apply gradient at proper resolution	In 256 × 224 array use gradient-directed, circular Hough method to find candidate sites; also detect large tumors	BOTTOM–UP
Pass III (Find nodule boundaries)	From 1024 × 896 array, extract 64 × 64 window about each candidate nodule site, then in window apply high-pass filter for edge enhancement; then apply gradient at proper resolution	In 64 × 64 full-resolution, pre-processed window, apply dynamic programming technique to find accurate nodule boundaries	TOP–DOWN

Fig. 10.19 A hierarchical tumor-detection algorithm. Technical details of the methods are found elsewhere in this volume. The processing proceeds in passes from top to bottom, and within each pass from left to right. The processing exhibits both top-down and bottom-up characteristics.

organization. Rather than a hierarchical structure (such as the military), one should imagine a community of cooperating and competing experts. They may be organized in their effort by a single executive, by a universal set of rules governing their behavior, or by an a priori system of ranking. If one can think of a task as consisting of many smaller subtasks, each requiring some expertise, and not necessarily performed globally in a fixed order, then the task could be suitable for heterarchical-like control structure.

The idea is to use, at any given time, the expert who can help *most* toward final task solution. The expert may be the most efficient, or reliable, or may give the most information; it is selected because according to some criterion its subtask is the best thing to do at that time. The criteria for selection are wide and varied, and several ideas have been tried. the experts may compute their own relevance, and the decision made on the basis of those individual local evaluations (as in PANDEMONIUM [Selfridge 1959]). They may be assigned a priori immutable rank, so that the highest-ranking expert that is applicable is always run (as in [Shirai 1975; Ambler et al. 1975]). A combination of empirically predetermined and dynamically situation-driven information can be combined to decide which expert applies.

The actual control structure of heterarchical programming can be quite simple; it can be a single iterative loop in which the best action to take is chosen, applied, and interpreted (Fig. 10.20).

10.4.3 Belief Maintenance and Goal Achievement

Belief maintenance and goal achievement are high-level processes that imply differing control styles. The former is concerned with maintaining a current state, the latter with a set of future states. Belief maintenance is an ongoing activity which can ensure that perceptions fit together in a coherent way. Goal achievement is the integration of vision into goal-directed activities such as searching for objects and navigation. There may be "unconscious" use of goal-seeking techniques (e.g., eye-movement control).

Belief Maintenance

An organism is presented with a rich visual input to interpret. Typically, it all makes sense: chairs and tables are supported by floors, objects have expected shapes and colors, objects appear to flow past as the organism moves, nearer objects obscure farther ones, and so on. However, every now and then something

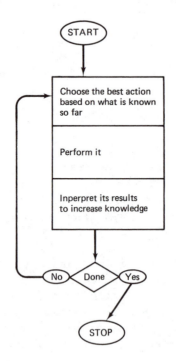

Fig. 10.20   A main executive control loop for heterarchical vision.

Ch. 10   Knowledge Representation and Use

enters the visual field that does not meet expectations. An unfamiliar object in a familiar environment or a sudden movement in the visual periphery can be "surprises" that do not fit in with our existing beliefs and thus have to be reckoned with.

It is sometimes impossible to ignore movements in our visual periphery, but if we are preoccupied it is easily possible to stay unconscious of small changes in our environment. How is it possible to notice some things and not others? The belief maintenance mechanism seems to be resource-limited. A certain amount of "computing resource" is allocated for the job. With this resource, only a limited amount of checking can be done. Checks to be made are ranked (somehow— responses to events in the periphery are like reflexes, or high-priority hard-wired interrupts) and those that cannot be done within the resource limit are omitted. Changes in our beliefs are often initiated in a *bottom-up* way, through unexpected inputs.

A second characteristic of belief maintenance is the almost total absence of sequential, simulation-based or "symbolic" planning or problem-solving activity. Our beliefs are "in the present"; manipulation of hypothetical worlds is not belief maintenance. "Truth maintenance" schemes have been discussed in various contexts [Doyle 1979; Stallman and Sussman 1977].

We conjecture that constraint-satisfaction (relaxation) mechanisms (Chapters 3, 7, and 12) are computationally suited to maintaining belief structures. They can operate in parallel, they seek to minimize inconsistency, they can tolerate "noise" in either input or axioms. Relaxation techniques are usually applied to low-level visual input where locally noisy parameters are combined into globally consistent intrinsic images. Chapter 12 is concerned with inference, in which constraint relaxation is applied to higher-level entities.

Characteristics of Goal Achievement

Goal achievement involves two related activities: planning and acting. Planning is a simulation of the world designed to generate a plan. A *plan* is a sequence of actions that, if carried out, should achieve a goal. *Actions* are the primitives that can modify the world. The motivation for planning is survival. By being able to simulate the effects of various actions, a human being is able to avoid dangerous situations. In an analogous fashion, planning can help machines with vision. For example, a Mars rover can plan its route so as to avoid steep inclines where it might topple over. The incline measurement is made by processing visual input. Since planning involves a sequence of actions, each of which if carried out could potentially change the world, and since planning does not involve actually making those changes, the difficult task of the planner is to keep track of all the different world states that could result from different action sequences.

Vision can clearly serve as an important information-gathering step in planning actions. Can planning techniques be of use directly to the vision process? Clearly so in "skilled vision," such as photointerpretation. Also, planning is a useful computational mechanism that need not be accompanied by conscious, cognitive behavior.

These inductive conclusions leading to the formation of our sense perceptions certainly do lack the purifying and scrutinizing work of conscious thinking. Nevertheless, in my opinion, by their particular nature they may be classed as *conclusions*, inductive conclusions unconsciously formed. [Helmholtz 1925]

The character of computations in goal achievement is related to the inference mechanisms studied in Chapter 11, only planning is distinguished by being dynamic through time. Inference (Chapter 12) is concerned with the knowledge base and deducing relations that logically follow from it. The primitives are *propositions*. In planning (Chapter 13) the primitives are *actions*, which are inherently more complex than propositions. Also, planning need not be a purely deductive mechanism; instead it can be integrated with visual "acting", or the interpretation of visual input. Often, a long deductive sequence may be obviated by using direct visual inspection. This raises a crucial point: Given the existence of plans, how does one choose between them? The solution is to have a method of scoring plans based on some measure of their effectiveness.

EXERCISES

10.1 (a) Diagram some networks for a simple dial telephone, at various levels of detail and with various complexities of relations.

(b) Now include in your network dial and pushbutton types.

(c) Embed the telephone frame into an office frame, describing where the telephone should be found.

10.2 Is a LISP vision program an analogical or propositional representation of knowledge?

10.3 Write a semantic net for the concept "leg," and use it to model human beings, tables, and spiders. Represent the fact "all tables have four legs." Can your "leg" model be shared between tables and spiders? Shared within spiders?

REFERENCES

AMBLER, A. P., H. G. BARROW, C. M. BROWN, R. M. BURSTALL, and R. J. POPPLESTONE. "A versatile system for computer controlled assembly." *Artificial Intelligence 6*, 2, 1975, 129–156.

ANDERSON, J. R. and G. H. BOWER. *Human Associative Memory.* New York: V. H. Winston & Sons, 1973.

BALLARD, D. H. *Hierarchic Recognition of Tumors in Chest Radiographs.* Basel: Birkhauser-Verlag (ISR-16), 1976.

BALLARD, D. H. "Model-directed detection of ribs in chest radiographs." TR11, Computer Science Dept., Univ. Rochester, March 1978.

BALLARD, D. H., U. SHANI, and R. B. SCHUDY. "Anatomical models for medical images." *Proc.*, 3rd COMPSAC, November 1979, 565-570.

BARROW, H. G. and J. M. TENENBAUM. "Computational vision." *Proc. IEEE 69*, 5, May 1981, 572-595.

BARTLETT, F. C. *Remembering: A Study in Experimental and Social Psychology.* Cambridge: Cambridge University Press, 1932.

BOBROW, D. G. and T. WINOGRAD. "An overview of KRL-0, a knowledge representation language." *Cognitive Science 1*, 1, 1977, 3–46.

BOBROW, D. G. and T. WINOGRAD. "KRL: another perspective." *Cognitive Science 3*, 1, 1979, 29–42.

BOLLES, R. C. "Verification vision for programmable assembly." *Proc.*, 5th IJCAI, August 1977, 569–575.

BRACHMAN, R. J. "What's in a concept? Structural foundations for semantic networks." Report 3433, Bolt, Beranek and Newman, October 1976.

BRACHMAN, R. J. "On the epistemological status of semantic networks." *In Associative Networks: Representation and Use of Knowledge by Computers*, N.V. Findler (Ed.). New York: Academic Press, 1979, 3–50.

BROOKS, R. A. and T. O. BINFORD. "Representing and reasoning about specified scenes." *Proc.*, DARPA IU Workshop, April 1980, 95–103.

BUCKHOUT, R. "Eyewitness testimony." *Scientific American*, December 1974, 23–31.

DOYLE, J. "A truth maintenance system." *Artificial Intelligence 12*, 3, 1979.

FAHLMAN, S. E. "A planning system for robot construction tasks." *Artificial Intelligence 5*, 1, 1974, 1–49.

FINDLER, N. V. (Ed.). *Associative Networks: Representation and Use of Knowledge by Computers*. New York: Academic Press, 1979.

FODOR, J. D., J. A. FODOR, and M. F. GARRETT. "The psychological unreality of semantic representations." *Linguistic Inquiry 4*, 1975, 515–531.

FREUDER, E. C. "A computer system for visual recognition using active knowledge." Ph.D. dissertation, MIT, 1975.

FUNT, B. V. "WHISPER: a problem-solving system utilizing diagrams." *Proc.*, 5th IJCAI, August 1977, 459–464.

GARVEY, J. D. "Perceptual strategies for purposive vision." Technical Note 117, AI Center, SRI International, 1976.

GELERNTER, H. "Realization of a geometry-theorem proving machine." In *Computers and Thought*, E. Feigenbaum and J. Feldman (Eds.). New York: McGraw-Hill, 1963.

GESCHWIND, N. "Neurological knowledge and complex behaviors." *Cognitive Science 4*, 2, April 1980, 185–193.

GOMBRICH, E. H. *Art and Illusion*. Princeton, NJ: Princeton University Press, 1972.

GREGORY, R. L. *The Intelligent Eye*. New York: McGraw-Hill, 1970.

HAYES, Patrick J. "The logic of frames." *In The Frame Reader*. Berlin: DeGruyter, 1980.

HAYES*, Philip J. "Some association-based techniques for lexical disambiguation by machine." Ph.D. dissertation, École polytechnique fédérale de Lausanne, 1977; also TR25, Computer Science Dept., Univ. Rochester, June 1977.

HELMHOLTZ, H. von. *Treatise on Physiological Optics* (translated by J. P. T. Sauthall). New York: Dover Publications, 1925.

HENDRIX, G. G. "Expanding the utility of semantic networks through partitions." *Proc.*, 4th IJCAI, September 1975, 115–121.

HENDRIX, G. G. "Encoding knowledge in partitioned networks." In *Associative Networks: Representation and Use of Knowledge by Computers*, N.V. Findler (Ed.). New York: Academic Press, 1979, 51–92.

HEWITT, C. "Description and theoretical analysis (using schemata) of PLANNER" (Ph.D. dissertation). AI-TR-258, AI Lab, MIT, 1972.

HINTON, G. E. "Relaxation and its role in vision." Ph.D. dissertation, Univ. Edinburgh, December 1977.

JOHNSON-LAIRD, P. N. "Mental models in cognitive science." *Cognitive Science 4*, 1, January–March 1980, 71–115.

KOSSLYN, S. M. and J. R. POMERANTZ. "Imagery, propositions and the form of internal representations." *Cognitive Psychology 9*, 1977, 52–76.

KOSSLYN, S. M. and S. P. SCHWARTZ. "A simulation of visual imagery." *Cognitive Science 1*, 3, July 1977, 265–295.

KOSSLYN, S. M. and S. P. SCHWARTZ. "Visual images as spatial representations in active memory." In *CVS*, 1978.

LEHNERT, W. and Y. WILKS. "A critical perspective on KRL." *Cognitive Science 3*, 1, 1979, 1–28.

LOZANO-PEREZ, T. and M. A. WESLEY. "An algorithm for planning collision-free paths among polyhedral obstacles." *Comm. ACM 22*, 10, October 1979, 560–570.

MCDERMOTT, D. "Artificial intelligence meets natural stupidity." *SIGART Newsletter 57*, April 1976, 4–9.

MINSKY, M. L. "A framework for representing knowledge." In *PCV*, 1975.

MOORE J. and A. NEWELL. "How can MERLIN understand?" In *Knowledge and Cognition*, L. Gregg (Ed.). Hillsdale, NJ: Lawrence Erlbaum Assoc., 1973.

MOORE, R. C. "Reasoning about knowledge and action." Techical Note 191, AI Center, SRI International, 1979.

NEISSER, U. *Cognitive Psychology*. New York: Appleton-Century-Crofts, 1967.

NEVATIA, R. and K.E. PRICE. "Locating structures in aerial images." USCIPI Report 800, Image Processing Institute, Univ. Southern California, March 1978, 41–58.

NILSSON, N. J. *Problem-Solving Methods in Artificial Intelligence*. New York: McGraw-Hill, 1971.

NILSSON, N. J. *Principles of Artificial Intelligence*. Palo Alto, CA: Tioga, 1980.

PALMER, S. E. "Visual perception and world knowledge: notes on a model of sensory-cognitive interaction." In *Explorations in Cognition*, D.A. Norman, D.E. Rumelhart, and the LNR Research Group (Eds.). San Francisco: W.H. Freeman, 1975.

PYLYSHYN, Z. W. "What the mind's eye tells the mind's brain; a critique of mental imagery." *Psychological Bulletin 80*, 1973, 1–24.

QUILLIAN, M. R. "Semantic memory." In *Semantic Information Processing*, M. Minsky (Ed.). Cambridge, MA: MIT Press, 1968.

ROBERTS, R. B. and I. P. GOLDSTEIN. "The FRL primer." AI Memo 408, AI Lab, MIT, 1977.

RUMELHART, D. E., P. H. LINDSAY, and D. A. NORMAN. "A process model for long-term memory." In *Organization of Memory*, E. Tulving and J. Donaldson (Eds.). New York: Academic Press, 1972.

RUSSELL, D. M. "Where do I look now?" *Proc.*, PRIP, August 1979, 175–183.

SCHANK, R. C. *Conceptual Information Processing*. Amsterdam: North-Holland, 1975.

SCHANK, R. C. and R. P. ABELSON. *Scripts, Plans, Goals and Understanding*. Hillsdale, NJ: Lawrence Erlbaum Assoc., 1977.

SCHUBERT, L. K. "Extending the expressive power of semantic networks." *Artificial Intelligence 7*, 2, 1976, 163–198.

SELFRIDGE, O. "Pandemonium, a paradigm for learning." In *Proc.*, Symp. on the Mechanisation of Thought Processes, National Physical Laboratory, Teddington, England, 1959.

SHEPARD, R. N. "The mental image." *American Psychologist 33*, 1978, 125–137.

SHIRAI, Y. "Analyzing intensity arrays using knowledge about scenes." In *PCV*, 1975.

SLOMAN, A. "Interactions between philosophy and artificial intelligence: the role of intuition and non-logical reasoning in intelligence." *Artificial Intelligence 2*, 3/4, 1971, 209–225.

STALLMAN, R. M. and G. J. SUSSMAN. "Forward reasoning and dependency-directed backtracking in a system for computer-aided circuit analysis." *Artificial Intelligence 9*, 2, 1977, 135–196.

STEFIK, M. "An examination of a frame-structured representation system." *Proc.*, 6th IJCAI, August 1979, 845–852.

SUSSMAN, G. J. and D. MCDERMOTT. "Why conniving is better than planning." AI Memo 255A, AI Lab, MIT, 1972.

WALTZ, D. and L. BOGGESS. "Visual analog representations for natural language understanding." *Proc.*, 6th IJCAI, August 1979, 226–234.

WINOGRAD, T. "Extended inference modes in reasoning by computer systems." *Proc.*, Conf. on Inductive Logic, Oxford Univ., August 1978.

WINOGRAD, T. "Frame representations and the declarative/procedural controversy." In *Representation and Understanding*, D. G. Bobrow and A. M. Collins (Eds.). New York: Academic Press, 1975, 185–210.

WINSTON, P. H. "Learning structural descriptions from examples." In *PCV*, 1975.

WINSTON, P. H. *Artificial Intelligence.* Reading, MA: Addison-Wesley, 1977.

WOODS, W. A. "What's in a link? Foundations for semantic networks." In *Representation and Understanding*, D. G. Bobrow and A. M. Collins (Eds.). New York: Academic Press, 1975.

Matching   11

11.1 ASPECTS OF MATCHING

11.1.1 Interpretation: Construction, Matching, and Labeling

Figure 10.1 shows a vision system organization in which there are several representations for visual entities. A complex vision system will at any time have several coexisting representations for visual inputs and other knowledge. Perception is the process of integrating the visual input with the preexisting representations, for whatever purpose. Recognition, belief maintenance, goalseeking, or building complex descriptions—all involve forming or finding relations between internal representations. These correspondences *match* ("model," "re-represent," "abstract," "label") entities at one level with those at another level.

Ultimately, matching "establishes an interpretation" of input data, where an interpretation is the correspondence between models represented in a computer and the external world of phenomena and objects. To do this, matching associates different representations, hence establishing a connection between their interpretations in the world. Figure 11.1 illustrates this point. Matching associates TOKNODE, a token for a linear geometric structure derived from image segmentation efforts with a model token NODE101 for a particular road. The token TOKNODE has the interpretation of an image entity; NODE101 has the interpretation of a particular road.

One way to relate representations is to *construct* one from the other. An example is the construction of an intrinsic image from raw visual input. Bottom-up construction in a complex visual system is for reliably useful, domain-independent, goal-independent processing steps. Such steps rely only on "compiled-in" ("hard-wired," "innate") knowledge supplied by the designer of the system. Matching becomes more important as the needed processing becomes more diverse and idiosyncratic to an individual's experience, goals, and

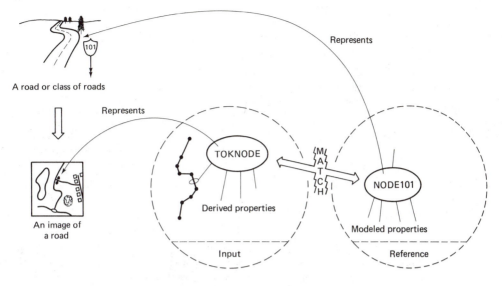

Fig. 11.1 Matching and interpretation.

knowledge. Thus as processing moves from "early" to "late," control shifts from bottom-up toward top-down, and existing knowledge begins to dominate perception.

This chapter deals with some aspects of matching, in which two already existing representations are put into correspondence. When the two representations are similar (both are images or relational structures, say), "matching" can be used in its familiar sense. When the representations are different (one image and one geometric structure, say), we use "matching" in an extended sense; perhaps "fitting" would be better. This second sort of matching usually has a top-down or expectation-driven flavor; a representation is being related to a preexisting one.

As a final extension to the meaning of matching, matching might include the process of checking a structure with a set of rules describing structural legality, consistency, or likelihood. In this sense a scene can be matched against rules to see if it is nonsense or to assign an interpretation. One such interpretation process (called *labeling*) assigns consistent or optimally likely interpretations (labels) at one level to entities of another level. Labeling is like matching a given structure with a possibly infinite set of acceptable structures to find the best fit. However, we (fairly arbitrarily) treat labeling in Chapter 12 as extended inference rather than here as extended matching.

11.1.2 Matching Iconic, Geometric, and Relational Structures

Chapter 3 presented various correlation techniques for matching *iconic* (image-like) structures with each other. The bulk of this chapter, starting in Section 11.2, deals with matching *relational* (semantic net) structures. Another important sort of matching between two dissimilar representations fits data to parameterized models (usually geometric). This kind of matching is an important part of computer vision. A typical example is shown in Fig. 11.2. A preexisting representation (here a straight line) is to be used to interpret a set of input data. The line that best "explains" the data is (by definition) the line of "best fit." Notice that the decision to use a line (rather than a cubic, or a piecewise linear template) is made at a higher level. Given the model, the fitting or matching means determining the *parameters* of the model that tailor it into a useful abstraction of the data.

Sometimes there is no parameterized mathematical model to fit, but rather a given geometric structure, such as a piecewise linear curve representing a shoreline in a map which is to be matched to a piece of shoreline in an image, or to another piecewise linear structure derived from such a shoreline. These geometric matching problems are not traditional mathematical applications, but they are similar in that the best match is defined as the one minimizing a measure of disagreement.

Often, the computational solutions to such geometric matching problems exhibit considerable ingenuity. For example, the shore-matching example above may proceed by finding that position for the segment of shore to be matched that minimizes some function (perhaps the square) of a distance metric (perhaps Euclidean) between input points on the iconic image shoreline and the nearest point on the reference geometric map shoreline. To compute the smallest distance between an arbitrary point and a piecewise linear point set is not a trivial task, and this calculation may have to be performed often to find the best match. The computation may be reduced to a simple table lookup by precomputing the metric in a "chamfer array," that contains the metric of disagreement for any point around the geometric reference shoreline [Barrow et al. 1978]. The array may be computed efficiently by symmetric axis transform techniques (Chapter 8) that "grow" the linear structure outward in contours of equal disagreement (distance) until a value has been computed for each point of the chamfer array.

Parameter optimization techniques can relate geometrical structures to lower-level representations and to each other through the use of a merit function measuring how well the relations match. The models are described by a vector of parameters $\mathbf{a} = (a_1, \ldots, a_n)$. The merit function M must rate each set of those parameters in terms of a real number. For example, M could be a function of both a, the parameters, and $f(x)$, the image. The problem is to find a such that

$$M(\mathbf{a}, f(\mathbf{x}))$$

Reference Input

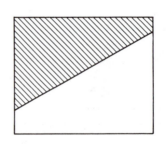

$Ax + By + C = 0$

Fig. 11.2 Matching or fitting a straight line model to data.

is maximized. Note that if **a** were some form of template function rather than a vector of parameters, the problem statement would encompass the iconic correlation techniques just covered. There is a vast literature on optimization techniques and we cannot do more than provide a cursory discussion of a few cases with examples.

Formally, the different techniques have to do with the form of the merit function M. A fundamental result from calculus is that if M is sufficiently well behaved (i.e., has continuous derivatives), then a condition for a local maximum (or minimum) is that

$$M_{a_j} = \frac{\partial M}{\partial a_j} = 0 \qquad \text{for } j = 1,\ldots,n \qquad (11.1)$$

This condition can be exploited in many different ways.

- Sometimes Eqs. (11.1) are sufficiently simple so that the a can be determined analytically, as in the least squares fitting, described in Appendix 1.
- An approximate solution a^0 can be iteratively adjusted by moving in the gradient direction or direction of maximum improvement:

$$a_j^k = a_j^{k-1} + cM_{a_j} \qquad (11.2)$$

where c is a constant. This is the most elementary of several kinds of *gradient (hill-climbing) techniques.* Here the gradient is defined with respect to M and does not mean edge strength.

- If the partial derivatives are expensive to calculate, the coefficients can be perturbed (either randomly or in a structured way) and the perturbations kept if they improve M:

  (1) $\mathbf{a}' := \mathbf{a} + \Delta \mathbf{a}$

  (2) $\mathbf{a} := \mathbf{a}'$ if $M(\mathbf{a}') > M(\mathbf{a})$

A program to fit three-dimensional image data with shapes described by spherical harmonics used these techniques [Schudy and Ballard 1978]. The details of the spherical harmonics shape representation appear in Chapter 9. The fitting proceeded by the third method above. A nominal expected shape was matched to boundaries in image data. If a subsequent perturbation in one of its parameters results in an improvement in fit it was kept; otherwise, a different perturbation was made. Figure 11.3 shows this fitting process for a cross section of the shape.

Though parameter optimization is an important aspect of matching, we shall not pursue it further here in view of the extensive literature on the subject.

11.2 GRAPH-THEORETIC ALGORITHMS

The remainder of this chapter deals with methods of matching relational structures. Chapter 10 showed how to represent a relational structure containing n-ary relations as a graph with labeled arcs. Recall that the labels can have values from a

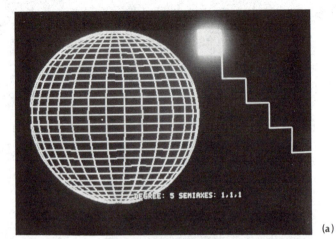

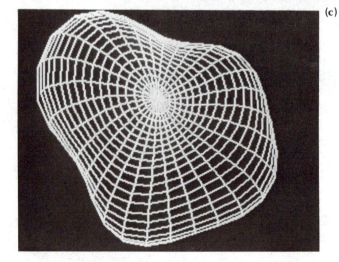

Fig. 11.3 An example of matching as parameter optimization. (a) Initial parameter set (displayed at left as three-dimensional surface (see Fig. 9.8) (b) Fitting process: iteratively adjust *a* based on M (see text). (c) Final parameter set yields this three-dimensional surface. (*See color inserts.*)

continuum, and that labeled arcs could be replaced by nodes to yield a directed graph with labeled nodes.

Depending on the attributes of the relational structure and of the correspondence desired, the definition of a match may be more or less elegant. It is always possible to translate powerful representations such as labeled graphs or *n*-ary relations into computational representations which are amenable to formal treatment (such as undirected graphs). However, when graph algorithms are to be implemented with computer data structures, the freedom and power of programming languages often tempts the implementer away from pure graph theory. He can replace elegant (but occasionally restrictive and impractical) graph-theoretic concepts and operations with arbitrarily complex data structures and algorithms.

One example is the "graph isomorphism" problem, a very pure version of relational structure matching. In it, all graph nodes and arcs are unlabeled, and graphs match if there is a 1:1 and onto correspondence between the arcs and nodes of the two graphs. The lack of expressive power in these graphs and the requirement that a match be "perfect" limits the usefulness of this pure model of matching in the context of noisy input and imprecise reference structures. In practice, graph nodes may have properties with continuous ranges of values, and an arbitrarily complex algorithm determines whether nodes or arcs match. The algorithm may even access information outside the graphs themselves, as long as it returns the answer "match" or "no match." Generalizing the graph-theoretic notions in this way can obscure issues of their efficiency, power, and properties; one must steer a course between the "elegant and unusable" and the "general and uncontrollable." This section introduces some "pure" graph-theoretic algorithms that form the basis for techniques in Sections 11.3 and 11.4.

11.2.1 The Algorithms

The following are several definitions of matching between graphs [Harary 1969; Berge 1976].

- *Graph isomorphism.* Given two graphs (V_1, E_1) and (V_2, E_2), find a 1:1 and onto mapping (an isomorphism) f between V_1 and V_2 such that for $v_1, v_2 \in V_1, V_2, f(v_1) = v_2$ and for each edge of E_1 connecting any pair of nodes v_1 and $v'_1 \in V_1$, there is an edge of E_2 connecting $f(v_1)$ and $f(v_1')$.

- *Subgraph isomorphism.* Find isomorphisms between a graph (V_1, E_1) and *subgraphs* of another graph (V_2, E_2). This is computationally harder than isomorphism because one does not know in advance which subsets of V_2 are involved in isomorphisms.

- *"Double" subgraph isomorphisms.* Find all isomorphisms between *subgraphs* of a graph (V_1, E_1) and *subgraphs* of another graph (V_2, E_2). This sounds harder than the subgraph isomorphism problem, but is equivalent.

- A match may not conform to strict rules of correspondence between arcs and nodes (some nodes and arcs may be "unimportant"). Such a matching criterion may well be implemented as a "computational" (impure) version of one of the pure graph isomorphisms.

Figure 11.4 shows examples of these kinds of matches.

One algorithm for finding graph isomorphism [Corneil and Gotlieb 1970] is based on the idea of separately putting each graph into a canonical form, from which isomorphism may easily be determined. For directed graphs (i.e., nonsymmetric relations) a backtrack search algorithm [Berztiss 1973] works on both graphs at once.

Two solutions to the subgraph isomorphism problem appear in [Ullman 1976]: The first is a simple enumerative search of the tree of possible matches between nodes. The second is more interesting; in it a process of "parallel-iterative" refinement is applied at each stage of the search. This process is a way of rejecting node pairs from the isomorphism and of propagating the effects of such rejections; one rejected match can lead to more matches being rejected. When the iteration converges (i.e., when no more matches can be rejected at the current stage), another step in the tree search is performed (one more matching pair is hypothesized). This mixing of parallel-iterative processes with tree search is useful in a variety of applications (Section 11.4.4, Chapter 12).

"Double" subgraph isomorphism is easily reduced to subgraph isomorphism via another well-known graph problem, the "clique problem." A *clique* of size N is a totally connected subgraph of size N (each node is connected to every other node in the clique by an arc). Finding isomorphisms between subgraphs of a graph A and subgraphs of a graph B is accomplished by forming an *association graph G* from the graphs A and B and finding cliques in G (for details, see Section 11.3.3). Clique

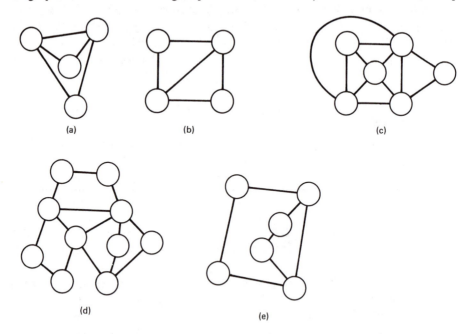

(a) (b) (c)

(d) (e)

Fig. 11.4 Isomorphisms and matches. The graph (a) has an isomorphism with (b), various subgraph isomorphisms with (c), and several "double" subgraph isomorphisms with (d). Several partial matches with (e) (and also (b), (c), and (d)), depending on which missing or extra nodes are ignored.

finding may be done with a subgraph isomorphism algorithm; hence the reduction. Several other clique-finding algorithms exist [Ambler et al. 1975; Knodel 1968; Bron and Kerbosch 1973; Osteen and Tou 1973].

11.2.2 Complexity

It is of some practical importance to be aware of the computational complexity of the matching algorithms proposed here; they may take surprising amounts of computer time. There are many accessible treatments of computational complexity of graph-theoretic algorithms [Reingold et al. 1977; Aho, Hopcroft and Ullman 1974]. Theoretical results usually describe worst-case or average time complexity. The state of knowledge in graph algorithms is still improving; some interesting worst-case bounds have not been established.

A "hard" combinatorial problem is one that takes time (in a usual model of computation based on a serial computer) proportional to an exponential function of the length of the input. "Polynomial-time" solutions are desirable because they do not grow as fast with the size of the problem. The time to find all the cliques of a graph is in the worst case inherently exponential in the size of the input graphs, because the output is an exponential number of graphs. Both the single subgraph isomorphism problem and the "clique problem" (does there exist a clique of size k?) are *NP-complete*; all known deterministic algorithms run (in the worst case) in time exponential in the length of the description of the graphs involved (which must specify the nodes and arcs). Not only this, but if either of these problems (or a host of other NP complete problems) could be solved deterministically in time polynomially related to the length of the input, it could be used to solve all the other NP problems in polynomial time.

Graph isomorphism, both directed and undirected, is at this writing in a netherworld (along with many other combinatorial problems). No polynomial-time deterministic algorithms are known to exist, but the relation of these problems to each other is not as clear-cut as it is between the NP-complete problem. In particular, finding a polynomial-time deterministic solution to one of them would not necessarily indicate anything about how to solve the other problems deterministically in polynomial time. These problems are not mutually reducible. Certain restrictions on the graphs, for instance that they are planar (can be arranged with their nodes in a plane and with no arcs crossing), can make graph isomorphism an "easy" (polynomial-time) problem.

The average-case complexity is often of more practical interest than the worst case. Typically, such a measure is impossible to determine analytically and must be approximated through simulation. For instance, one algorithm to find isomorphisms of randomly generated graphs yields an average time that seems not exponential, but proportional to N^3, with N the number of nodes in the graph [Ullman 1976]. Another algorithm seems to run in average time proportional to N^2 [Corneil and Gotlieb 1970].

All the graph problems of this section are in NP. That is, a *non*deterministic algorithm can solve them in polynomial time. There are various ways of visualizing nondeterministic algorithms; one is that the algorithm makes certain significant "good guesses" from a range of possibilities (such as correctly guessing which subset of nodes from graph B are isomorphic with graph A and then only having to worry about the arcs). Another way is to imagine *parallel* computation; in the clique problem, for example, imagine multiple machines running in parallel, each with a different subset of nodes from the input graph. If any machine discovers a totally connected subset, it has, of course, discovered a clique. Checking whether N nodes are all pairwise connected is at most a polynomial-time problem, so all the machines will terminate in polynomial time, either with success or not. Several interesting processes can be implemented with parallel computations. Ullman's algorithm uses a refinement procedure which may run in parallel between stages of his tree search, and which he explains how to implement in parallel hardware [Ullman 1976].

11.3 IMPLEMENTING GRAPH-THEORETIC ALGORITHMS

11.3.1 Matching Metrics

Matching involves *quantifiable similarities*. A match is not merely a correspondence, but a correspondence that has been quantified according to its "goodness." This measure of goodness is the *matching metric*. Similarity measures for correlation matching are lumped together as one number. In relational matching they must take into account a relational, structured form of data [Shapiro and Haralick 1979].

Most of the structural matching metrics may be explained with the physical analogy of "templates and springs" [Fischler and Elschlager 1973]. Imagine that the reference data comprise a structure on a transparent rubber sheet. The matching process moves this sheet over the input data structure, distorting the sheet so as to get the best match. The final goodness of fit depends on the individual matches between elements of the input and reference data, and on the amount of work it takes to distort the sheet. The continuous deformation process is a pretty abstraction which most matching algorithms do not implement. A computationally more tractable form of the idea is to consider the model as a set of rigid "templates" connected by "springs" (see Fig. 11.5). The templates are connected by "springs" whose "tension" is also a function of the relations between elements. A spring function can be arbitrarily complex and nonlinear; for example the "tension" in the spring can attain very high or infinite values for configurations of templates which cannot be allowed. Nonlinearity is good for such constraints as: in a picture of a face the two eyes must be essentially in a horizontal line and must be within fixed limits of distance. The quality of the match is a function of the goodness of fit of the templates locally and the amount of "energy" needed to stretch the springs to force the input onto the reference data. Costs may be imposed for missing or extra elements.

The template match functions and spring functions are general procedures, thus the templates may be more general than pure iconic templates. Further,

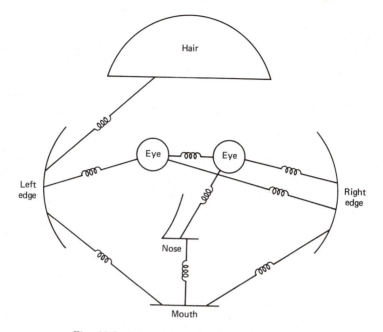

Fig. 11.5 A templates and springs model of a face.

matches may be defined not only between nodes and other nodes, but between nodes and image data directly. Thus the template and springs formalism is workable for "cross-representational" matching. The mechanism of minimizing the total cost of the match can take several forms; more detailed examples follow in Section 11.4.

Equation 11.3 a general form of the template-and-springs metric. TemplateCost measures dissimilarity between the input and the templates, and SpringCost measures the dissimilarity between the matched input elements' relations and the reference relations between the templates. MissingCost measures the penalties for missing elements. $F(\cdot)$ is the mapping from templates of the reference to elements of the input data. F partitions the reference templates into two classes, those found {FoundinRefer} and those not found {MissinginRefer} in the input data. If the input data are symbolic they may be similarly partitioned. The general metric is

$$
\begin{aligned}
\text{Cost} = &\sum_{d \,\in\, \{\text{FoundinRefer}\}} \text{TemplateCost}(d, F(d)) \\
+ &\sum_{(d, e) \,\in\, \{\text{FoundinRefer} \,\times\, \text{FoundinRefer}\}} \text{SpringCost}(F(d), F(e)) \qquad (11.3) \\
+ &\sum_{c \,\in\, \{\text{MissinginRefer}\} \,\cup\, \{\text{MissinginInput}\}} \text{MissingCost}\,(c)
\end{aligned}
$$

Equation 11.3 may be written as one sum of generalized SpringCosts in which the template properties are included (as 1-ary relations), as are "springs" involving missing elements.

As with correlation metrics, there are normalization issues involved with structural matching metrics. The number of elements matched may affect the ultimate magnitude of the metric. For instance, if springs always have a finite cost, then the more elements that are matched, the higher the total spring energy must be; this should probably not be taken to imply that a match of many elements is worse than a match of a few. Conversely, suppose that relations which agree are given positive "goodness" measures, and a match is chosen on the basis of the total "goodness." Then unless one is careful, the sheer number of possibly mediocre relational matches induced by matching many elements may outweigh the "goodness" of an elegant match involving only a few elements. On the other hand, a small, elegant match of a part of the input structure with one particular reference object may leave much of the search structure unexplained. This good "submatch" may be less helpful than a match that explains more of the input. To some extent the general metric (Eq.11.3) copes with this by acknowledging the "missing" category of elements.

If the reference templates actually contain iconic representations of what the input elements should look like in the image, a TemplateCost can be associated with a template and a location in the image by

$$\text{TemplateCost(Template, Location)}$$
$$= (1 - \text{normalized correlation metric between}$$
$$\text{template shape and input image at the location}).$$

If the match is, for instance, to match reference descriptions of a chair with an input data structure, a typical "spring" might be that the chair seat must be supported by its legs. Thus if F is the association function mapping reference elements such as LEG or TABLETOP to input elements,

$$\text{SpringCost}_1(F(\text{LEG}), F(\text{TABLETOP})$$
$$= \begin{cases} 0 & \text{if } F(\text{LEG}) \text{ appears to support } F(\text{TABLETOP}), \\ 1 & \text{if } F(\text{LEG}) \text{ does not appear to support } F(\text{TABLETOP}). \end{cases}$$

For quantified relations, one might have

$$\text{MatchCost} = \text{number of standard deviations from the}$$
$$\text{canonical mean value for this relation.}$$

Another version of MatchCost is used in the following expression for total match cost by [Barrow and Popplestone 1971].

$$\text{Cost of Match} = \frac{\text{SpringCosts of properties (unary) and binary relations}}{\text{total number of unary and binary springs}} \quad (11.4)$$
$$+ \frac{\text{Empirical Constant}}{\text{Total number of reference elements matched}}$$

The first term measures the average badness of matches between properties (unary relations) and relations between regions. The second term is inversely proportional to the number of regions that are matched, effectively increasing the cost of matches that explain less of the input.

11.3.2 Backtrack Search

Backtrack search is a generic name for a type of potentially exhaustive search organized in stages; each processing stage attempts to extend a partial solution derived in the previous stage. Should the attempt fail, the search "backtracks" to the most recent partial solution, from which a new extension is attempted. The technique is basic, amounting to a depth-first search through a tree of partial solutions (Fig. 11.6). Backtracking is a pervasive control structure in artificial intelli-

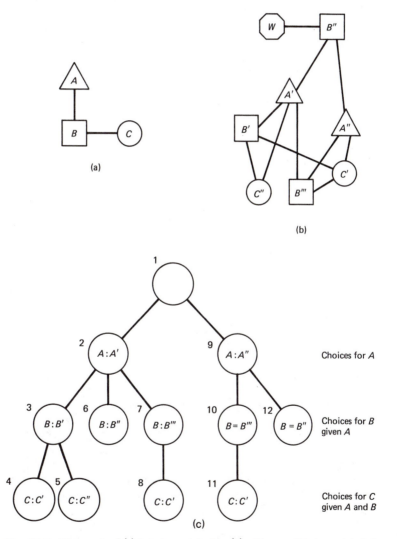

Fig. 11.6 The graph of (a) is to be matched in (b) with arcs all being unlabeled but nodes having properties indicated by their shapes, (c) is the tree of solutions built by a backtrack algorithm.

gence, and through the years several general classes of techniques have evolved to make the basic, brute-force backtrack search more efficient.

Example: Graph Isomorphisms

Given two graphs,

$$X = (V_X, E_X)$$
$$Y = (V_Y, E_Y),$$

without loss of generality, let $V_X = V_Y = \{1, 2, \ldots, n\}$, and let X be the reference graph, Y the input graph. The isomorphism is given by: If $i \in V_X$, the corresponding node under the isomorphism is $F(i) \in V_Y$.

In the algorithm, S is the set of nodes accounted for in Y by a partial solution. k gives the current level of the search in the tree of partial solutions, the number of nodes in the current partial solution, and the node of X whose match in Y is currently being sought. v is a node of Y currently being considered to extend the current partial solution. As written, the algorithm finds all isomorphisms. It is easily modified to quit after finding the first.

Algorithm 11.1   Backtrack Search for Directed Graph Isomorphism

Recursive Procedure DirectedGraphIsomorphisms(S,k);
begin
  if $S = V_Y$ *then* ReportAsIsomorphism(F)
  else
   forall $v \in (V_Y - S)$
    do
    if Match(k, v)
    then
     begin
      $F(k) := v$;
      DirectedGraphIsomorphisms $(S \cup \{v\}, k+1)$;
     end;
end;

ReportAsIsomorphism could print or save the current value of F, the global structure recording the current solution. Match(k, v) is a procedure that tests whether $v \in V_Y$ can correspond to $k \in V_X$ under the isomorphism so far defined by F. Let X_k be the subgraph of X with vertices $\{1, 2, \ldots, k\}$. The procedure "Match" must check for $i < k$, whether (i, k) is an edge of X_k iff $(F(i), v)$ is an edge of Y *and* whether (k, i) is an edge of X_k iff $(v, F(i))$ is an edge of Y.

Improving Backtrack Search

Several techniques are useful in improving the efficiency of backtrack search [Bittner and Reingold 1975]:

1. *Branch pruning.* All techniques of this variety examine the current partial solution and prune away descendents that are not viable continuations of the solution. Should none exist, backtracking can take place immediately.

2. *Branch merging.* Do not search branches of the solution tree isomorphic with those already searched.

3. *Tree rearrangement and reordering.* Given pruning capabilities, more nodes are likely to be eliminated by pruning if there are fewer choices to make early in the search (partial solution nodes of low degree should be high in the search tree). Similarly, search first those extensions to the current solution that have the fewest alternatives.

4. *Branch and bound.* If a cost may be assigned to solutions, standard techniques such as heuristic search and the A* search algorithm [Nilsson 1980] (Section 4.4) may be employed to allow the search to proceed on a "best-first" rather than a "depth-first" basis.

For extensions of these techniques, see [Haralick and Elliott 1979].

11.3.3 Association Graph Techniques

Generalized Structure Matching

A general relational structure "best match" is less restricted than graph isomorphism, because nodes or arcs may be missing from one or the other graph. Also, it is more general than subgraph isomorphism because one structure may not be exactly isomorphic to a substructure of the other. A more general match consists of a set of nodes from one structure and a set of nodes from the other and a 1:1 mapping between them which preserves the compatibilities of properties and relations. In other words, corresponding nodes (under the node mapping) have sufficiently similar properties, and corresponding sets under the mapping have compatible relations.

The two relational structures may have a complex makeup that falls outside the normal purview of graph theory. For instance, they may have parameterized properties attached to their nodes and edges. The definition of whether a node matches another node and whether two such node matches are mutually compatible can be determined by arbitrary procedures, unlike the much simpler criteria used in pure graph isomorphism or subgraph isomorphism, for example. Recall that the various graph and subgraph isomorphisms rely heavily on a 1:1 match, at least locally, between arcs and nodes of the structures to be matched. However, the idea of a "best match" may make sense even in the absence of such perfect correspondences.

The *association graph* defined in this section is an auxiliary data structure produced from two relational structures to be matched. The beauty of the association graph is that it *is* a simple, pure graph-theoretic structure which is amenable to pure graph-theoretic algorithms such as clique finding. This is useful for several reasons.

- It takes relational structure matching from the ad hoc to the classical domain.

- It broadens the base of people who are producing useful algorithms for structure matching. If the rather specialized relational structure matching enterprise is reducible to a classical graph-theoretical problem, then everyone working on the classical problem is also working indirectly on structure matching.

- Knowledge about the computational complexity of classical graph algorithms illuminates the difficulty of structure matching.

Clique Finding for Generalized Matching

Let a relational structure be a set of elements V, a set of properties (or more simply unary predicates) P defined over the elements, and a set of binary relations (or binary predicates) R defined over pairs of the elements. An example of a graph representation of such a structure is given in Fig. 11.7.

Given two structures defined by (V_1, P, R) and (V_2, P, R), say that "similar" and "compatible" actually mean "the same." Then we construct an association graph G as follows [Ambler et al. 1975]. For each v_1 in V_1 and v_2 in V_2, construct a node of G labeled (v_1, v_2) if v_1 and v_2 have the same properties [$p(v_1)$ iff $p(v_2)$ for each p in P]. Thus the nodes of G denote assignments, or pairs of nodes, one each from V_1 and V_2, which have similar properties. Now connect two nodes (v_1, v_2) and (v'_1, v'_2) of G if they represent *compatible* assignments according to R, that is, if the pairs satisfy the same binary predicates [$r(v_1, v'_1)$ iff $r(v_2, v'_2)$ for each r in R].

A match between (V_1, P, R) and (V_2, P, R), the two relational structures, is just a set of assignments that are all mutually compatible. The "best match" could well be taken to be the largest set of assignments (node correspondences) that were all mutually compatible under the relations. But this in the association graph G is just the largest totally connected (completely mutually compatible) — set of nodes. It is a *clique*. A clique to which no new nodes may be added without destroying the clique properties is a *maximal* clique. In this formulation of matching, larger cliques are taken to indicate better matches, since they account for more nodes.

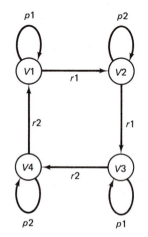

Fig. 11.7 A graph representation of a relational structure. Elements (nodes) v_1 and v_3 have property p1, v_2 and v_4 have property p2, and the arcs between nodes indicate that the relation r1 holds between v_1 and v_2 and between v_2 and v_3, and r2 holds between v_3 and v_4 and between v_4 and v_1.

Thus the best matches are determined by the largest maximal cliques in the association graph. Figure 11.8 shows an example: Certain subfeatures of the objects have been selected as "primitive elements" of the objects, and appear as nodes (elements) in the relational structures. To these nodes are attached properties, and between them can exist relations. The choice of primitives, properties, and relations is up to the designer of the representation. Here the primitives of the representation correspond to edges and corners of the shape.

The association graph is shown in 11.8e. Its nodes correspond to pairs of nodes, one each from A and B, whose properties are similar. [Notice that there is no node in the association graph for (6,6')]. The arcs of the association graph indicate that the endpoints of the arc represent compatible associations. Maximal cliques in the association graph (shown as sets of nodes with the same shape) indicate sets of consistent associations. The largest maximal clique provides the node pairings of the "best match."

In the example construction, the association graph is formed by associating nodes with exactly the same properties (actually unary predicates), and by allowing as compatible associations only those with exactly the same relations (actually binary predicates). These conditions are easy to state, but they may not be exactly what is needed. In particular, if the properties and relations may take on ranges of values greater than the binary "exists" and "does not exist," then a measure of similarity must be introduced to define when node properties are similar enough for association, and when relations are similar enough for compatibility. Arbitrarily complex functions can decide whether properties and relations are similar. As long as the function answers "yes" or "no," the complexity of its computations is irrelevant to the matching algorithm.

The following recursive clique-finding algorithm builds up cliques a node at a time [Ambler et al. 1975]. The search tree it generates has states that are ordered pairs (set of nodes chosen for a clique, set of nodes available for inclusion in the clique). The root of the tree is the state ($\varnothing$, all graph nodes), and at each branch a choice is made whether to include or not to include an eligible node in the clique. (If a node is eligible for inclusion in clique X, then *each* clique including X must either include the node or exclude it).

Algorithm 11.2: Clique-Finding Algorithm

Comment *Nodes* is the set of nodes in the input graph.

Comment
    Cliques (X, Y) takes as arguments a clique X, and Y, a set of nodes that includes
                X. It returns all cliques that include X and are included in Y.
    Cliques ($\varnothing$,Nodes) finds all cliques in the graph.
Cliques$(X, Y) :=$
    if no node in $Y-X$ is connected to all elements of X
      then $\{X\}$
      else
          Cliques$(X \bigcup \{y\}, Y) \bigcup$ Cliques $(X, Y-\{y\})$
        where y is connected to all elements of X.

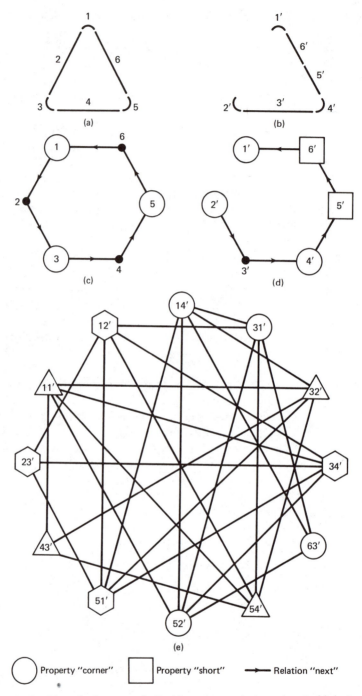

Fig. 11.8 Clique-finding example. Entities to be matched are given in (a) (reference) and (b) (input). The relational structures corresponding to them are shown in (c) and (d). The resulting association graph is shown in (e) with its largest cliques indicated by node shapes.

Extensions

Modifications to the clique-finding algorithm extend it to finding maximal cliques and finding largest cliques. To find largest cliques, perform an additional test to stop the recursion in *Cliques* if the size of X plus the number of nodes in $Y-X$ connected to all of X becomes less than k, which is initially set to the size of the largest possible clique. If no cliques of size k are found, decrement k and run *Cliques* with the new k.

To find maximal cliques, at each stage of *Cliques*, compute the set

$$Y' = \{z \in \text{Nodes: } z \text{ is connected to each node of Y}\}.$$

Since any maximal clique must include Y', searching a branch may be terminated should Y' not be contained in Y, since Y can then contain no maximal cliques.

The association graph may be searched not for cliques, but for r-connected components. An *r-connected* component is a set of nodes such that each node is connected to at least r other nodes of the set. A clique of size n is an $n-1$-connected component. Fig. 11.9 shows some examples.

The r-connected components generalize the notion of clique. An r-connected component of N nodes in the association graph indicates a match of N pairs of nodes from the input and reference structures, as does an N-clique. Each matching pair has similar properties, and each pair is compatible with at least r other matches in the component.

Whether or not the r-connected component definition of a match between two structures is useful depends on the semantics of "compatibility." For instance, if all relations are either compulsory or prohibited, clearly a clique is called for. If the relations merely give some degree of mutual support, perhaps an r-connected component is the better definition of a match.

11.4 MATCHING IN PRACTICE

This section illustrates some principles of matching with examples from the computer vision literature.

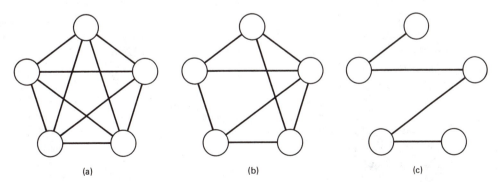

Fig. 11.9   *r*-connected components. (a) A 5-clique (which is 4-connected). (b) A 3-connected set of 5 nodes. (c) A 1-connected set of 5 nodes.

11.4.1 Decision Trees

Hierarchical decision-tree matching with ad hoc metrics is a popular way to identify input data structures as instances of reference models and thus classify the input instances [Nevatia 1974; Ambler et al. 1975; Winston 1975]. Decision trees are indicated when it is predictable that certain features are more reliably extracted than others and that certain relations are either easier to sense or more necessary to the success of a match.

Winston and Nevatia both compare matches with a "weighted sums of difference" metric that reflects cumulative differences between the parameters of corresponding elements and relations in the reference and input data. In addition, Nevatia does parameter fitting; his reference information includes geometrical information.

Matching Structural Descriptions

Winston is interested in matching such structures as appear in Fig. 11.10B. The idea is to recognize instances of structural concepts such as "arch" or "house," which are relational structures of primitive blocks (Fig.11.10A) [Winston 1975]. An important part of the program learns the concept in the first place—only the matching aspect of the program is discussed here. His system has the pleasant property of representational uniqueness: reference and input data structures that are identical up to the resolution of the descriptors used by the program have identical representations. Matching is easy in this case. Reflections of block structures can be recognized because the information available about relations (such as LEFT-OF and IN-FRONT-OF) includes their OPPOSITE (i.e., RIGHT-OF and BEHIND). The program thus can recognize various sorts of symmetry by replacing all input data structure relations by their relevant opposite, then comparing with the reference.

The next most complicated matching task after exact or symmetric matches is to match structures in isolation. Here the method is sequentially to match the input data against the whole reference data catalog of structures and determine which match is best (which difference description is most inconsequential). Easily computed scene characteristics can rule out some candidate models immediately.

The models contain arcs such as MUST-BE and MUST-NOT, expressing relations mandatory or forbidden relations. A match is not allowed between a description and a model if one of the strictures is violated. For instance, the program may reject a "house" immediately as not being a "pedestal," "tent," or "arch," since the pedestal top must be a parallelepiped, both tent components must be wedges, and the house is missing a component to support the top piece that is needed in the arch. These outright rejections are in a sense easy cases; it can also happen that more than one model matches some scene description. To determine the best match in this case, a weighted sum of differences is made to express the amount of difference.

The next harder case is to match structures in a complex scene. The issue here is what to do about evidence that is missing through obscuration. Two heuristics help:

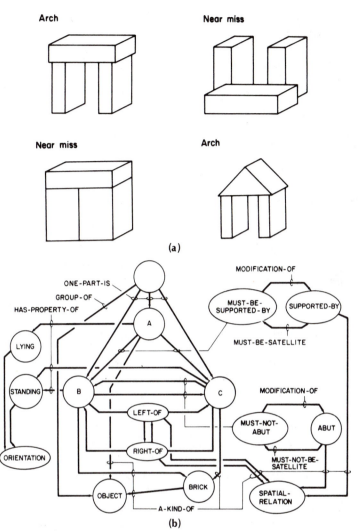

Fig. 11.10 (a) Several arches and non-arches. (b) The computer-generated arch description to be used for matching.

1. Objects that seem to have been stacked and could be the same type are of the same type.
2. Essential model properties may be hidden in the scene, so the match should not be aborted because of missing essential properties (though the presence of forbidden properties is enough to abort a match).

This latter rule is equivalent to Nevatia's rules about connectivity difference and missing input instance parts (see below). In terms of the general structure metric introduced earlier, neither Winston or Nevatia penalize the match for missing elements or relations in the reference data. One result of this is that the best match is sometimes missed in favor of the first possible match. Winston suggests that complex scenes be analyzed by identifying subscenes and subtracting out the identified parts, as was done by Roberts.

Winston's program can learn shortcuts in matching strategy by itself; it builds for itself a similarity network relating models whose differences are slight. If a reference model does not quite fit an input structure, the program can make an intelligent choice of the next model to try. A good choice is a model that has only minor differences with the first. This self-organization and cataloging of the models according to their mutual differences is a powerful way to use matching work that is already performed to guide further search for a good match.

Backtrack Search

Nevatia addresses a domain of complex articulated biological-like forms (hands, horses, dolls, snakes) [Nevatia 1974]. His strategy is to segment the objects into parts with central axes and "cross section" (not unlike generalized cylinders, except that they are largely treated in two dimensions). The derived descriptions of objects contain the connectivity of subparts, and descriptions of the shape and joint types of the parts. Matching is needed to compare descriptions and find differences, which can then be explained or used to abort the match. Partial matches are important (as in most real-world domains) because of occlusions, noise, and so on.

A priori ideas as to the relative importance of different aspects of structures are used to impose a hierarchical order on the matching decision tree. Nevatia finds this heuristic approach more appealing than a uniform, domain-independent one such as clique finding. His system knows that certain parts of a structure are more important than others, and uses them to index into the reference data catalog containing known structures. Thus relevant models for matching may be retrieved efficiently on the basis of easily-computed functions of the input data. The models are generated by the machine by the same process that later extracts descriptions of the image for recognition. Several different models are stored for the same view of the same object, because his program has no idea of model equivalence, and cannot always extract the same description.

The matching process is basically a depth-first tree search, with initial choices being constrained by "distinguished pieces." These are important pieces of image which first dictate the models to be tried in the match, and then constrain the possible other matches of other parts.

There is a topological and a geometrical component to the match. The topological part is based on the connectivity of the "stick figure" that underlies the representation. The geometrical part matches the more local characteristics of individual pieces. Consider Nevatia's example of matching a doll with stored reference descriptions, including those of a doll and a horse.

By a process not concerning us here, the doll image is segmented into pieces as shown in Fig. 11.11. From this, before any matching is done, a connection graph of pieces is formed, as shown in Fig. 11.12.

This connection graph is topologically the same as the reference connection graph for the doll, which looks as one would expect. In both reference and input, "distinguished pieces" are identified by their large size. During reference forma-

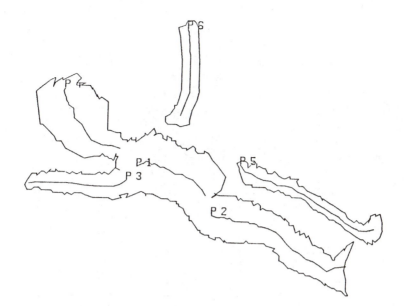

Fig. 11.11 A view of a doll, with derived structure.

tion time, the two largest pieces were the head and the trunk, and these are the distinguished pieces in the reference. There are the same pieces picked as distinguished in the instance to be matched consistent with the hierarchical decision-tree style, distinguished pieces are matched first.

Because of noise, connections at joints may be missed; because of the nature of the objects, extra joints are hardly ever produced. Thus there is a domain-dependent rule that an input piece with two other pieces connected at a single joint (a "two-ended piece") cannot match a one-ended reference piece, although the reverse is possible.

On the basis of the distinguished pieces in the input instance, the program decides that the instance could be a doll or a horse. Both these possibilities are evaluated carefully; Fig. 11.13 shows a schematic view of the process. Piece-match evaluation must be performed at the nodes of the tree to determine which pieces at a joint should be made to correspond.

The final best match between the doll input and the horse reference model is diagrammed in Fig. 11.14. This match is not as good as the match between the doll input and the doll reference.

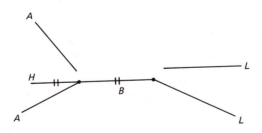

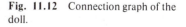

Fig. 11.12 Connection graph of the doll.

The final choice of matches is made with a version of the general relational structure matching metric (Eq. 11.3). It takes into account the connectivity relations, which are the same in this case, and the quality of the individual piece matches. In the doll-horse match, more reference parts are missing, but this can happen if parts are hidden in a view, and do not count against the match. The doll-doll match is preferred on the basis of piece matching, but both matches are considered possible.

In summary, the selection of best match proceeds roughly as follows: unacceptable differences are first sought (not unlike the Winston system). The number of input pieces not matched by the reference is an important number (not vice versa, because of the possibility of hidden input parts). Only elongated, large, parts

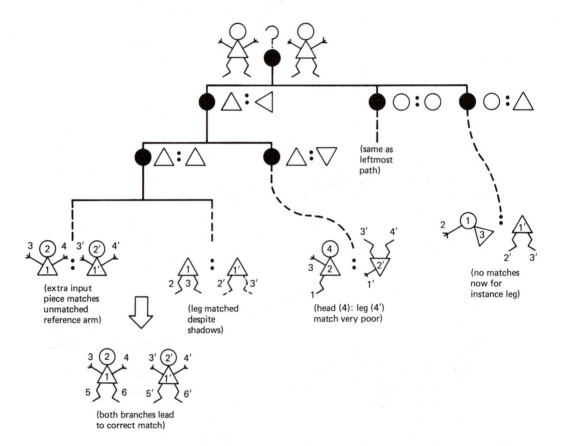

Fig. 11.13 A pictorial guide to the combinations tried by the matcher establishing the best correspondence of the doll input with the doll reference. The graphic shapes are purely pedagogical: the program deals with symbolic connectivity information and geometric measurements. Inferences and discoveries made by the program while matching are given in the diagram. A:B means that structure A is matched with structure B, with the numbered substructures of A matching their primed counterparts in B.

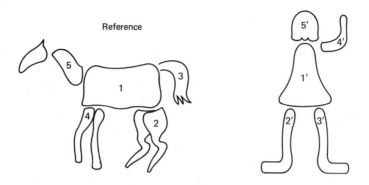

Fig. 11.14 The best match of the doll input with the horse reference model. One doll arm is unmatched, as is the horse head and two legs.

are considered for this determination, to eliminate small "noise" patches. The match with fewest unmatched input pieces is chosen.

If no deciding structural differences exist, the quality of piece matches determines the quality of the match. These correspond to the template cost term in Eq. (11.3). If a "significant" difference in match error exists, the better match is exclusively selected; if the difference is not so great as that, the better match is merely preferred.

Piece matching is a subprocess of joint matching. The difference in the number of pieces attached at the ends of the piece to be matched is the *connectivity difference*. If the object piece has more pieces connected to it than the model piece, the match is a poor one; since pieces may not be visible in a view, the converse is not true. If one match gives fewer excess input pieces, it is accepted at this point. If not, the goodness of the match is computed as a weighted sum of width difference, length-to-width ratio difference, and difference in acuteness of the generalized cylinders (Chapter 9) forming the pieces. The weighted sum is thresholded to yield a final "yes or no" match result. Shadowing phenomena are accommodated by allowing the input piece to be narrower than the reference model piece with no penalty. The error function weights are derived empirically; one would not expect the viewing angle to affect seriously the width of a piece, for example, but it could affect its length. Piece axis shapes (what sort of space curve they are) are not used for domain-dependent reasons, nor are cross section functions (aside from a measure of "acuteness" for cone-like generalized cylinders).

11.4.2 Decision Tree and Subgraph Isomorphism

A robotics program for versatile assembly [Ambler et al. 1975] uses matching to identify individual objects on the basis of their boundaries, and to match several individual blobs on a screen with a reference model containing the known location of multiple objects in the field of view. In both cases the best subgraph isomorphism between input and reference data structures is found when necessary by the clique-finding technique (Algorithm 11.2).

The input data to the part recognizer consist of silhouettes of parts with outlines of piecewise linear and circular segments. A typical set of shapes to be recognized might be stored in terms of boundary primitives as shown in Fig. 11.15a, with matchable and unmatchable scenes shown in Fig. 11.15b.

Generally, the matching process works on hierarchical structures which capture increasing levels of detail about the objects of interest. The matching works its way down the hierarchy, from high-level, easily computable properties such as size down to difficult properties such as the arrangements of linear segments in a part outline. After this decision tree pre-processing, all possible matches are computed by the clique-finding approach to subgraph isomorphism. A scene can be assigned a number of interpretations, including those of different views of the same part. The hierarchical organization means that complicated properties of the scene are not computed unless they are needed by the matcher. Once computed they are never recomputed, since they are stored in accessible places for later retrieval if needed. Each matching level produces multiple interpretations; ambiguity is treated with backtracking. The system recognizes rotational and translational invariance, but must be taught different views of the same object in its different gravitationally stable states. It treats these different states basically as different objects.

11.4.3 Informal Feature Classification

The domain of this work is one of small, curved tabletop objects, such as a teacup (Fig. 11.16) [Barrow and Popplestone 1971]. The primitives in models and image descriptions are regions which are found by a process irrelevant here. The regions have certain properties (such as shape or brightness), and they have certain parameterized relations with other regions (such as distance, adjacency, "aboveness"). The input and reference data are both relational structures. The properties and relations are the following:

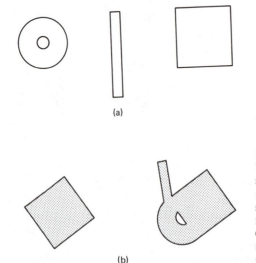

(a)

(b)

Fig. 11.15 A small catalog of part boundaries (a) and some sample silhouettes (b). The "heap" will not match any part very well, while the square can match the square model in four different ways, through rotations. Gross properties such as area may be used cheaply to reject matches such as the square with the axle.

Fig. 11.16 An object for recognition by relational matching.

1. *Region Properties*

   Shape 1–Shape 6: the first six root mean square amplitudes of the Fourier components of the $\phi(s)$ curve [Chapter 8].

2. *Relations between Regions A and B*

   Bigger: Area(A)/Area(B)

   Adjacency: Fraction of A's boundary which also is a boundary of B.

   Distance: Distance between centroids divided by the geometric mean of average radii. The average radius is twice the area over the perimeter. Distance is scale, rotation, translation, reflection invariant.

   Compactness: $4 * \pi * area / perimeter^2$

   Above, Beside: Vertical and horizontal distance between centroids, normalized by average radius. Not rotation invariant.

The model that might be derived for the cup of Fig. 11.16 is shown in Fig. 11.17.

   The program works on objects such as spectacles, pen, cup, or ball. During training, views and their identifications are given to the program, and the program forms a relational structure with information about the mean and variance of the values of the relations and properties. After training, the program is presented with a scene containing one of the learned objects. A relational structure is built describing the scene; the problem is then to match this input description with a reference description from the set of models.

   One approximation to the goodness of a match is the number of successes provided by a region correspondence. A one-region object description has 7 relations to check, a two-region object has 28, a three-region one has 63. Therefore, the "successes" criterion could imply the choice of a terrible three-region interpretation over a perfect one-region match. The solution adapted in the matching evaluation is first to grade failures. A failure weight is assigned to a trial match according to how many standard deviations σ from the model mean the relevant

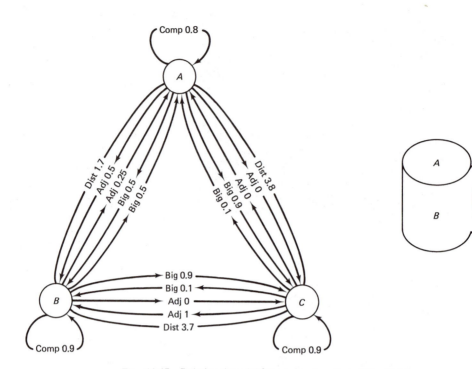

Fig. 11.17 Relational model for cups such as that of Fig. 11.16.

parameter is. From zero to three σ imply a success, or a failure weight of 0; from three to six σ, a failure weight of 1; from six to nine σ, failure weight of 2, and so on. Then the measure "trials–cumulative failure weight" is an improvement on just "successes." On the other hand, simple objects are often found as subparts of complex ones, and one does not want to reject a good interpretation as a complex object in favor of a less explanatory one as a simple object. The final evaluation function adapted is

$$\text{Cost of Match} = \frac{1 - \text{(tries-failure weight)}}{\text{number of relations}} \qquad (11.5)$$

$$+ \frac{K}{\text{number of regions in view description}}$$

As in Eq. (11.4), the first term measures the average badness of matches between properties (unary relations) and relations between regions. The second term is inversely proportional to the number of regions that are matched, effectively increasing the cost of matches that explain less of the input.

11.4.4 A Complex Matcher

A program to match linear structures like those of Fig. 11.18 is described in [Davis 1976]. This matcher presents quite a diversity of matching techniques incorporated into one domain-dependent program.

The matching metric is very close to the general metric of Eg. (11.3). The match is characterized by a structural match of reference and input elements and a geometrical transformation (found by parameter fitting) which accounts for the spatial relations between reference and input. Davis forms an association graph between reference and input data. This graph is reduced by parallel-iterative relaxation (see Section 12.4) using the "spring functions" to determine which node associations are too costly. Eliminating one node–node match may render others

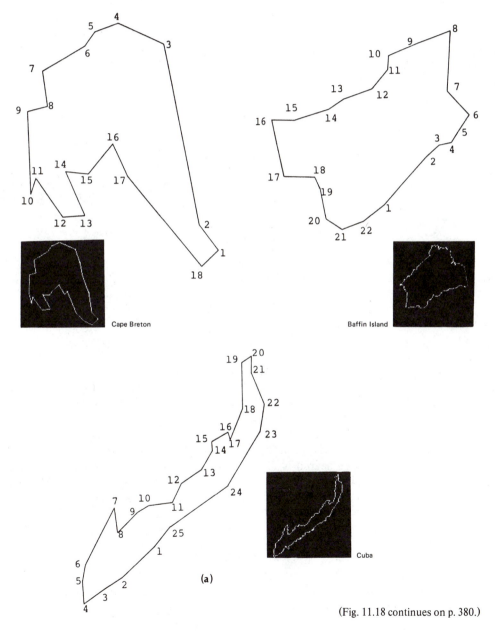

Cape Breton

Baffin Island

Cuba

(a)

(Fig. 11.18 continues on p. 380.)

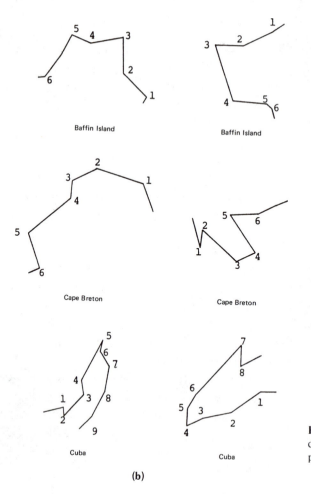

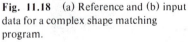

Baffin Island

Baffin Island

Cape Breton

Cape Breton

Cuba

Cuba

(b)

Fig. 11.18 (a) Reference and (b) input data for a complex shape matching program.

more unlikely, so the node-pruning process iterates until no more nodes are eliminated. What remains is something like an *r*-connected component of the graph, which specifies an approximate match supported by some amount of consistent relations between nodes.

After the process of constraint relaxation, there are still in general several locally consistent interpretations for each component of the input structure. Next, therefore, a tree search is used to establish global consistency and therefore the best match. The tree search is the familiar "best first" heuristic search through the partial match space, with pruning taking place between each stage of search again by using the parallel-iterative relaxation technique.

EXERCISES

11.1 Relational structures *A* and *B* are to be matched by the association-graph, clique-finding method.

Relational structure A: entities u, v, w, x, y, z.
relations $P(u)$, $P(w)$, $P(y)$, $R(v)$, $R(x)$, $R(z)$,
$F(u, v)$, $F(v, w)$, $F(w, x)$, $F(x, y)$, $F(y, z)$, $F(z, u)$

Relational structure B: entities a, b, c, d, e, f.
relations $P(a)$, $P(b)$, $P(d)$, $Q(e)$, $Q(f)$, $R(c)$
$F(b, c)$, $F(c, d)$, $F(d, e)$, $F(e, f)$, $F(f, a)$.

(a) Construct graph structures corresponding to the structures A and B. Label the nodes and arcs.

(b) Construct the association graph of structures A and B.

(c) Visually find the largest maximal cliques in the association graph and thus the best matches between A and B. (There are three.)

11.2 Suppose in a geometric match that two input points on the xy plane are identified with two others taken to correspond with two reference points. It is known that the input data comes about only through rotation and translation of the reference data. Given the two input points (x_1, y_1) and (x_2, y_2) and the two reference points (x'_1, y'_1) and (x'_2, y'_2), one way to find the transformation from reference to input is to solve the equation

$$\sum_{i=1}^{2} [x_i - (ax'_i + by'_i + c)]^2 + [y_i - (bx'_i + ay'_i + d)]^2 = 0$$

The resulting values of a, b, c, and d represent the desired transformation. Solve the equation analytically to get expressions for a, b, c, and d in terms of the reference and input coordinates. What happens if the reference and input data are not related by simple rotation and translation?

11.3 What are the advantages and disadvantages of a uniform method (such as subgraph isomorphism algorithm approach) to matching as compared to an ad hoc (such as a decision-tree approach with various empirically derived metrics) one?

11.4 In the worst case, for graphs of n nodes, how many partial solutions total will Algorithm 11.1 have to proceed through? Construct "worst case" graphs X and Y (label their nodes $1, \ldots, n$, of course), assuming that nodes of Y are selected in ascending order at any stage.

11.5 Find out something about the state of associative memories in computers. How do they work? How are they used? Would anything like this technology be useful for computer vision? Introspect about familiar phenomena of visual recall, recognition, and memory. Do you have a theory about how human visual memory could possibly work?

11.6 What graph of N nodes has the maximum number of maximal cliques? How many does it have?

11.7 Think about reasoning by analogy and find out something about programs that do analogical reasoning. In what sense can analogical process be used for computer vision, and technically do the matching techniques necessary provide any insight?

11.8 Compare Nevatia's structure matching with Hinton's relaxation-based puppet recognition (Chapter 12).

11.9 Verify the observation made in Section 11.4.3 about the number of relations that must be checked between regions (one region, 7; two regions, 28; three regions, 63; etc.).

REFERENCES

AHO, A. V., J. E. HOPCROFT and J. D. ULLMAN. *The Design and Analysis of Algorithms*. Reading, MA: Addison-Wesley, 1974.

AMBLER, A. P., H. G. BARROW, C. M. BROWN, R. M. BURSTALL, and R. J. POPPLESTONE. "A versatile computer-controlled assembly system." *Artificial Intelligence 6*, 2, 1975, 129–156.

BARROW, H. G. and R. J. POPPLESTONE. "Relational descriptions in picture processing." In *MI6*, 1971.

BARROW, H. G., J. M. TENENBAUM, R. C. BOLLES, and H. C. WOLF. "Parametric correspondence and chamfer matching: two new techniques for image matching." *Proc.*, DARPA IU Workshop, May 1978, 21-27.

BERGE, C. *Graphs and Hypergraphs* 2nd rev. ed.. New York: American Elsevier, 1976.

BERZTISS, A. T. "A backtrack procedure for isomorphism of directed graphs." *J. ACM 20*, 3, July 1973, 365-377.

BITTNER, J. R. and E. M. REINGOLD. "Backtrack programming techniques." *Comm. ACM 18*, 11, November 1975, 651-656.

BRON, C. and J. KERBOSCH. "Algorithm 457: finding all cliques in an undirected graph (H)." *Comm. ACM 16*, 9, September 1973, 575-577.

CORNEIL, D. G. and C. C. GOTLIEB. "An efficient algorithm for graph isomorphism." *J. ACM 17*, 1, January 1970, 51-64.

DAVIS, L. S. "Shape matching using relaxation techniques." *IEEE-PAMI 1*, 1, January 1979, 60-72.

FISCHLER, M. A. and R. A. ELSCHLAGER. "The representation and matching of pictorial structures." *IEEE Trans. Computers 22*, 1, January 1973, 67-92.

HARALICK, R. M. and G. L. ELLIOTT. "Increasing tree search efficiency for constraint satisfaction problems." *Proc.*, 6th IJCAI, August 1979, 356-364.

HARARY, F. *Graph Theory*. Reading, MA: Addison-Wesley, 1969.

KNODEL, W. "Bestimmung aller maximalen vollstandigen Teilgraphen eines Graphen G nach Stoffers." *Computing 3*, 3, 1968, 239–240 (and correction in *Computing 4*, p. 75).

NEVATIA, R. "Structured descriptions of complex curved objects for recognition and visual memory." AIM-250, Stanford AI Lab, October 1974.

NILLSON, N.J. *Principles of Artificial Intelligence*. Palo Alto, CA: Tioga, 1980.

OSTEEN, R. E. and J. T. TOU. "A clique-directed algorithm based on neighbourhoods in graphs." *International J. Computer and Information Science 2*, 4, December 1973, 257–268.

REINGOLD, E. M., J. NIEVERGELT, and N. DEO. *Combinatorial Algorithm Theory and Practice*. Englewood Cliffs,N. J.: Prentice-Hall, 1977.

SCHUDY, R. B. and D. H. BALLARD. "Model-directed detection of cardiac chambers in ultrasound images." TR12, Computer Science Dept., Univ. Rochester, November 1978.

SHAPIRO, L. G. and R. M. HARALICK. "Structural descriptions and inexact matching." Technical Report CS79011-R, Computer Science Dept., Virginia Polytechnic Institute, November 1979.

ULLMAN, J. R. "An algorithm for a subgraph isomorphism." *J. ACM 23*, 1, January 1976, 31-42.

WINSTON, P. H. "Learning structural descriptions from examples." In *PCV*, 1975.

Inference   12

Classical and Extended Inference

This chapter explores *inference,* the process of deducing facts from other known facts. Inference is useful for belief maintenance and is a cornerstone of rational thought. We start with *predicate logic*, and then explore *extended inference* systems—production systems, relaxation labeling, and active knowledge (procedures).

Predicate logic (Section 12.1) is a system for expressing propositions and for deriving consequences of facts. It has evolved over centuries, and many clear accounts describe predicate logic in its various forms [Mendelson 1964; Robinson 1965]. It has good formal properties, a nontrivial but automatable inference procedure, and a history of study in artificial intelligence. There are several "classical" extensions (modal logics, higher-order logics) which are studied in well-settled academic disciplines of metamathematics and philosophy. *Extended inference* (Section 12.2) is possible in automated systems, and is interesting technically and from an implementational standpoint.

A *production system* (Section 12.3) is a general rewriting system consisting of a set of *rewriting rules* ($A \rightarrow BC$ could mean "rewrite A as BC") and an executive program to apply rewrites. More generally, the rules can be considered "situation–action" pairs ("in situation A, do B and C"). Thus production systems can be used to control computational activities. Production systems, like semantic nets, embody powerful notions that can be used for extended inference.

Labeling schemes (Section 12.4) are unlike most inference mechanisms in that they often involve mathematical optimization in continuous spaces and can be implemented with parallel computation. Labeling is like inference because it establishes consistent "probability-like" values for "hypotheses" about the interpretation of entities.

Active knowledge (Section 12.5) is an implementation of inference in which each chunk of knowledge is a program. This technique goes far in the direction of "proceduralizing" the implementation of propositions. The design issues for such a system include the vocabulary of system primitives and their actions, mechanisms for implementing the flow of control, and overall control of the action of the system.

12.1 FIRST ORDER PREDICATE CALCULUS

Predicate logic is in many ways an attractive knowledge representation and inference system. However, despite its historical stature, important technical results in automated inference, and much research on inference techniques, logic has not dominated all aspects of mechanized inference. Some reasons for this are presented in Sections 12.1.6 and 12.2. The logical system that has received the most study is *first order predicate logic*. General theorem provers in this calculus are cumbersome for reasons which we shall explore. Furthermore, there is some controversy as to whether this logical system is adequate to express the reasoning processes used by human beings [Hayes 1977; Collins 1978; Winograd 1978; McCarthy and Hayes 1969]. We briefly describe some aspects of this controversy in Section 12.1.6. Our main purpose is to give the flavor of predicate calculus-based methods by describing briefly how automated inference can proceed with the formulae of predicate calculus expressed in the convenient *clause form*. Clause form is appealing for two reasons. First, it can be represented usefully in relational n-tuple or semantic network notation (Section 12.1.5). Second, the predicate calculus clause and inference system may be easily compared to production systems (Section 12.3).

12.1.1 Clause-Form Syntax (Informal)

In this section we describe the syntax of clause-form predicate calculus sentences. In the next, a more standard nonclausal syntax is described, together with a method for assigning meaning to grammatical logical expressions. Next, we show briefly how to convert from nonclausal to clausal syntax.

A *sentence* is a set of *clauses*. A clause is an ordered pair of sets of *atomic formulae*, or *atoms*. Clauses are written as two (possibly null) sets separated by an arrow, pointing from the *hypotheses* or *conditions* of the clause to its *conclusion*. The *null clause,* whose hypotheses and conclusion are both null, is written □. For example, a clause could appear as

$$A_1, \ldots, A_n \rightarrow B_1, \ldots, B_m$$

where the A's and B's are atoms. An atom is an expression

$$P(t_1, \ldots, t_j),$$

where P is a predicate symbol which "expects j arguments," each of which must be a *variable, constant symbol,* or a *term*. A term is an expression

$$f(t_1, \ldots, t_k)$$

where f is a *function symbol* which "expects k arguments," each of which may be a term. It is convenient to treat constant symbols alone as terms.

A careful (formal) treatment of the syntax of logic must deal with technical issues such as keeping constant and term symbols straight, associating the number of expected arguments with a predicate or function symbol, and assuring an infinite supply of symbols.

For example, the following are sentences of logic.

$\rightarrow$ Obscured(Backface(Block1))
Visible(Kidney) $\rightarrow$
Road(x), Unpaved(x) $\rightarrow$ Narrow(x)

12.1.2 Nonclausal Syntax and Logic Semantics (Informal)

Nonclausal Syntax

Clause form is a simplified but logically equivalent form of logic expressions which are perhaps more familiar. A brief review of non-clausal syntax follows.

The concepts of constant symbols, variables, terms, and atoms are still basic. A set of *logical connectives* provides unary and binary operators to combine atoms to form *well-formed formulae* (wffs). If A and B are atoms, then A is a wff, as is $\tilde{\ }A$ ("not A") $A \Longrightarrow B$ ("A implies B," or "if A then B"), $A \bigvee B$ ("A or B"), $A \bigwedge B$ ("A and B"), $A \Longleftrightarrow B$ ("A is equivalent to B," or "A if and only if B"). Thus an example of a wff is

Back(Face) $\bigvee$ (Obscured(Face)) $\Longrightarrow \tilde{\ }$ (Visible(Face))

The last concept is that of *universal* and *existential quantifiers*, the use of which is illustrated as follows.

$(\forall x)$ (wff using "x" as a variable).
$(\exists$ thing) (wff using "thing" as a variable).

A universal quantifier $\forall$ is interpreted as a conjunction over all domain elements, and an existential quantifier $\exists$ as a disjunction over all domain elements. Hence their usual interpretation as "for each element . . ." and "there exists an element . . . ."

Since a quantified wff is also a wff, quantifiers may be iterated and nested. A quantifier quantifies the "dummy" variable associated with it (x and thing in the examples above). The wff within the *scope* of a quantifier is said to have this quantified variable *bound* by the quantifier. Typically only wffs or clauses all of whose variables are bound are allowed.

Semantics

How does one assign meaning to grammatical clauses and formulae? The semantics of logic formulae (clauses and wffs alike) depends on an *interpretation* and on the meaning of connectives and quantifiers. An interpretation specifies the following.

1. A *domain* of individuals
2. A particular domain element is associated with each constant symbol
3. A function over the domain (mapping k individuals to individuals) is associated with each function symbol.
4. A relation over the domain (a set of ordered k-tuples of individuals) is associated with each predicate symbol.

The interpretation establishes a connection between the symbols in the representation and a domain of discourse (such as the entities one might see in an office or chest x-ray). To establish the truth or falsity of a clause or wff, a value of TRUE or FALSE must be assigned to each atom. This is done by checking in the world of the domain to see if the terms in the atom satisfy the relation specified by the predicate of the atom. If so, the atom is TRUE; if not, it is FALSE. (Of course, the terms, after evaluating their associated functions, ultimately specify individuals). For example, the atom

GreaterThan$(5,\pi)$

is true under the obvious interpretation and false with domain assignments such that

GreaterThan means "Is the author of"
5 means the book *Gone With the Wind*
π means Rin-Tin-Tin.

After determining the truth values of atoms, wffs with connectives are given truth values by using the *truth tables* of Table 12.1, which specify the semantics of the logical connectives. The relation of this formal semantics of connectives with the usual connectives used in language (especially "*implies*") is interesting, and one must be careful when translating natural language statements into predicate calculus.

The semantics of clause form expressions is now easy to explain. A sentence is the *conjunction* of its clauses. A clause

$$A_1, \ldots, A_n \rightarrow B_1, \ldots, B_m$$

with variables $x_1, \ldots, x_k$ is to be understood

Table 12.1

A	B	$\bar{A}$	$A \wedge B$	$A \vee B$	$A \Longrightarrow B$	$A \Longleftrightarrow B$
T	T	F	T	T	T	T
T	F	F	F	T	F	F
F	T	T	F	T	T	F
F	F	T	F	F	T	T

$$\forall\, x_1, \ldots, x_k,\ (A_1 \wedge \ldots \wedge A_n) \implies (B_1 \vee \ldots \vee B_m).$$

The null clause is to be understood as a contradiction. A clause with no conditions is an assertion that at least one of the conclusions is true. A clause with null conclusion is a denial that the conditions (hypotheses) are true.

12.1.3 Converting Nonclausal Form to Clauses

The conversion of nonclausal to clausal form is done by applying straightforward rewriting rules, based on logic identities (ultimately the truth tables). There is one trick necessary, however, to remove existential quantifiers. *Skolem functions* are used to replace existentially quantified variables, according to the following reasoning.

Consider the wff

$$(\forall\, x)\,((\exists\, y)\,(\text{Behind } (y, x))).$$

With the proper interpretation, this wff might correspond to saying "For any object x we consider, there is another object y which is behind x." Since the $\exists$ is within the scope of the $\forall$, the particular y might depend on the choice of x. The Skolem function trick is to remove the existential quantifier and use a function to make explicit the dependence on the bound universally quantified variable. The resulting wff could be

$$(\forall\, x)\ (\text{Behind}(\text{SomethingBehind}(x), x))$$

which might be rendered in English: "Any object x has another object behind it; furthermore, some Skolem function we choose to call SomethingBehind determines which object is behind its argument." This is a notational trick only; the existence of the new function is guaranteed by the existential quantification; both notations are equally vague as to the entity the function actually produces.

In general, one must replace each occurrence of an existentially quantified variable in a wff by a (newly created Skolem) function of all the universally quantified variables whose scope includes the existential quantifier being eliminated. If there is no universal quantifier, the result is a new function of no arguments, or a new constant.

$$(\exists\, x)\,(\text{Red}(x)),$$

which may be interpreted "Something is red," is rewritten as something like

$$\text{Red}(\text{RedThing})$$

or

"Something is red, and furthermore let's call it RedThing."

The conversion from nonclausal to clausal form proceeds as follows (for more details, see [Nilsson 1971]). Remove all implication signs with the identity $(A \implies B) \iff ((\tilde{\ } A) \vee B)$. Use DeMorgan's laws (such as $\tilde{\ }(A \vee B) \iff ((\tilde{\ } A) \wedge (\tilde{\ } B))$), and the extension to quantifiers, together with cancellation of double negations, to force negations to refer only to single predicate letters. Rewrite variables to give each quantifier its own unique dummy variable. Use Skolem functions to remove existential quantifiers. Variables are all now universally quantified, so eliminate the quantifier symbols (which remain implicitly), and rearrange the expression into conjunctive normal form (a conjunction of disjunctions.) The $\bigwedge$'s now connect disjunctive *clauses* (at last!). Eliminate the $\bigwedge$'s, obtaining from the original expression possibly several clauses.

At this point, the original expression has yielded multiple disjunctive clauses. Clauses in this form may be used directly in automatic theorem provers [Nilsson 1971]. The disjunctive clauses are not quite in the clause form as defined earlier, however; to get clauses into the final form, convert them into implications. Group negated atoms, reexpanding the scope of negation to include them all and converting the $\bigvee$ of $\tilde{\ }$'s into a $\tilde{\ }$ of $\bigwedge$'s. Reintroduce one implication to go from

$$B_1 \bigvee B_2 \ldots \bigvee B_m \bigvee (\tilde{\ }(A_1 \bigwedge A_2 \ldots \bigwedge A_n))$$

to

$$A_1 \bigwedge \ldots \bigwedge A_n \rightarrow B_1 \bigvee B_2 \ldots \bigvee B_m$$

To obtain the final form, replace the connectives (which remain implicitly) with commas.

12.1.4 Theorem Proving

Good accounts of the basic issues of automated theorem proving are given in [Nilsson 1971; Kowalski 1979; Loveland 1978]. The basic ideas are as follows. A sentence is *inconsistent*, or *unsatisfiable*, if it is false in every interpretation. Some trivially inconsistent sentences are those containing the null clause, or simple contradictions such as the same clause being both unconditionally asserted and denied. A sentence that is true in all interpretations is *valid*. Validity of individual clauses may be checked by applying the truth tables unless quantifiers are present, in which case an infinite number of formulae are being specified, and the truth status of such a clause is not algorithmically decidable. Thus it is said that first order predicate calculus is *undecidable*. More accurately, it is *semidecidable*, because any valid wff can be established as such in some (generally unpredictable) finite time. The validation procedure will run forever on invalid formulae; the rub is that one can never be sure whether it is running uselessly, or about to terminate in the next instant.

The notion of a *proof* is bound up with the notion of logical entailment. A clause C *logically follows* from a set of clauses S (we take S to *prove* C) if every interpretation that makes S true also makes C true. A formal proof is a sequence of inferences which establishes that C logically follows from S. In nonclausal predicate logic, inferences are rewritings of axioms and previously established formulae in accordance with *rules of inference* such as

Modus Ponens: From (A) and $(A \Longrightarrow B)$ infer (B)
Modus Tollens: From $(\tilde{\ }B)$ and $(A \Longrightarrow B)$ infer $(\tilde{\ }A)$
Substitution: e.g. From $(\forall x)(\text{Convex}(x))$ infer $(\text{Convex}(\text{Region31}))$
Syllogisms,

and so forth.

Automatic clausal theorem provers usually try to establish that a clause C logically follows from the set of clauses S. This is accomplished by showing the *unsatisfiability* of S and $(\neg C)$ taken together. This rather backward approach is a technical effect of the way that theorem provers usually work, which is to derive a contradiction.

The fundamental and surprising result that all true theorems are provable in finite time, and an algorithmic (but inefficient) way to find the proof, is due to Herbrand [Herbrand 1930]. The crux of the result is that although the domain of individuals who might participate in an interpretation may be infinite, only a finite number of interpretations need be investigated to establish unsatisfiability of a set of clauses, and in each only a finite number of individuals must be considered. A computationally efficient way to perform automatic inference was discovered by Robinson [Robinson 1965]. In it, a single rule of inference called *resolution* is used. This single rule preserves the *completeness* of the system (all true theorems are provable) and its *correctness* (no false theorems are provable).

The rule of resolution is very simple. Resolution involves matching a condition of one clause A with a conclusion of another clause B. The derived clause, called the *resolvent*, consists of the unmatched conditions and conclusions of A and B instantiated by the matching substitution. *Matching* two atoms amounts to finding a substitution of terms for variables which if applied to the atoms would make them identical.

Theorem proving now means resolving clauses with the hope of producing the empty clause, a contradiction.

As an example, a simple resolution proof goes as follows. Say it is desired to prove that a particular wastebasket is invisible. We know that the wastebasket is behind Brian's desk and that anything behind something else is invisible (we have a simpleminded view of the world in this little example). The givens are the wastebasket location and our naive belief about visibility:

$$\rightarrow \text{Behind}(\text{WasteBasket}, \text{DeskOf}(\text{Brian})) \qquad (12.1)$$
$$\text{Behind}(\text{object}, \text{obscurer}) \rightarrow \text{Invisible}(\text{object}) \qquad (12.2)$$

Here Behind and Invisible are predicates, DeskOf is a function, Brian and WasteBasket are constants (denote particular specific objects), and object and obscurer are (universally quantified) variables. The negation of the conclusion we wish to prove is

$$\text{Invisible}(\text{WasteBasket}) \rightarrow \qquad (12.3)$$

or, "Asserting the wastebasket is invisible is contradictory." Our task is to show this set of clauses is inconsistent, so that the invisibility of the wastebasket is proved. The resolution rule consists of matching clauses on opposite sides of the arrow which can be unified by a substitution of terms for variables. A substitution that works is:

Substitute WasteBasket for object and DeskOf(Brian) for obscurer in (12.2).

Then a cancellation can occur between the right side of (12.1) and the left side of (12.2). Another cancellation can then occur between the right side of (12.2) and the left side of (12.3), deriving the empty clause (a contradiction), Quod Erat Demonstrandum.

Anyone who has ever tried to do a nontrivial logic proof knows that there is searching involved in finding which inference to apply to make the proof terminate. Usually human beings have an idea of "what they are trying to prove," and can occasionally call upon some domain semantics to guide which inferences make sense. Notice that at no time in a resolution proof or other formal proof of logic is a specific interpretation singled out; the proof is about all possible interpretations. If deductions are made by appealing to intuitive, domain-dependent, semantic considerations (instead of purely syntactic rewritings), the deduction system is *informal*. Almost all of mathematics is informal by this definition, since normal proofs are not pure rewritings.

Many nonsemantic heuristics are also possible to guide search, such as trying to reduce the differences between the current formulae and the goal formula to be proved. People use such heuristics, as does the Logic Theorist, an early nonclausal, nonresolution theorem prover [Newell et al. 1963].

A basic resolution theorem prover *is* guaranteed to terminate with a proof if one exists, but usually resource limitations such as time or memory place an upper limit on the amount of effort one can afford to let the prover spend. As all the resolvents are added to the set of clauses from which further conclusions may be derived, the question of selecting which clauses to resolve becomes quite a vital one. Much research in automatic theorem proving has been devoted to reducing the search space of derivations for proofs [Nilsson 1980; Loveland 1970]. This has usually been done through heuristics based on formal aspects of the deductions (such as: make deductions that will not increase drastically the number of active clauses). Guidance from domain-dependent knowledge is not only hard to implement, it is directly against the spirit of resolution theorem proving, which attempts to do all the work with a uniform inference mechanism working on uninterpreted symbol strings. A moderation of this view allows the "intent" of a clause to guide its application in the proof. This can result in substantial savings of effort; an example is the treatment of "frame axioms" recommended by Kowalski (Section 13.1.4). Ad hoc, nonformalizable, domain-dependent methods are not usually welcome in automatic theorem-proving circles; however, such heuristics only guide the activity of a formal system; they do not render it informal.

12.1.5 Predicate Calculus and Semantic Networks

Predicate calculus theorem proving may be assisted by the addition of more relational structure to the set of clauses. The structure in a semantic net comes from *links* which connect *nodes*; nodes are accessed by following links, so the availability of information in nodes is determined by the link structure. Links can thus help by providing quick access to relevant information, given that one is "at" a particular node.

Although there are several ways of representing predicate calculus formulae in networks, we adopt here that of [Kowalski 1979; Deliyanni and Kowalski 1979]. The steps are simple:

1. Use a partition to represent the clause.

2. Convert all atoms to binary predicate atoms.

3. Distinguish between conditions and conclusions.

Recall that in Chapter 10, a partition is defined as a set of nodes and arcs in a graph. The internal structure of the partition cannot be determined from outside it. Partitioning extends the structure of a semantic net enough to allow unambiguous representations of all of first order predicate calculus.

The first step in developing the network representation for clauses is to convert each relation to a binary one. We distinguish between conditions and conclusions by using an additional bit of information for each arc. Diagrammatically, an arc is drawn with a double line if it is a condition and a single line if it is a conclusion. Thus the earlier example $S = \{(12.1), (12.2), (12.3)\}$ can be transformed into the network shown in Fig. 12.1.

This figure hints at the advantages of the network embedding for clauses: It is an indexing scheme. This scheme does not indicate which clauses to resolve next but can help reduce the possibilities enormously. If the most recent resolution involved a given clause with a given set of terms, other clauses which also have those terms will be represented by explicit arcs nearby in the network (this would *not* be true if the clauses were represented as a set). Similarly, other clauses involving the same predicate symbols are also nearby being indexed by those symbols. Again, this would not be true in the set representation. Thus the embedded network

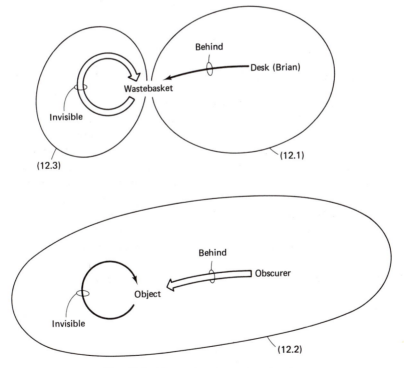

Fig. 12.1 Converting clauses to networks.

representation contains argument indices and predicate indices which can be extremely helpful in the inference process.

A very simple example illustrates the foregoing points. Suppose that S consists of the set of clauses

$$\text{SouthOf(river2},x), \text{NorthOf(river1},x) \rightarrow \text{Between(river1, river2, }x) \quad (12.4)$$
$$\rightarrow \text{SouthOf }(u, \text{silo30}) \quad (12.5)$$
$$\rightarrow \text{NorthOf (river1, silo30)} \quad (12.6)$$

Clause (12.5) might arise when it is determined that "silo30" is south of some feature in the image whose identity is not known. *Bottom up inference* derives new assertions from old ones. Thus in the example above the variable substitutions

$$u = \text{river2} \qquad x = \text{silo30}$$

match assertion (12.5) with the general clause (12.4) and allow the inference

$$\text{NorthOf(river1, silo30)}$$
$$\rightarrow \text{Between(river1, river2, silo30)} \quad (12.7)$$

Consequently, use (12.6) and (12.7) to assert

$$\rightarrow \text{Between(river1, river2, silo 30)} \quad (12.8)$$

Suppose that this was not the case: that is, that

$$\text{Between(river1, river2, silo30)} \rightarrow \quad (12.9)$$

and that $S = \{(12.4), (12.9)\}$. One could then use *top-down inference*, which infers new denials from old ones. In this case

$$\text{NorthOf(river1,silo30), SouthOf(river2,silo30)} \rightarrow \quad (12.10)$$

follows with the variable substitution $x = \text{silo30}$. This can be interpreted as follows: "If x is really silo30, then it is neither north of river1 or south of river2." Figure 12.2 shows two examples using the network notation.

Now suppose the goal is to prove that (12.8) logically follows from (12.4) through (12.6) and the substitutions. The strategy would be to negate (12.8), add it to the data base, and show that the empty clause can be derived. Negating an assertion produces a denial, in this case (12.9), and now the set of axioms (including the denial) consists of $\{(12.4), (12.5), (12.6), (12.9)\}$. It is easy to repeat the earlier steps to the point where the set of clauses includes (12.8) and (12.9), which resolve to produce the empty clause. Hence the theorem is proved.

12.1.6 Predicate Calculus And Knowledge Representation

Pure predicate calculus has strengths and weaknesses as a knowledge representation system. Some of the seeming weaknesses can be overcome by technical "tricks." Some are not inherent in the representation but are a property of the common interpreters used on it (i.e., on state-of-the-art theorem provers). Some problems are relatively basic, and the majority opinion seems to be that first order

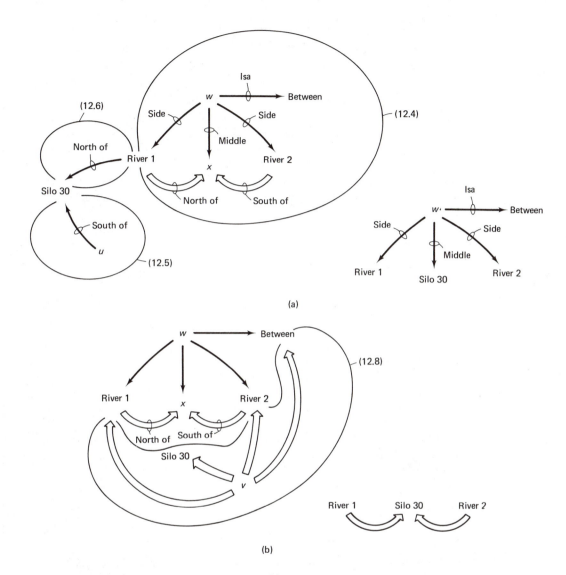

Fig. 12.2 Resolution using networks. (a) Bottom-up inference as a result of substitutions u = river2, x = silo30. (b) Top-down inference as a result of substitutions w = v, x = silo30.

predicate logic must be extended in order to become a representation scheme that is satisfactorily matched to the power of the deductive methods applied by human beings. Opinion is divided on the technical aspects of such enhancements. Predicate calculus has several strengths, some of which we list below.

1. Predicate logic is a well-polished gem, having been refined and studied for several generations. It was designed to represent knowledge and inference. One knows what it means. Its model theory and proof theory are explicit and lucid [Hayes 1977; 1980].

2. Predicate logic can be considered a language with a machine-independent semantics; the meaning of the language is determined by the laws of logic, not the actual programming system upon which the logic is "executed."

3. Predicate calculus clauses with only one conclusion atom (Horn clauses) may be considered as "procedures," with the single conclusion being the name of the procedure and the conditions being the procedure body, which itself is made up of procedure calls. This view of logic leads to the development of predicate logic-based programming languages (such as PROLOG [Warren et al. 1977; McDermott 1980]). These programs exhibit nondeterminism in several interesting ways; the order of computations is not specified by the proof procedure (and is not restricted by it, either). Several resolutions are in general possible for any clause; the combinations determine many computations and several distinguishable forms of nondeterminism [Kowalski 1974].

4. Predicate logic may be interpreted as a problem-reduction system. Then a (Horn) clause of the form

$$\rightarrow B$$

represents a solved problem. One of the form

$$A_1, \ldots, A_n \rightarrow$$

with variables $x_1, \ldots, x_k$ is a goal statement, or command, which is to find the x's that solve the problems $A_1, \ldots, A_n$. Finding the x's solves the goal. A clause

$$A_1, \ldots, A_n \rightarrow B$$

is a solution method, which reduces the solution of B to a combination of solutions of A's. This interpretation of Horn clauses maps cleanly into a standard and–or goal tree formulation of problem solving.

5. Resolutions may be performed on the left or right of clauses, and the resulting derivation trees correspond, in the problem-solving interpretation of predicate calculus, to top-down and bottom-up versions of problem solving. This duality is very important in conceptualizing aspects of problem solving.

6. There is a uniform proof procedure for logic which is guaranteed to prove in finite time any true theorem (logic is semidecidable and complete). No false theorems are provable (logic is correct). These and other good formal properties are important when establishing formally the properties of a knowledge representation system.

Predicate calculus is not a favorite of everyone, however; some of the (perceived) disadvantages are given below, together with ways they might be countered.

1. Sometimes the axioms necessary to implement relatively common concepts are not immediately obvious. A standard example is "equality." These largely technical problems are annoying but not basic.

2. The "first order" in first order predicate calculus means that the system

does not allow clauses with variables ranging over an infinite number of predicates, functions, assertions and sentences (e.g., "All unary functions are boring" cannot be stated directly). This problem may be ameliorated by a notational trick; the situations under which predicates are true are indicated with a Holds predicate. Thus instead of writing On(block1, surface, situation1), write Holds (On(block1,surface), situation1). This notation allows inferences about many situations with only one added axiom. The "situational calculus" reappears in Section 12.3.1. Another useful notational trick is a Diff relation, which holds between two terms if they are syntactically different. There are infinitely many axioms asserting that terms are different; the actual system can be made to incorporate them implicitly in a well-defined way. The Diff relation is also used in Section 12.3.1.

3. The *frame problem* (so called for historical reasons and not related to the frames described in Section 10.3.1) is a classic bugbear of problem-solving methods including predicate logic. One aspect of this problem is that for technical reasons, it must be explicitly stated in axioms that describe actions (in a general sense a visual test is an action) that almost all assertions were true in a world state *remain* true in the new world state after the action is performed. The addition of these new axioms causes a huge increase in the "bureaucratic overhead" necessary to maintain the state of the world. Currently, no really satisfactory way of handling this problem has been devised. The most common way to attack this aspect of the frame problem is to use explicit "add lists" and "delete lists" ([Fikes 1977], Chapter 13) which attempt to specify exactly what changes when an action occurs. New true assertions are added and those that are false after an action must be deleted. This device is useful, but examples demonstrating its inadequacy are readily constructed. More aspects of the frame problem are given in Chapter 13.

4. There are several sorts of reasoning performed by human beings that predicate logic does not pretend to address. It does not include the ability to describe its own formulae (a form of "quotation"), the notion of defaults, or a mechanism for plausible reasoning. Extensions to predicate logic, such as modal logic, are classically motivated. More recently, work on extensions addressing the topics above have begun to receive attention [McCarthy 1978; Reiter 1978; Hayes 1977]. There is still active debate as to whether such extensions can capture many important aspects of human reasoning and knowledge within the model-theoretic system. The contrary view is that in some reasoning, the very *process* of reasoning itself is an important part of the semantics of the representation. Examples of such extended inference systems appear in the remainder of this chapter, and the issues are addressed in more detail in the next section.

12.2 COMPUTER REASONING

Artificial intelligence in general and computer vision in particular must be concerned with *efficiency* and *plausibility* in inference [Winograd 1978]. Computer-based knowledge representations and their accompanying inference processes often sacrifice classical formal properties for gains in control of the inference process and for flexibility in the sorts of "truth" which may be inferred.

Automated inference systems usually have inference methods that achieve efficiency through implementational, computation-based, inference criteria. For example, truth may be defined as a successful lookup in a data base, falsity as the failure to find a proof with a given allocation of computational resources, and the establishment of truth may depend on the order in which deductions are made.

The semantics of computer knowledge representations is intimately related to the inference process that acts on them. Therefore, it is possible to define knowledge representations and interpreters in computers whose properties differ fairly radically from those of classical representations and proof procedures, such as the first-order predicate calculus. For instance, although the systems are deterministic, they may not be formally consistent (loosely, they may contain contradictory information). They may not be complete (they cannot derive all true theorems from the givens); it may be possible to prove P from Q but ^-P from Q and R. The set of provable theorems may not be recursively enumerable [Reiter 1978]. Efforts are being made to account for the "extended inference" needed by artificial intelligence using more or less classical logic [McCarthy 1978; Reiter 1978; Hayes 1977; 1978a; 1978b; Kowalski 1974, 1979]. In each case, the classical view of logic demands that the deductive process and the deducible truths be independent. On the other hand, it is reasonable to devote attention to developing a nonclassical semantics of these inference processes; this topic is in the research stage at this writing.

Several knowledge representations and inference methods using them are "classical" in the artificial intelligence world; that is, they provide paradigmatic methods of dealing with the issues of computational inference. They include STRIPS [Fikes and Nilsson 1971], the situational calculus [McCarthy and Hayes 1969], PLANNER and CONNIVER [Hewitt 1972; Sussman and McDermott 1972], and semantic net representations [Hendrix 1979; Brachman 1979].

To illustrate the issue of consistency, and to illustrate how various sorts of propositions can be represented in semantic nets, we address the question of how the order of inference can affect the set of provable theorems in a system.

Consider the semantic net of Fig. 12.3. The idea is that in the absence of specific information to the contrary, one should assume that railroad bridges are narrow. There are exceptions, however, such as Bridge02 (which has a highway bridge above the rail bridge, say). The network is clearly inconsistent, but trouble is avoided if inferences are made "from specific to general." Such ordering implies that the system is incomplete, but in this case incompleteness is an advantage.

Simple ordering constraints are possible only with simple inferential powers in the system [Winograd 1978]. Further, there is as yet little formal theory on the effects of ordering rules on computational inference, although this has been an active topic [Reiter 1978].

12.3 PRODUCTION SYSTEMS

The last section explored why the process of inference itself could be an important part of the semantics of a knowledge representation system. This idea is an impor-

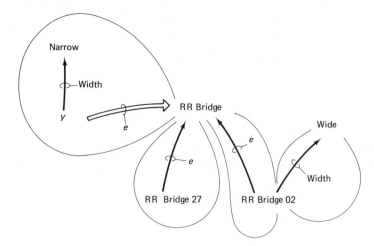

Fig. 12.3 An inconsistent network.

tant part of production systems. Perceived limitations in logic inference mechanisms and the seductive power of arbitrary algorithmic processes for inference has spawned the development of *rule-based* systems which differ from first-order logic in the following respects:

- Arbitrary additions and deletions to the clausal data base are allowed.
- An interpreter that controls the inference process in special ways is usually an integral part of the system.

Early examples of systems with the first addition are STRIPS [Fikes and Nilsson 1971] and PLANNER [Hewitt 1972]. Later examples of systems with both additions are given in [Waterman and Hayes-Roth 1978]. The virtues of trying to control inferences may be appreciated after our brief introduction to clausal automatic theorem proving, where there are no very good semantic heuristics to guide inferences. However, the price paid for restricting the inference process is the loss of formal properties of consistency and correctness of the system, which are not guaranteed in rule-based systems. We shall look in some detail at a particular form of rule-based inference system called production systems.

A *production system* supports a general sort of "inference." It has in common with resolution that matching is needed to identify which inference to make. It is different in that the action upon finding a matching data item is less constrained. Actions of arbitrary complexity are allowed. A production system consists of an explicit set of situation–action nodes, which can be applied against a data base of situations. For example, in a very constrained visual domain the rule

$$(\text{Green (Region } X)) \rightarrow (\text{Grass (Region } X)) \tag{12.11}$$

could infer directly the interpretation of a given region. Segmentation rules can also be developed; the following example merges two adjacent green regions into a single region.

$(\text{Green}(\text{Region } X))\wedge(\text{Green}(\text{Region } Y))\wedge$
$(\text{Adjacent}(\text{Region } X), (\text{Region } Y))$

$\rightarrow (\text{Green}(\text{Region } Z))\wedge((\text{Region } Z) :=$
$(\text{Union}(\text{Region } X, \text{Region } Y)))$

These examples highlight several points. The first is that basic idea of production systems is simple. The rules are easy to "read" by both the programmer and his program and new rules are easily added. Although it is imaginable that "situations" could extend down to the pixel level, and production systems could be used (for instance) to find lines, the system overhead would render such an approach impractical. In the visual domain, the production system usually operates on the segmented image (Chapters 4 and 5) or with the high-level internal model. In the rules above, X and Y are variables that must be bound to specific instances of regions in a data base. This process of binding variables or matching can become very complex, and is one of the two central issues of this kind of inference. The other is how to choose rules from a set all of whose situations match the current situation to some degree.

12.3.1 Production System Details

In its simplest form a production system has three basic components:

1. A data base
2. A set of rules
3. An interpreter for the rules

The vision data base is usually a set of facts that are known about the visual environment. Often the rules are considered to be themselves a manipulable part of the data base. Examples of some visual facts may be

$$(\text{ABOVE } (\text{Region } 5) \ (\text{Region } 10))$$

$$(\text{SIZE } (\text{Region } 5) \ 300)$$

$$(\text{SKY } (\text{Region } 5)) \qquad\qquad (12.12)$$

$$(\text{TOP } (\text{Region } 5) \ 255)$$

The data base is the sole storage medium for all state variables of the system. In particular, unlike procedurally oriented languages, there is no provision for separate storage of control state information—no separate program counter, pushdown stack, and so on [Davis and King 1975].

A rule is an ordered pair of *patterns* with a left-hand side and a right-hand side. A pattern may involve only data base primitives but usually will have variables and special forms as subpatterns which are matched against the data base by the interpreter. For example, applying the following rule to a data base which includes (12.12),

$$\text{(TOP (Region X) (GreaterThan 200))}$$

$$\rightarrow \qquad\qquad (12.13)$$

$$\text{(SKY (Region X))}$$

region 5 can be inferred to be sky. The left-hand side matches a set of data-base facts and this causes (SKY (Region 5)) to be added to the data base. This example shows the kinds of matching that the interpreter must do: (1) the primitive TOP in the data base fact matches the same symbol in the rule, (2) (Region X) matched (Region 5) and X is bound to 5 as a side effect, and (3) (GreaterThan 200) matches 255. Naturally, the user must design his own interpreter to recognize the meaning of such operational subpatterns.

However, even the form of the rules outlined so far is relatively restrictive. There is no reason why the right-hand side cannot do almost arbitrary things. For instance, the application of a rule may result in various productions being deleted or added from the set of productions; the data base of productions and assertions thus can be adaptive [Waterman and Hayes-Roth 1978]. Also, the right-hand side may specify programs to be run which can result in facts being asserted into the data base or actions performed.

Control in a basic production system is relatively simple: Rules are applied until some condition in the data base is reached. Rules may be applied in two distinct ways: (1) a match on the left-hand side of a rule may result in the addition of the consequences on the right-hand side to the data base, or (2) a match on the right-hand side may result in the addition of the antecedents in the left-hand side to the data base. The order of application of rules in the first case is termed *forward chaining* reasoning, where the objective is to see if a set of consequences can be derived from a given set of initial facts. The second case is known as *backward chaining*; the objective is to determine a set of facts that could have produced a particular consequence.

12.3.2 Pattern Matching

In the process of matching rules against the data base, several problems occur:

- Many rule situations may match data base facts
- Rules designed for a specific context may not be appropriate for larger context
- The pattern matching process may become very expensive
- The data base or set of rules may become unmanageably large.

The problem of multiple matches is important. Early systems simply resolved it by scanning the data base in a linear fashion and choosing the first match, but this is an ineffective strategy for large data bases, and has conceptual problems as well. Accordingly, strategies have evolved for dealing with these conflicts. Like most inference-controlling heuristics, their effectiveness can be domain-dependent, they can introduce incompleteness into the system, and so on.

On the principle of *least commitment*, when there are many chances of errors, one strategy is to apply the most general rule, defined by some metric on the components of the pattern. One simple such metric is the number of elements in a pattern. Antithetical to this strategy is the heuristic of applying the *most specific pattern*. This may be appropriate where the likelihood of making a false inference is small, and where specific actions may be indicated (match (MAD DOG) with (MAD DOG), not with (DOG)). Another popular but inelegant technique is to exercise control over the *order of production application* by using state markers which are inserted into the data base by right-hand sides and looked for by left-hand sides.

1. $A \rightarrow B \wedge <$marker 1$>$.
2. $A \rightarrow B \wedge <$marker 2$>$.
3. $B \wedge <$marker 1$> \rightarrow C$.
4. $B \wedge <$marker 2$> \rightarrow D$.

Here if rule 1 is executed, "control goes to rule 3," i.e., rule 3 is now executable, whereas if rule 2 is applied, "control goes to rule 4." Similarly, such control paradigms as subroutining, iteration and co-routining may be implemented with production sytems [Rychner 1978].

The use of connectives and special symbols can make matching become arbitrarily complex. Rules might be interpreted as allowing all partial matches in their antecedent clauses [Bajcsy and Joshi 1978]. Thus

$$(A \ B \ C) \rightarrow (D)$$

is interpreted as

$$(ABC) \bigvee (BC) \bigvee (AB) \bigvee (AC) \bigvee (A) \bigvee (B) \bigvee (C) \rightarrow (D)$$

where the leftmost actual match is used to compare the rule to others in the case of conflicts.

The problem of large data bases is usually overcome by structuring them in some way so that the interpreter applies the rules only to a subset of the data base or uses a subset of the rules. This structuring undermines a basic principle of pure rule-based systems: Control should be dependent on the contents of the data base alone. Nevertheless, many systems divide the data base into two parts: an active smaller part which functions like the original data base but is restricted in size, and a larger data base which is inaccessible to the rule set in the active smaller part. "Meta-rules" have actions that move situation-action rules and facts from the smaller data base to the larger one and vice versa. The incoming set of rules and facts is presumably that which is applicable in the context indicated by the situation triggering the meta-rule. This two-level organization of rules is used in "blackboard" systems, such as Hearsay for speech-understanding [Erman and Lesser 1975]. The meta-rules seem to capture our idea of "mental set," or "context," or "frame" (Section 10.3.1, [Minsky 1975]). The two data bases are sometimes referred to as short-term memory and long term memory, in analogy with certain models of human memory.

12.3.3 An Example

We shall follow the actions of a production system for vision [Sloan 1977; Sloan and Bajcsy 1979]. The intent here is to avoid a description of all the details (which may be found in the References) and concentrate on the performance of the system as reflected by a sample of its output. The program uses a production system architecture in the domain of outdoor scenes. The goal is to determine basic features of the scene, particularly the separation between sky and ground. The interpreter is termed the "observer" and the memory has a two-tiered structure: (1) short term memory (STM) and (2) long term memory (LTM), a data base of all facts ever known or established, structured to prefer access to the most recently used facts. The image to be analyzed is shown in Fig. 12.4, and the action may be followed in Fig. 12.5. The analysis starts with the initialization command

*(look 100000 100 nil)

This command directs the Observer to investigate all regions that fall in the size range 100 to 100000, in decreasing order of size. The LTM is initialized to NIL.

our first look at (region 11)

x	y	r-g	y-b	w-b	size	top	bottom	left	right
35	2	24	29	6	2132	35	97	2	127

This report is produced by an image-processing procedure that produces assertions about (region 11). This region is shown highlighted in Fig. 12.5c.

————— Progress Report —————

regions on this branch:
(11)
context stack:

Fig. 12.4 Outdoor scene to be analyzed with production system.

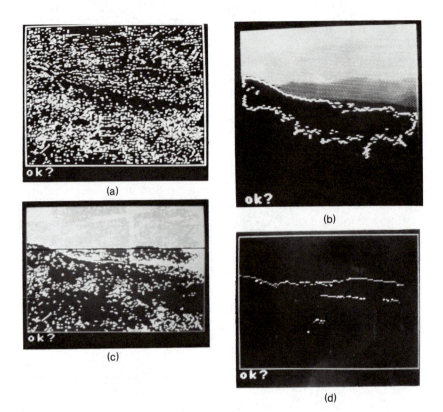

Fig. 12.5 Images corresponding to steps in production system analysis. (a) Texture in the scene. (b) Region 11 outlined. (c) Sky-Ground separation. (d) Skyline.

nil
contents of short term memory:
((far-left (region 11)) (far-right (region 11))
(right (region 11) 127) (left (region 11) 2)
(bottom (region 11) 97) (top (region 11) 35)
(w-b (region 11) minus) (y-b (region 11) zero)
(r-g (region 11) zero) (size (region 11) 2132))

_____ end of progress report _____

Note that gray-level information is represented as a vector in opponent color space (Chapter 2), where the components axes are WHITE-BLACK (w-b), RED-GREEN (r-g), and YELLOW-BLUE (y-b). Three values (plus, zero, minus) are used for each component. The display above is generated once after every iteration of the Observer. The report shows that (REGION 11) is being investigated; there is no known context for this investigation; the information about (REGION 11) created by the image-processing apparatus has been placed in STM. The context stack is for information only, and shows a trace of activated sets of rules.

i think that (far-left (region 11))
i think that (far-right (region 11))
i think that (right (region 11) 127)
i think that (left (region 11) 2)
i think that (bottom (region 11) 97)
i think that (top (region 11) 35)
i think that (size (region 11) 2132)

This portion of the trace shows assertions moving from STM to LTM. They are reported because this is the first time they have been REMEMBERed (a special procedure in the Observer).

_____ Progress Report _____

regions on this branch:
(11)
context stack:
nil
contents of short term memory:
((color (region 11) black))

_____ end of progress report _____

The assertions created from the region data structure have been digested, and lead only to the conclusion that (REGION 11) is BLACK, based on a production that looks like:

$(w\text{-}b$ (region x) minus) $\bigwedge$ $(r\text{-}w$ (region x) zero)
$\bigwedge$ $(b\text{-}w$ (region x) zero) $\rightarrow$ (color (region x) black)

_____ Progress Report _____

regions on this branch:
(11)
context stack:
nil
contents of short term memory:
((ground (region 11)) (shadow (region 11)))

_____ end of progress report _____

The observer knows that things that are black are GROUND and SHADOW. The facts it deduces about region 11 are again stored in the LTM.

Having discovered a piece of ground, the Observer has activated the GROUND-RULES, and changed context. It now investigates the neighbors of (REGION 11).

our first look at (region 16)

x	y	$r\text{-}g$	$y\text{-}b$	$w\text{-}b$	size	top	bottom	left	right
58	2	23	30	3	1833	57	119	2	127

(REGION 16) is a neighbor of (REGION 11), and the observer is trying to determine whether or not they are sufficiently similar, in both color and texture, to justify merging them.

————————— Progress Report —————————

regions on this branch:
(16 11)
context stack:
(ground)
contents of short term memory:
((texture-difference (region 16) (region 11)))
(color-similar (region 16) (region 11))
(distance (region 16) near) (ground (region 16))
(color (region 16 black))

————————— end of progress report —————————

The Observer decides that (REGION 16) is ground because it is at the bottom of the picture.

The ground-growing process continues, until finally one of the neighbors of a ground region is a piece of sky. The Observer will not immediately recognize this region as sky, but will see that a depth discontinuity exists and that the border between these two regions represents a section of three dimensional skyline.

our first look at (region 8)

x	y	r-g	y-b	w-b	size	top	bottom	left	right
27	2	13	13	33	394	15	38	2	57

————————— Progress Report —————————

regions on this branch:
(8 13 16 11)
context stack:
(ground ground ground)
contents of short term memory:
((new-neighbor (region 800) (far-left (region 8))
(right (region 8) 57) (left (region 8) 2) (bottom (region 8) 38)
(top (region 8) 15) (w-b (region 8) zero) (y-b (region 8) minus)
(r-g (region 8) minus) (size (region 8) 394))

————————— end of progress report —————————

texture descriptors for (region 8) are (54 50)
texture descriptors for (region 13) are (44 51)

Texture measurement is appropriate in the context of ground areas.

Ch. 12   Inference

regions on this branch:
(8 13 16 11)
context stack:
(ground ground ground)
contents of short term memory:
((texture-similar (region 8) (region 13)) (color-difference
(region 8) (region 13)) (color (region 8) blue-green))

_____ end of progress report _____

(REGION 8) passes the texture similarity test, but fails the color match.

_____ Progress Report _____

regions on this branch:
(8 13 16 11)
context stack:
(ground ground ground)
contents of short term memory:
((darker (region 13) (region 8)) (brighter (region 8) (region
13))
(yellower (region 13) (region 8)) (bluer (region 8) (region 13))
(redder (region 13) 13)
(below (region 13) (region 8)) (above (region 8) (region 13)))

_____ end of progress report _____

checking the border between (region 13) and (region 8)

_____ Progress Report _____

regions on this branch:
(8 13 16 11)
context stack:
(skyline ground ground ground)
contents of short term memory:
((segments built) (skyline-segment ((117 42))) (region 13)
(region 8)) (skyline-segment ((14 40) (13 40))) (region 13)
(region 8)))

_____ end of progress report _____

_____ Progress Report _____

regions on this branch:
(8 13 16 11)
context stack:
(skyline ground ground ground)

contents of short term memory:
((peak (14 40)) (peak (17 42)))

——————— end of progress report ———————

Two local maxima have been discovered in the skyline. On the basis of a depth judgment, these peaks are correctly identified as treetops.

The analysis continues until all the major regions have been analyzed. The sky-ground separation is shown in Fig. 12.5a and skyline in Fig. 12.5e.

In most cases, complete analysis of the image follows from the context established by the first (largest) region. This implies that initial scanning of such scenes can be quite coarse, and very simple ideas about gross context are enough to get started. Once started, inferences about local surroundings lead the Observer's attention over the entire scene, often returning many times to the same part of the image, each time with a bit more knowledge.

12.3.4 Production System Pros and Cons

In their pure form, the productions of production systems are completely "modular," and are themselves independent of the control process. The data base of facts, or situations, is unordered set accessed in undetermined order to find one matching some rule. The rule is applied, and the system reports the search for a matching situation and situation-action pair (rule). This completely unstructured organization of knowledge could be a model for the human learning of "facts" which become available for use by some associative mechanism that finds relevant facts in our memories. The hope for pure production systems is that performance will degrade noncatastrophically from the deletion of rules or facts, and that the rules can interact in synergistic and surprising ways. A learning curve may be simulated by the addition of productions. Thus one is encouraged to experiment with how knowledge may best be broken up into disjoint fragments that interact to produce intelligent behavior.

Together with the modularity of productions in a simple system, there is a corresponding simplicity in the overall control program. The pure controller simply looks at the data base and somehow finds a matching situation (left-hand side) among the productions, applies the rule, and cycles. This simple structure remains constant no matter how the rules change, so any nondeterminism in the performance arises from the matcher, which may find different left hand side matches for sets of assertions in the data base.

The productions usually have a syntax that is machine-readable. Their semantics is similarly constrained, and so it begins to seem hopeful that a program (perhaps fired up by a production) could reason about the rules themselves, add them, modify them, or delete them. This is in contrast to the situation with *procedurally embedded knowledge* (Section 10.1.3), because it is difficult or impossible for programs to answer general questions about other programs. Thus the claim is that a production system can more easily reason about itself than can many other knowledge representation systems.

Productions often interact in ways that are not foreseen. This can be an advantage or a drawback, depending on the behavior desired. The pattern-matching control structure allows knowledge to be used whenever it is relevant, not only when the original designer thought that it might be. Symbiotic interaction of knowledge may also produce unforeseen insights. Production systems are a primary tool of *knowledge engineering*, an enterprise that attempts to encode and use expert knowledge at such tasks as medical diagnosis and interpretation of mass spectrograms [Lindsay et al. 1980; Buchanan and Mitchell 1978; Buchanan and Feigenbaum 1978; Shortliffe 1976; Aikins 1980].

There are many who are not convinced that production systems really offer the advantages they initially seem to. They use the following sorts of arguments.

The pure form of production system is almost never seen doing anything useful. In particular, the production system is most naturally a forward-chaining inference system, and one must exercise restraints and guidelines on it to keep it from running away and deducing lots of irrelevant facts instead of doing useful work. Of course, production systems may be written to do backward chaining by hypothesizing a RHS and seeing which LHS must be true for the desired RHS to occur (the process may be iterated to any depth). In practical systems based on production systems, there is implicit or explicit ordering of production rules so the matcher tries them in some order. Often the ordering is determined in a rather complex and dynamic manner, with groups of related rules being more likely to be applied together, the most recently used rule not allowed to be reapplied immediately, and so on. In fact, many production systems's controllers have all the control structure tricks mentioned above (and more) built into them; the simple and elegant "bag of rules" ideal is inadequate for realistic examples. When the rules are explicitly written with an idiosyncratic control structure in mind, the system can become unprincipled and inexplicable.

On the same lines, notice how difficult it is to specify a time-ordered sequence of actions by a completely modular set of rewriting rules. It is unnatural to force knowledge about processes that may contain iteration, tests, and recursion into the form of independent situation–action rules. A view that is more easily defensible is that knowledge about procedures for perception should be encoded as (embedded in) computer procedures, not assertions or rules. The causal chain that dictates that some actions are best performed before others is implicit in the sequential execution of procedures, and the language constraints, such as iterate and test, test and branch, or subroutine invocation, are all fairly natural ways to think about solving certain problems. Production systems can in fact be made to perform all these procedural-like functions, but only through an abrogation of the ideal of modular, unordered, matching-oriented rule invocation which is the production system ideal. The question turns into one of aesthetics; how to use productions in a good style, and to work with their philosophy instead of against it.

To summarize the previous two objections: Production-based knowledge systems may in practice be no more robust, easily modified, modular, extensible, understandable, or self-understanding than any other (say, procedural) system unless great care is taken. After a certain level of complexity is reached, they are likely to be as opaque as any other scheme because of the control-structuring methods that must be imposed on the pure production system form.

12.4 SCENE LABELING AND CONSTRAINT RELAXATION

The general computational problem of assigning labels consistently to objects is sometimes called the "labeling problem," and arises in many contexts, such as graph and automata homomorphism, graph coloring, Latin square generation, and of course, image understanding [Davis and Rosenfeld 1976; Zucker 1976; Haralick and Shapiro 1979]. "Relaxation labeling," "constraint satisfaction," and "cooperative algorithms" are natural implementations for labeling, and their potential parallelism has been a very influential development in computer vision. As should any important development, the relaxation paradigm has had an impact on the conceptualization as well as on the implementation of processes.

Cooperating algorithms to solve the labeling problem are useful in low level vision (e.g., line finding, stereopsis) and in intermediate-level vision (e.g., line-labeling, semantics-based region growing). They may also be useful for the highest-level vision programs, those that maintain a consistent set of beliefs about the world to guide the vision process.

Section 12.4.1 presents the main concepts in the labeling problem. Section 12.4.2 outlines some basic forms that "discrete labeling" algorithms can take. Section 12.4.3 introduces a continuing example, that of labeling lines in a line drawing, and gives a mathematically well-behaved probabilistic "linear operator" labeling method. Section 12.4.4 modifies the linear operator to be more in accord with our intuitions, and Section 12.4.5 describes relaxation as linear programming and optimization, thereby gaining additional mathematical rigor.

12.4.1 Consistent and Optimal Labelings

All labeling problems have the following notions.

1. A set of *objects*. In vision, the objects usually correspond to entities to be labeled, or assigned a "meaning."

2. A finite set of *relations* between objects. These are the sorts of relations we saw in Chapter 10; in vision, they are often geometric or topological relations between segments in a segmented image. Properties of objects are simply unary relations. An input scene is thus a relational structure.

3. A finite set of *labels*, or symbols associated with the "meanings" mentioned above. In the simplest case, each object is to be assigned a single label. A *labeling* assigns one or more labels to (a subset of) the objects in a relational structure. Labels may be weighted with "probabilities"; a (label, weight) pair can indicate something like the "probability of an object having that label."

4. *Constraints*, which determine what labels may be assigned to an object and what sets of labels may be assigned to objects in a relational structure.

A basic labeling problem is then: Given a finite input scene (relational structure of objects), a set of labels, and a set of constraints, find a "consistent labeling." That is, assign labels to objects without violating the constraints. We saw this problem in Chapter 11, where it appeared as a matching problem. Here we shall start with the discrete labeling of Chapter 11 and proceed to more general labeling schemes.

As a simple example, consider the indoor scene of Fig. 12.6. The segmented office scene is to have its regions labeled as Door, Wall, Ceiling, Floor, and Bin, with the obvious interpretation of the labels. Here are some possible constraints, informally stated. Note that these particular constraints are in terms of the input relational structure, not the world from which the structure arose. A more complex (but reasonable) situation arises if scene constraints must be derived from rules about the three dimensional domain of the scene and the imaging process. Unary constraints use object properties to constrain labels; n-ary constraints force sets of label assignments to be compatible.

Unary constraints

1. The Ceiling is the single highest region in the image.

2. The Floor must be checkered.

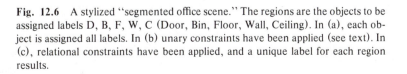

Fig. 12.6 A stylized "segmented office scene." The regions are the objects to be assigned labels D, B, F, W, C (Door, Bin, Floor, Wall, Ceiling). In (a), each object is assigned all labels. In (b) unary constraints have been applied (see text). In (c), relational constraints have been applied, and a unique label for each region results.

3. A Wall is adjacent to the Floor and Ceiling.

4. A Door is adjacent to the Floor and a Wall.

5. A Bin is adjacent to a Floor.

6. A Bin is smaller than a Door.

Obviously, there are many constraints on the appearance of segments in such a scene; which ones to use depends on the available sensors, the ease of computation of the relations and their power in constraining the labeling. Here the application of the constraints (Fig. 12.6) results in a unique labeling. Although the constraints of this example are purely for illustration, a system that actually performs such labeling on real office scenes is described in [Barrow and Tenenbaum 1976].

Labelings may be characterized as *inconsistent* or *consistent*. A weaker notion is that of an *optimal* labeling. Each of these adjectives reflects a formalizable property of the labeling of a relational structure and the set of constraints. If the constraints admit of only completely compatible or absolutely incompatible labels, then a labeling is consistent if and only if all its labels are mutually compatible, and inconsistent otherwise. One example is the line labels of Section 9.5; line drawings that could not be consistently labeled were declared "impossible." Such a black-and-white view of the scene interpretation problem is convenient and neat, but it is sometimes unrealistic. Recall that one of the problems with the line-labeling approach of Chapter 9 is that it does not cope gracefully with missing lines; strictly, missing lines often mean "impossible" line drawings. Such an uncompromising stance can be modified by introducing constraints that allow more degrees of compatibility than two (wholly compatible or strictly incompatible). Once this is done, both consistent and inconsistent labelings may be ranked on compatibility and likelihood. It is possible that a formally inconsistent labeling may rank better than a consistent but unlikely labeling.

Some examples are shown in Fig. 12.7. In 12.7b, the "inconsistent" labels are not nonsensical, but can only arise from (a very unlikely) accidental alignment of convex edges with three of the six vertices of a hexagonal hole in an occluding surface. The vertices that arise are not all included in the traditional catalog of legal vertices, hence the "inconsistent" labeling. The "floating cube" interpretation is consistent, but the "sitting cube" interpretation may be more likely if support and gravity are important concepts in the system. In Fig.12.7c, the scene with a missing line cannot be consistent according to the traditional vertex catalog, but the "inconsistent" labels shown are still the most likely ones. Labelings are only "consistent," "inconsistent," or "optimal" with respect to a given relational structure of objects (an input scene) and a set of constraints. These examples are meant to be illustrative only.

12.4.2 Discrete Labeling Algorithms

Let us consider the problem of finding a consistent set of labels, taken from a discrete finite set. This problem may be placed in an abstract algebraic context [Haralick and Kartus 1978; Haralick 1978; Haralick et al. 1978]. Perhaps the sim-

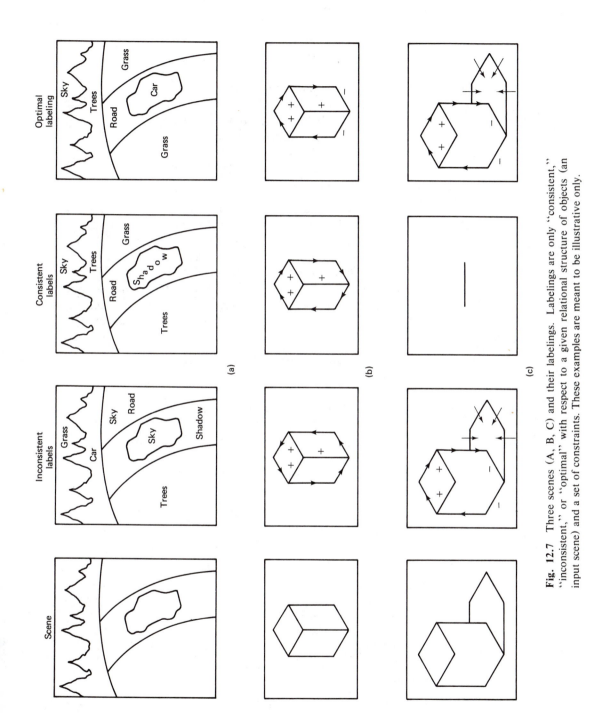

Fig. 12.7 Three scenes (A, B, C) and their labelings. Labelings are only "consistent," "inconsistent," or "optimal" with respect to a given relational structure of objects (an input scene) and a set of constraints. These examples are meant to be illustrative only.

plest way to find a consistent labeling of a relational structure (we shall often say "labeling of a scene") is to apply a depth-first *tree search* of the labeling possibilities, as in the backtracking algorithm (11.1).

Label an object in accordance with unary constraints.

Iterate until a globally consistent labeling is found:

> Given the current labeling, label another object
> consistently—in accordance with all constraints.
>
> If the object cannot be labeled consistently, backtrack
> and pick a new label for a previously labeled object.

This labeling algorithm can be computationally inefficient. First, it does not prune the search tree very effectively. Second, if it is used to generate all consistent labelings, it does not recognize important independences in the labels. That is, it does not notice that conclusions reached (labels assigned) in part of the tree search are usable in other parts without recomputation.

In a *serial relaxation*, the labels are changed one object at a time. After each such change, the new labeling is used to determine which object to process next. This technique has proved useful in some applications [Feldman and Yakimovsky 1974].

Assign all possible labels to each object in accordance with unary constraints.

Iterate until a globally consistent labeling is found:

> Somehow pick an object to be processed.
>
> Modify its labels to be consistent with the current
> labeling.

A *parallel iterative* algorithm adjusts all object labels at once; we have seen this approach in several places, notably in the "Waltz filtering algorithm" of Section 9.5.

Assign all possible labels to each object in accordance with unary constraints.

Iterate until a globally consistent labeling is found:

> In parallel, eliminate from each object's label set
> those labels that are inconsistent with the current
> labels of the rest of the relational structure.

A less structured version of relaxation occurs when the iteration is replaced with an *asynchronous interaction* of labeled objects. Such interaction may be implemented with multiple cooperating processes or in a data base with "demons" (Appendix 2). This method of relaxation was used in MSYS [Barrow and Tenenbaum 1976]. Here imagine that each object is an active process that knows its own label set and also knows about the constraints, so that it knows about its relations with other objects. The program of each object might look like this.

> If I have just been activated, and my label set is not
> consistent with the labels of other objects in the
> relational structure, then I change my label set to be
> consistent, else I suspend myself.

> Whenever I change my label set, I activate other objects
> whose label set may be affected, then I suspend myself.

To use such a set of active objects, one can give each one all possible labels consistent with the unary constraints, establish the constraints so that the objects know where and when to pass on activity, and activate all objects.

Constraints involving arbitrarily many objects (i.e., constraints of arbitrarily high *order*) can efficiently be relaxed by recording acceptable labelings in a graph structure [Freuder 1978]. Each object to be labeled initially corresponds to a node in the graph, which contains all legal labels according to unary constraints. Higher order constraints involving more and more nodes are incorporated successively as new nodes in the graph. At each step the new node constraint is *propagated*; that is, the graph is checked to see if it is consistent with the new constraint. With the introduction of more constraints, node pairings that were previously consistent may be found to be inconsistent. As an example consider the following graph coloring problem: color the graph in Fig. 12.8 so that neighboring nodes have different colors. It is solved by building constraints of increasingly higher order and propagating them. The node constraints are given explicitly as shown in Fig. 12.8a, but the higher-order constraints are given in functional implicit form; prospective colorings must be tested to see if they satisfy the constraints. After the node constraints are given, order two constraints are synthesized as follows: (1) make a node for each node pairing; (2) add all labelings that satisfy the constraint. The result is shown in Fig. 12.8b. The single constraint of order three is synthesized in the same way, but now the graph is inconsistent: the match " Y,Z : Red,Green" is ruled out by the third order legal label set (RGY,GRY). To restore consistency the constraint is propagated through node (Y,Z) by deleting the inconsistent labelings. This means that the node constraint for node Z is now inconsistent. To remedy this, the constraint is propagated again by deleting the inconsistency, in this case the labeling $(Z:G)$. The change is propagated to node (X,Z) by deleting $(X,Z$: Red,Green) and finally the network is consistent.

In this example constraint propagation did not occur until constraints of order three were considered. Normally, some constraint propagation occurs after every order greater than one. Of course it may be impossible to find a consistent graph. This is the case when the labels for node Z in our example are changed from (G, Y) to (G, R). Inconsistency is then discovered at order three.

It is quite possible that a discrete labeling algorithm will not yield a unique label for each object. In this case, a consistent labeling exists using each label for the

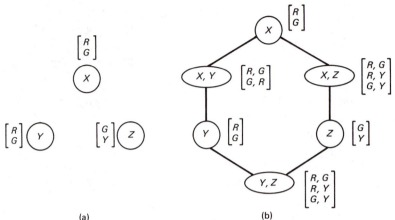

(a)   (b)

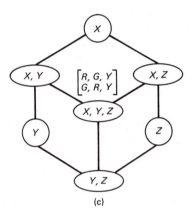

(c)

Fig. 12.8 Coloring a graph by building constraints of increasingly higher order.

object. However, which of an object's multiple labels goes with which of another object's multiple labels is not determined. The final enumeration of consistent labelings usually proceeds by tree search over the reduced set of possibilities remaining after the relaxation.

Convergence properties of relaxation algorithms are important; convergence means that in some finite time the labeling will "settle down" to a final value. In discrete labeling, constraints may often be written so that the label adjustment phase always reduces the number of labels for an object (inconsistent ones are eliminated). In this case the algorithm clearly must converge in finite time to a consistent labeling, since for each object the label set must either shrink or stay stable. In schemes where labels are added, or where labels have complex structure (such as real number "weights" or "probabilities"), convergence is often not guaranteed mathematically, though such schemes may still be quite useful. Some probabilistic labeling schemes (Section 12.4.3) have provably good convergence properties.

It is possible to use relaxation schemes without really considering their mathematical convergence properties, their semantics (What is the semantics of weights attached to labels—are they probabilities?), or a clear definition of what exactly the relaxation is to achieve (What is a good set of labels?). The fact that some schemes can be shown to have unpleasant properties (such as assigning nonzero weights to each of two inconsistent hypotheses, or not always converging to a solution), does not mean that they cannot be used. It only means that their behavior is not formally characterizable or possibly even predictable. As relaxation computations become more common, the less formalizable, less predictable, and less conceptually elegant forms of relaxation computations will be replaced by better behaved, more thoroughly understood schemes.

12.4.3 A Linear Relaxation Operator and a Line Labeling Example

The Formulation

We now move away from discrete labeling and into the realm of continuous *weights* or *supposition values* on labels. In Sections 12.4.3 and 12.4.4 we follow closely the development of [Rosenfeld et al. 1976]. Let us require that the sum of label weights for each object be constrained to sum to unity. Then the weights are reminiscent of probabilities, reflecting the "probability that the label is correct." When the labeling algorithm converges, a label emerges with a high weight if it occurs in a probable labeling of the scene. Weights, or supposition values, are in fact hard to interpret consistently as probabilities, but they are suggestive of likelihoods and often can be manipulated like them.

In what follows p refers to probability-like weights (supposition values) rather than to the value of a probability density function. Let a relational structure with n objects be given by a_i, $i = 1, ..., n$, each with m discrete labels $\lambda_1, ..., \lambda_m$. The shorthand $p_i(\lambda)$ denotes the weight, or (with the above caveats) the "probability" that the label λ (actually λ_k for some k) is correct for the object a_i. Then the probability axioms lead to the following constraints,

$$0 \leqslant p_i(\lambda) \leqslant 1 \tag{12.14}$$

$$\sum_\lambda p_i(\lambda) = 1 \tag{12.15}$$

The labeling process starts with an initial assignment of weights to all labels for all objects [consistent with Eqs. (12.14) and (12.15)]. The algorithm is parallel iterative: It transforms all weights at once into a new set conforming to Eqs. (12.14) and (12.15), and repeats this transformation until the weights converge to stable values.

Consider the transformation as the application of an operator to a vector of label weights. This operator is based on the *compatibilities* of labels, which serve as constraints in this labeling algorithm. A compatibility p_{ij} looks like a conditional probability.

$$\sum_\lambda p_{ij}(\lambda|\lambda') = 1 \qquad \text{for all } i, j, \lambda' \tag{12.16}$$

$$p_{ii} \, (\lambda \,|\, \lambda') = 1 \qquad iff \; \lambda = \lambda', \quad \text{else } 0. \qquad (12.17)$$

The $p_{ij} \, (\lambda \,|\, \lambda')$ may be interpreted as the conditional probability that object a_i has label λ given that another object a_j has label λ'. These compatibilities may be gathered from statistics over a domain, or may reflect a priori belief or information.

The operator iteratively adjusts label weights in accordance with other weights and the compatibilities. A new weight $p_i(\lambda)$ is computed from old weights and compatibilities as follows.

$$p_i(\lambda) := \sum_j c_{ij} \left\{ \sum_{\lambda'} p_{ij} \, (\lambda \,|\, \lambda') p_j(\lambda') \right\} \qquad (12.18)$$

The c_{ij} are coefficients such that

$$\sum_j c_{ij} = 1 \qquad (12.19)$$

In Eq. (12.18), the inner sum is the expectation that object a_i has label λ, given the evidence provided by object a_j. $p_i(\lambda)$ is thus a weighted sum of these expectations, and the c_{ij} are the weights for the sum.

To run the algorithm, simply pick the p_{ij} and c_{ij}, and apply Eq. (12.18) repeatedly to the p_i until they stop changing. Equation (12.18) is in the form of a matrix multiplication on the vector of weights, as shown below; the matrix elements are weighted compatibilities, the $c_{ij}p_{ij}$. The relaxation operator is thus a matrix; if it is partitioned into several *component* matrices, one for each set of non-interacting weights, linear algebra yields proofs of convergence properties [Rosenfeld et al. 1976]. The iteration for the reduced matrix for each component does converge, and converges to the weight vector that is the eigenvector of the matrix with eigenvalue unity. This final weight vector is independent of the initial assignments of label weights; we shall say more about this later.

An Example

Let us consider the input line drawing scene of Fig. 12.9a used in [Rosenfeld et.al. 1976]. The line labels given in Section 9.5 allow several consistent labels as shown in Fig. 12.9b-e, each with a different physical interpretation.

In the discrete labelling "filtering" algorithm presented in Section 9.5 and outlined in the preceding section, the relational structure is imposed by the neighbor relation between vertices induced by their sharing a line. Unary constraints are imposed through a catalog of legal combinations of line labels at vertices, and the binary constraint is that a line must not change its label between vertices. The algorithm eliminates inconsistent labels.

Let us try to label the sides of the triangle a_i, a_2, and a_3 in Fig. 12.9 with the solid object edge labels $\{>, <, +, -\}$. To do this requires some "conditional probabilities" for compatibilities $p_{ij}(\lambda \,|\, \lambda')$, so let us use those that arise if all eight interpretations of Fig. 12.9 are equally likely. Remembering that

$$p(X \,|\, Y) = \frac{p(X, Y)}{p(Y)} \qquad (12.20)$$

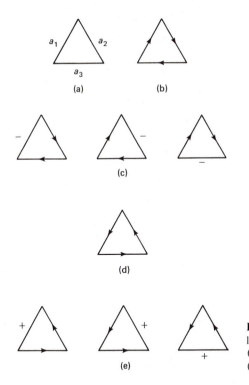

Fig. 12.9 A triangle and its possible labels. (a) Edge names. (b) Floating. (c) Flap folded up. (d) Triangular hole. (e) Flap folded down.

and taking $p(X, Y)$ to mean the probability that labels X and Y occur consecutively in clockwise order around the triangle, one can derive Table 12.2. Of course, we could choose other compatibilities based on any considerations whatever as long as Eqs. (12.16) and (12.17) are preserved.

Table 12.2 shows that there are two noninteracting components, $\{-,>\}$ and $\{+,<\}$. Consider the first component that consists of the weight vector

$$[p_1(>), \, p_1(-), \, p_2(>), \, p_2(-), \, p_3(>), \, p_3(-)] \qquad (12.21)$$

The second is treated similarly. This vector describes weights for the subpopulation of labelings given by Fig. 12.9b and c. The matrix M of compatibilities has columns of weighted p_{ij}.

$$M = \begin{bmatrix} c_{11}p_{11}(>|>) & c_{21}p_{21}(>|>) & \cdots \\ c_{11}p_{11}(>|-) & c_{21}p_{21}(>|-) & \cdots \\ c_{12}p_{12}(>|>) & c_{22}p_{22}(>|>) & \cdots \\ c_{12}p_{12}(>|-) & c_{22}p_{22}(>|-) & \cdots \\ c_{13}p_{13}(>|>) & c_{23}p_{23}(>|>) & \cdots \\ c_{13}p_{13}(>|-) & c_{23}p_{23}(>|-) & \cdots \end{bmatrix} \qquad (12.22)$$

Table 12.2

λ_1	λ_2	$p(\lambda_1, \lambda_2)$	$p(\lambda_1 \mid \lambda_2)$
>	>	$1/4$	$2/3$
>	−	$1/8$	1
−	>	$1/8$	$1/3$
−	−	0	0
>	<	0	0
>	+	0	0
−	<	0	0
−	+	0	0
<	>	0	0
+	>	0	0
<	−	0	0
+	−	0	0
<	<	$1/4$	$2/3$
<	+	$1/8$	1
+	<	$1/8$	$1/3$
+	+	0	0

If we let $c_{ij} = 1/3$ for all i, j, then

$$M = \frac{1}{3} \begin{bmatrix} 1 & 0 & 2/3 & 1/3 & 2/3 & 1/3 \\ 0 & 1 & 1 & 0 & 1 & 0 \\ 2/3 & 1/3 & 1 & 0 & 2/3 & 1/3 \\ 1 & 0 & 0 & 1 & 1 & 0 \\ 2/3 & 1/3 & 2/3 & 1/3 & 1 & 0 \\ 1 & 0 & 1 & 0 & 0 & 1 \end{bmatrix} \qquad (12.23)$$

An analytic eigenvector calculation (Appendix 1) shows that the M of Eq. (12.23) yields (for any initial weight vector) the final weight vector of

$$[3/4, 1/4, 3/4, 1/4, 3/4, 1/4] \qquad (12.24)$$

Thus each line of the population in the component we chose (Fig. 12.9b and c) has label > with "probability" $3/4$, −with "probability" $1/4$. In other words, from an initial assumption that all labelings in Fig. 12.9b and c were equally likely, the system of constraints has "relaxed" to the state where the "most likely" labeling is that of Fig. 12.9b, the floating triangle.

This relaxation method is a crisp mathematical technique, but it has some drawbacks. It has good convergence properties, but it converges to a solution entirely determined by the compatibilities, leaving no room for preferences or local scene evidence to be incorporated and affect the final weights. Further, the algorithm perhaps does not exactly mirror the following intuitions about how relaxation should work.

1. Increase $p_i(\lambda)$ if high probability labels for other objects are compatible with assignment of λ to a_i.

2. Decrease $p_i(\lambda)$ if high probability labels are incompatible with the assignment of λ to a_i.

3. Labels with low probability, compatible or incompatible, should have little influence on $p_i(\lambda)$.

However, the operator of this section decreases $p_i(\lambda)$ the most when other labels have both low compatibility and low probability. Thus it accords with (1) above, but not with (2) or (3). Some of these difficulties are addressed in the next section.

12.4.4 A Nonlinear Operator

The Formulation

If compatibilities are allowed to take on both positive and negative values, then we can express strong incompatibility better and obtain behavior more like (1), (2), and (3) just above. Denote the compatibility of the event "label λ on a_i" with the event "label λ on a_j" by $r_{ij}(\lambda, \lambda')$. If the two events occur together often, r_{ij} should be positive. If they occur together rarely, r_{ij} should be negative. If they are independent, r_{ij} should be 0. The *correlation coefficient* behaves like this, and the compatibilities of this section are based on correlations (hence the the notation r_{ij} for compatibilities). The correlation is defined using the covariance.

$$\text{cov}(X, Y) = p(X, Y) - p(X)p(Y)$$

Now define a quantity σ which is like the standard deviation

$$\sigma(X) = [p(X^2) - (p(X))^2]^{1/2} \qquad (12.25)$$

then the correlation is the normalized covariance

$$\text{cor}(X, Y) = \frac{\text{cov}(X, Y)}{\sigma(X)\sigma(Y)} \qquad (12.26)$$

This allows the formulation of an expression precisely analogous to Eq. (12.18), only that r_{ij} instead of p_{ij} is used to obtain a means of calculating the positive or negative change in weights.

$$q_i^{(k)}(\lambda) = \sum_j c_{ij} \left[\sum_{\lambda'} r_{ij}(\lambda, \lambda') p_j^{(k)}(\lambda') \right] \qquad (12.27)$$

In Eqs. (12.27)–(12.29) the superscripts indicate iteration numbers. The weight change (Eq. 12.27) could be applied as follows,

$$p_i^{(k+1)}(\lambda) = p_i^{(k)}(\lambda) + q_i^{(k)}(\lambda) \qquad (12.28)$$

but then the resultant label weights might not remain nonnegative. Fixing this in a straightforward way yields the iteration equation

$$p_i^{(k+1)}(\lambda) = \frac{p_i^{(k)}(\lambda)[1 + q_i^{(k)}(\lambda)]}{\sum_\lambda p_i^{(k)}(\lambda)[1 + q_i^{(k)}(\lambda)]} \qquad (12.29)$$

The convergence properties of this operator seem to be unknown, and like the linear operator it can assign nonzero weights to maximally incompatible labelings. However, its behavior can accord with intuition, as the following example shows.

An Example

Computing the covariances and correlations for the set of labels of Fig. 12.9b-e yields Table 12.3.

Figure 12.10 shows the nonlinear operator of Eq. (12.29) operating on the example of Fig. 12.9. Figure 12.10 shows several cases.

1. Equal initial weights: convergence to apriori probabilities $(\frac{3}{8}, \frac{3}{8}, \frac{1}{8}, \frac{1}{8})$.
2. Equal weights in the component $\{>, -\}$: convergence to "most probable" floating triangle labeling.
3. Slight bias toward a flap labeling is not enough to overcome convergence to the "most probable" labeling, as in (2).
4. Like (3), but greater bias elicits the "improbable" labeling.
5. Contradictory biases toward "improbable" labelings: convergence to "most probable" labeling instead.
6. Like (5), but stronger bias toward one "improbable" labeling elicits it.
7. Bias toward one of the components $\{>, -\}$, $\{<, +\}$ converges to most probable labeling in that component.
8. Like (7), only biased to less probable labelling in a component.

12.4.5 Relaxation as Linear Programming

The Idea

Linear programming (LP) provides some useful metaphors for thinking about relaxation computations, as well as actual algorithms and a rigorous basis [Hummel and Zucker 1980]. In this section we follow the development of [Hinton 1979].

Table 12.3

λ_1	λ_2	$\text{cov}(\lambda_1, \lambda_2)$	$\text{cor}(\lambda_1, \lambda_2)$
>	>	$\frac{7}{64}$	$\frac{7}{15}$
>	−	$\frac{5}{64}$	$5/\sqrt{105}$
−	>	$\frac{5}{64}$	$5/\sqrt{105}$
−	−	$-\frac{1}{64}$	$-\frac{1}{7}$
>	<	$-\frac{9}{64}$	$-\frac{3}{5}$
.	.	.	.
.	.	.	.
.	.	.	.

(a)

	>	<	-	+
a_1	$p_1(>)$	.	.	.
a_2	.	.	.	.
a_3	.	.	.	$p_3(+)$

(b)

Case	Initial weights				After 2 to 3 iterations				After 20 to 30 iterations				Limit			
	>	<	-	+	>	<	-	+	>	<	-	+	>	<	-	+
(1)	0.25	0.25	0.25	0.25	0.3	0.3	0.2	0.2	0.33	0.33	0.17	0.17	0.37	0.37	0.13	0.13
	0.25	0.25	0.25	0.25	0.3	0.3	0.2	0.2	0.33	0.33	0.17	0.17	0.37	0.37	0.13	0.13
	0.25	0.25	0.25	0.25	0.3	0.3	0.2	0.2	0.33	0.33	0.17	0.17	0.37	0.37	0.13	0.13
(2)	0.5	0	0.5	0	0.8	0	0.2	0	0.98	0	0.2	0	1	0	0	0
	0.5	0	0.5	0	0.8	0	0.2	0	0.98	0	0.2	0	1	0	0	0
	0.5	0	0.5	0	0.8	0	0.2	0	0.98	0	0.2	0	1	0	0	0
(3)	0.5	0	0.5	0	0.62	0	0.37	0	1	0	0	0	1	0	0	0
	0.4	0	0.6	0	0.49	0	0.51	0	0.97	0	0.03	0	1	0	0	0
	0.5	0	0.5	0	0.62	0	0.37	0	1	0	0	0	1	0	0	0
(4)	0.5	0	0.5	0	0.64	0	0.36	0	1	0	0	0	1	0	0	0
	0.3	0	0.7	0	0.36	0	0.64	0	0.07	0	0.93	0	0	0	1	0
	0.5	0	0.5	0	0.64	0	0.36	0	1	0	0	0	1	0	0	0
(5)	0.3	0	0.7	0	0.5	0	0.5	0	0.95	0	0.05	0	1	0	0	0
	0.3	0	0.7	0	0.5	0	0.5	0	0.95	0	0.05	0	1	0	0	0
	0.5	0	0.5	0	0.84	0	0.16	0	1	0	0	0	1	0	0	0
(6)	0.2	0	0.8	0	0.3	0	0.7	0	0.06	0	0.94	0	0	0	1	0
	0.3	0	0.7	0	0.51	0	0.49	0	1	0	0	0	1	0	0	0
	0.5	0	0.5	0	0.83	0	0.17	0	1	0	0	0	1	0	0	0
(7)	0.3	0.2	0.3	0.2	0.41	0.13	0.32	0.14	0.98	0	0.02	0	1	0	0	0
	0.3	0.2	0.3	0.2	0.41	0.13	0.32	0.14	0.98	0	0.02	0	1	0	0	0
	0.3	0.2	0.3	0.2	0.41	0.13	0.32	0.14	0.98	0	0.02	0	1	0	0	0
(8)	0.3	0.2	0.3	0.2	0.38	0.17	0.29	0.16	1	0	0	0	1	0	0	0
	0.25	0.25	0.25	0.25	0.35	0.20	0.25	0.20	1	0	0	0	1	0	0	0
	0.2	0.2	0.4	0.2	0.23	0.16	0.45	0.16	0.2	0	0.8	0	0	0	1	0

(c)

Fig. 12.10 The nonlinear operator produces labelings for the triangle in (a). (b) shows how the label weights are displayed, and (c) shows a number of cases (see text).

To put relaxation in terms of linear programming, we use the following translations.

- LABEL WEIGHT VECTORS $\Longrightarrow$ POINTS IN EUCLIDEAN N-SPACE. Each possible assignment of a label to an object is a *hypothesis*, to which a weight (supposition value) is to be attached. With N hypotheses, an N-vector of weights describes a labeling. We shall call this vector a (hypothesis or label) *weight vector*. For *m* labels and *n* objects, we need at most Euclidean *nm*-space.

- CONSTRAINTS $\Longrightarrow$ INEQUALITIES. Constraints are mapped into *linear inequalities* in hypothesis weights, by way of various identities like those of "fuzzy logic" [Zadeh 1965]. Each inequality determines an infinite half-space. The weight vectors within this half-space satisfy the constraint. Those outside do not. The convex solid that is the set intersection of all the half-spaces includes those weight vectors that satisfy all the constraints: each represents a "consistent" labeling. In linear programming terms, each such weight vector is a *feasible solution*. We thus have the usual geometric interpretation of the linear programming problem, which is to find the best (optimal) consistent (feasible) labeling (solution, or weight vector). Solutions should have *integer*-valued (1- or 0-valued) weights indicating convergence to actual labelings, not probabilistic ones such as those of Section 12.4.3, or the one shown in Fig. 12.10c, case 1.

- HYPOTHESIS PREFERENCES $\Longrightarrow$ PREFERENCE VECTOR. Often some hypotheses (label assignments) are preferred to others, on the basis of a priori knowledge, image evidence, and so on. To express this preference, make an *N*-dimensional *preference vector*, which expresses the relative importance (preference) of the hypotheses. Then

  - The *preference of a labeling* is the dot product of the preference vector and the weight vector (it is the sum for all hypotheses of the weight of each hypothesis times its preference).

  - The preference vector defines a *preference direction* in *N*-space. The optimal feasible solution is that one "farthest" in the preference direction. Let $\mathbf{x}$ and $\mathbf{y}$ be feasible solutions; they are *N*-dimensional weight vectors satisfying all constraints. If $\mathbf{z} = \mathbf{x} - \mathbf{y}$ has a component in the positive preference direction, then $\mathbf{x}$ is a better solution than $\mathbf{y}$, by the definition of the preference of a labeling.

It is helpful for our intuition to let the preference direction define a "downward" direction in N-space as gravity does in our three-space. Then we wish to pick the lowest (most preferred) feasible solution vector.

- LABELING $\Longrightarrow$ OPTIMAL SOLUTION. The relaxation algorithm must solve the linear programming problem—find the best consistent labeling. Under the conditions we have outlined, the best solution vector occurs generally at a vertex of the *N*-space solid. This is so because usually a vertex will be the "lowest" part of the convex solid in the preference direction. It is a rare coincidence that the solid "rests on a face or edge," but when it does a whole edge or face of the solid contains equally preferred solutions (the preference direction is normal to the edge or face). For integer solutions, the solid should be the convex hull of integer solutions and not have any vertices at noninteger supposition values.

The "simplex algorithm" is the best known solution method in linear programming. It proceeds from vertex to vertex, seeking the one that gives the optimal solution. The simplex algorithm is not suited to parallel computation, however, so here we describe another approach with the flavor of hill-climbing optimization. Basically, any such algorithm moves the weight vector around in N-space, iteratively adjusting weights. If they are adjusted one at a time, serial relaxation is taking place; if they are all adjusted at once, the relaxation is parallel iterative. The feasible solution solid and the preference vector define a "cost function" over all N-space, which acts like a potential function in physics. The algorithm tries to reach an optimum (minimum) value for this cost function. As with many optimization algorithms, we can think of the algorithm as trying to simulate (in N-space) a ball bearing (the weight vector) rolling along some path down to a point of minimum gravitational (cost) potential. Physics helps the ball bearing find the minimum; computer optimization techniques are sometimes less reliable.

Translating Constraints to Inequalities

The supposition values, or hypothesis weights, may be encoded into the interval $[0, 1]$, with 0 meaning "false," 1 meaning "true." The extension of weights to the whole interval is reminiscent of "fuzzy logic," in which truth values may be continuous over some range [Zadeh 1965]. As in Section 12.4.3, we denote supposition values by $p(\cdot)$; H, A, B, and C are label assignment events, which may be considered as hypotheses that the labels are correctly assigned. $\tilde{}$, $\vee$, $\wedge$, $\Longrightarrow$ and $\Longleftrightarrow$ are the usual logical connectives relating hypotheses. The connectives allow the expression of complex constraints. For instance, a constraint might be "Label x as 'y' if and only if z is labeled 'w' or q is labelled 'v'." This constraint relates three hypotheses: h_1: (x is "y"), h_2: (z is "w"), h_3: (q is "v"). The constraint is then $h_1 \Longleftrightarrow (h_2 \vee h_3)$.

Inequalities may be derived from constraints this way.

1. *Negation.* $p(H) = 1 - p(\tilde{}(H))$.

2. *Disjunction.* The sums of weights of the disjunct are greater than or equal to one. $p(A \vee B \vee \ldots \vee C)$ gives the inequality $p(A) + p(B) + \ldots + p(C) \geqslant 1$.

3. *Conjunction.* These are simply separate inequalities, one per conjunct. In particular, a conjunction of disjunctions may be dealt with conjunct by conjunct, producing one disjunctive inequality per conjunct.

4. *Arbitrary expressions.* These must be put into conjunctive normal form (Chapter 10) by rewriting all connectives as $\wedge$'s and $\vee$'s. Then (3) applies.

As an example, consider the simple case of two hypotheses A and B, with the single constraint that $A \Longrightarrow B$. Applying rules 1 through 4 results in the following five inequalities in $p(A)$ and $p(B)$; the first four assure weights in $[0, 1]$. The fifth arises from the logical constraint, since $A \Longrightarrow B$ is the same as $B \vee \tilde{}(A)$.

$$0 \leqslant p(A)$$
$$p(A) \leqslant 1$$
$$0 \leqslant p(B)$$
$$p(B) \leqslant 1$$
$$p(B) + (1 - p(A)) \geqslant 1 \quad \text{or} \quad p(B) \geqslant p(A)$$

These inequalities are shown in Fig. 12.11. As expected from the $\Longrightarrow$ constraint, optimal feasible solutions exist at: (1,1) or (A,B); (0,1) or $(\tilde{}(A),B)$; (0,0) or $(\tilde{}(A), \tilde{}(B))$. Which of these is preferred depends on the preference vector. If both its components are positive, (A,B) is preferred. If both are negative, $(\tilde{}(A), \tilde{}(B))$ is preferred, and so on.

A Solution Method

Here we describe (in prose) a search algorithm that can find the optimal feasible solution to the linear programming problem as described above. The description makes use of the mechanical analogy of an N-dimensional solid of feasible solutions, oriented in N-space so that the preference vector induces a "downward" direction in space. The algorithm attempts to move the vector of hypothesis weights to the point in space representing the feasible solution of maximum preference. It should be clear that this is a point on the surface of the solid, and unless the preference vector is normal to a face or edge of the solid, the point is a unique "lowest" vertex.

To establish a potential that leads to feasible solutions, one needs a measure of the infeasibility of a weight vector for each constraint. Define the amount a vector violates a constraint to be zero if it is on the feasible side of the constraint hyperplane. Otherwise the violation is the normal distance of the vector to the hyperplane. If $\mathbf{h}_i$ is the coefficient vector of the ith hyperplane (Appendix 1) and $\mathbf{w}$ the weight vector, this distance is

$$d_i = \mathbf{w} \cdot \mathbf{h}_i \tag{12.30}$$

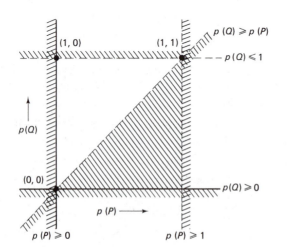

Fig. 12.11 The feasible region for two hypotheses A and B and the constraint A B. Optimal solutions may occur at the three vertices. The preferred vertex will be that one farthest in the direction of the preference vector, or lowest if the preference vector defines "down."

If we then define the infeasibility as

$$I = \sum_i \frac{d_i^2}{2} \tag{12.31}$$

then $\partial I/\partial d_i = d_i$ is the rate the infeasibility changes for changes in the violation. The force exerted by each constraint is proportional to the normal distance from the weight vector to the feasible region defined by that constraint, and tends to pull the weight vector onto the surface of the solid.

Now add a weak "gravity-like" force in the preference direction to make the weight vector drift to the optimal vertex. At this point an optimization program might perform as shown in Fig. 12.12.

Figure 12.12 illustrates a problem: The forces of preference and constraints will usually dictate a minimum potential outside the solid (in the preference direction). Fixes must be applied to force the weight vector back to the closest (presumably the optimum) vertex. One might round high weights to 1 and low ones to 0, or add another local force to draw vectors toward vertices.

Examples

An algorithm based on the principles outlined in the preceeding section was successfully used to label scenes of "puppets" such as Fig. 12.13 with body parts [Hinton 1979].

The discrete, consistency-oriented version of line labeling may be extended to incorporate the notion of optimal labelings. Such a system can cope with the explosive increase in consistent labelings that occurs if vertex labels are included for cases of missing lines, accidental alignment, or "two-dimensional" objects such as folded paper. It allows modeling of the fact that human beings do not "see" all possible interpretations of scenes with accidental alignments. If labelings are given

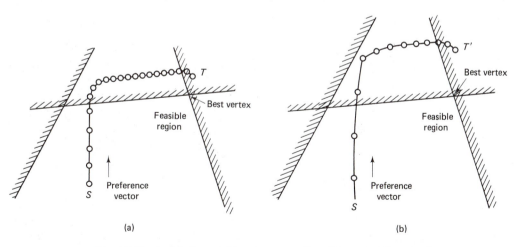

Fig. 12.12 In (a), the weight vector moves from S to rest at T, under the combined influence of the preferences and the violated constraints. In (b), convergence is speeded by making stronger preferences, but the equilibrium is farther away from the optimal vertex.

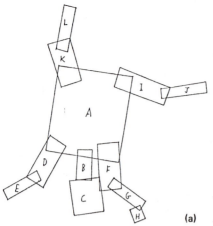

```
!.bestset;
A1  BOT TRUNK   NECK B1  UPPERARM D2 F2   THIGH I3 K2
B1  BOT NECK    HEAD C1   TRUNK A1
C1  BOT HEAD    NECK B1
D2  TOP UPPERARM  TRUNK A1   LOWERARM E4
E4  TOP LOWERARM  UPPERARM D2   HAND -
F2  TOP UPPERARM  TRUNK A1   LOWERARM G2
G2  TOP LOWERARM  UPPERARM F2   HAND H2
H2  TOP HAND   LOWERARM G2
I3  TOP THIGH  TRUNK A1   CALF J4
J4  BOT CALF   THIGH I3   FOOT -
K2  BOT THIGH  TRUNK A1   CALF L4
L4  BOT CALF   THIGH K2   FOOT -
```

(a)

```
!trytointerpret [trunk as upright importance=1];
!trytointerpret [thigh as upright importance=1];

!.bestset;
A2  TOP TRUNK   NECK -   UPPERARM I2 K1   THIGH D3 F3
B1  BOT NECK    HEAD C1   TRUNK -
C1  BOT HEAD    NECK B1
D3  TOP THIGH   TRUNK A2   CALF E3
E3  TOP CALF    THIGH D3   FOOT -
F3  TOP THIGH   TRUNK A2   CALF E3
G3  TOP CALF    THIGH F3   FOOT H1
H1  TOP FOOT    CALF G3
I2  TOP UPPERARM  TRUNK A2   LOWERARM J3
J3  BOT LOWERARM  UPPERARM I2   HAND -
K1  BOT UPPERARM  TRUNK A2   LOWERARM L3
L3  BOT LOWERARM  UPPERARM K1   HAND -
```

(b)

```
!.bestset;
A1  TOP HEAD    NECK B1
B1  TOP NECK    HEAD A1   TRUNK C2
C2  TOP TRUNK   NECK B1   UPPERARM H1 J1   THIGH D3 F3
D3  TOP THIGH   TRUNK C2   CALF E3
E3  TOP CALF    THIGH D3   FOOT -
F3  TOP THIGH   TRUNK C2   CALF G3
G3  TOP CALF    THIGH F3   FOOT-
H1  TOP UPPERARM  TRUNK C2   LOWERARM I1
I1  TOP LOWERARM  UPPERARM H1   HAND -
J1  TOP LOWERARM  TRUNK C2   LOWERARM K4
K4  BOT LOWERARM  UPPERARM J1   HAND L6
L6  BOT HAND   LOWERARM K4
```

(c)

Fig. 12.13 Puppet scenes interpreted by linear programming relaxation. (a) shows an upside down puppet. (b) is the same input along with preferences to interpret the trunk and thighs as upright; these result in an interpretation with trunk and neck not connected. In (c), the program finds only the "best" puppet, since it was only expecting one.

Ch. 12 Inference

costs, then one can include labels for missing lines and accidental alignment as high-cost labels, rendering them usable but undesirable. Also, in a scene-analysis system using real data, local evidence for edge appearance can enhance the a priori likelihood that a line should bear a particular label. If such preferences can be extracted along with the lines in a scene, the evidence can be used by the line labeling algorithm.

The inconsistency constraints for line labels may be formalized as follows. Each line and vertex has one label in a consistent labeling; thus for each line L and vertex J,

$$\sum_{\text{all line labels}} p\,(L \text{ has label LLABEL}) = 1 \qquad (12.32)$$

$$\sum_{\text{all vertex labels}} p\,(J \text{ has label VLABEL}) = 1 \qquad (12.33)$$

Of course, the VLABELS and LLABELS in the above constraints must be forced to be compatible (if L has LLABEL, JLABEL must agree with it). For a line L and a vertex J at its end,

$$p\,(L \text{ has LLABEL}) = \sum_{\substack{\text{all VLABELS} \\ \text{giving LLABEL to} L}} p\,(J \text{ has label VLABEL}) \qquad (12.34)$$

This constraint also enforces the coherence rule (a line may not change its label between vertices).

Using these constraints, linear programming relaxation labeled the triangle example of Fig. 12.7 as shown in Fig. 12.14, which shows three cases.

1. Preference 0.5 for each of the three junction label assignments (hypotheses) corresponding to the floating triangle, 0 preference for all other junction and line label hypotheses: converges to floating triangle.
2. Like (1), but with equal preferences given to the junction labels for the triangular hole interpretation, 0 to all other preferences.
3. Preference 3 to the convex edge label for a 2 overrides the three preferences of 1/2 for the floating triangle of case (1). All preferences but these four were 0.

Some Extensions

The translation of constraints to inequalities described above does not guarantee that they produce a set of half-spaces whose intersection is the convex hull of the feasible integer solutions. They can produce "noninteger optima," for which supposition values are not forced to 1 or 0. This is reminiscent of the behavior of the linear relaxation operator of Section 12.4.3, and may not be objectionable. If it is, some effort must be expended to cope with it. Here is an example

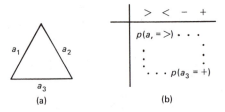

(a)   (b)

Case	After 10 iterations				After 20 iterations				After 30 to 40 iterations				
(1)	0.65	0.22	0.01	0.14	0.90	0.07	0	0.04	0.99	0	0	0	
	0.65	0.22	0.01	0.14	0.90	0.07	0	0.04	0.99	0	0	0	
	0.65	0.22	0.01	0.14	0.90	0.07	0	0.04	0.99	0	0	0	
(2)	0.39	0.89	0	0	0.14	0.95	0	0	0	0.99	0	0	
	0.39	0.89	0	0	0.14	0.95	0	0	0	0.99	0	0	
	0.39	0.89	0	0	0.14	0.95	0	0	0	0.99	0	0	
(3)	0.56	0.48	0	0.05	0.81	0.23	0	0	0.99	0	0	0	
	0	0.34	0	0.99	0	0.15	0	0.99	0	0	0	0.99	
	0.56	0.48	0	0.05	0.81	0.23	0	0	0.99	0	0	0	

(c)

Fig. 12.14 As in Fig. 12.10, the triangle of (a) is to be assigned labels, and the changing label weights are shown for three cases in (c) using the format of (b). Supposition values for junction labels were used as well, but are not shown. All initial supposition values were 0.

of the problem. Assume three logical constraints, $\bar{}(A \wedge B)$, $\bar{}(B \wedge C)$, and $\bar{}(C \wedge A)$. Suppose A, B, and C have equal preferences of unity (the preference vector is $(1, 1, 1)$). Translating the constraints yields

$$p(A) + p(B) \leqslant 1$$
$$p(B) + p(C) \leqslant 1 \tag{12.35}$$
$$p(C) + p(A) \leqslant 1$$

The best feasible solution has a total preference of $1\frac{1}{2}$, and is

$$p(A) = p(B) = p(C) = \frac{1}{2} \tag{12.36}$$

Here the "best" solution is outside the convex hull of the integer solutions (Fig. 12.15).

The basic way to ensure integer solutions is to use stronger constraints than those arising from the simple rules given above. These may be introduced at first, or when some noninteger optimum has been reached. These stronger constraints are called *cutting planes*, since they cut off the noninteger optima vertices. In the example above, the obvious stronger constraint is

$$p(A) + p(B) + p(C) \leqslant 1 \tag{12.37}$$

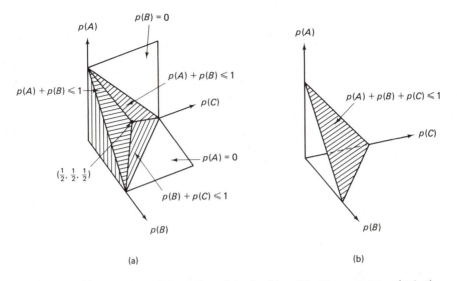

Fig. 12.15 (a) shows part of the surface of the feasible solid with constraints $\neg (A \& B)$, $\neg (B \& C)$, $\neg (C \& A)$, and the non-integer vertex where the three halfspaces intersect. (b) shows a cutting plane corresponding to the constraint "at most one of A, B, or C" that removes the non-integer vertex.

which says that at most one of A, B, and C is true (this is a logical consequence of the logical constraints). Such cutting planes can be derived as needed, and can be guaranteed to eliminate all noninteger optimal vertices in a finite number of cuts [Gomory 1968; Garfinkel and Nemhauser 1972]. Equality constraints may be introduced as two inequality constraints in the obvious way: This will constrain the feasible region to a plane.

Suppose that one desires "weak rules," which are usually true but which can be broken if evidence demands it? For each constraint arising from such a rule, add a hypothesis to represent the situation where the rule is broken. This hypothesis is given a negative preference depending on the strength of the rule, and the constraint enhanced to include the possibility of the broken rule. For example, if a weak rule gives the constraint $P \lor Q$, create a hypothesis H equivalent to $\tilde{\ }(P \lor Q) = (\tilde{\ }(P) \land \tilde{\ }(Q))$, and replace the constraint with $P \lor Q \lor H$. Then by "paying the cost" of the negative preference for H, we can have neither P nor Q true.

Hypotheses can be created as the algorithm proceeds by having demon-like "generator hypotheses." The demon watches the supposition value of the generator, and when it becomes high enough, runs a program that generates explicit hypotheses. This is clearly useful; it means that all possible hypotheses do not need to be generated in advance of any scene investigation. The generator can be given a preference equal to that of the best hypotheses that it can generate.

Relaxation sometimes should determine a real number (such as the slope of a line) instead of a truth value. A generator-like technique can allow the method to refine the value of real-valued hypotheses. Basically, the idea is to assign a (Boolean-valued) generator hypothesis to a range of values for the real value to be determined. When this generator triggers, more hypotheses are generated to get a finer partition of the range, and so on.

The enhancements to the linear programming paradigm of relaxation give some idea of the flexibility of the basic idea, but also reveal that the method is not at all cut-and-dried, and is still open to basic investigation. One of the questions about the method is exactly how to take advantage of parallel computation capabilities. Each constraint and hypothesis can be given its own processor, but how should they communicate? Also, there seems little reason to suppose that the optimization problems for this form of relaxation are any easier than they are for any other multidimensional search, so the method will encounter the usual problems inherent in such optimization. However, despite all these technical details and problems of implementation, the linear programming paradigm for the relaxation computation is a coherent formalization of the process. It provides a relatively "classical" context of results and taxonomy of problems [Hummel and Zucker 1980].

12.5 ACTIVE KNOWLEDGE

Active knowledge systems [Freuder 1975] are characterized by the use of procedures as the elementary units of knowledge (as opposed to propositions or data base items, for instance). We describe how active knowledge might work, because it is a logical extreme of the procedural implementation of propositions. In fact, this style of control has not proven influential; some reasons are given below.

Active knowledge is notionally parallel and heterarchical. Many different procedures can be active at the same time depending on the input. For this reason active knowledge is more easily applied to belief maintenance than to planning; it is very difficult to organize sequential activity within this discipline. Basically, each procedure is responsible for a "chunk" of knowledge, and knows how to manage it with respect to different visual inputs. Control in an active knowledge system is completely distributed. Active knowledge can also be viewed as an extension of the constraint relaxation problem; powerful procedures can make arbitrary detailed tests of the consistency between constraints.

Each piece of active knowledge (program module) knows which other modules it depends on, which depend on it, which it can complain to, and so forth. Thus the choice of "what to do next" is contained in the modules and is not made by an exterior executive.

We describe HYPER, a particular active knowledge system design which illustrates typical properties of active knowledge [Brown 1975]. HYPER provides a less structured mechanism for construction and exploration of hypotheses than does LP-relaxation. Using primitive control functions of the system, the user may write programs for establishing hypotheses and for using the conclusions so reached. The programs are "procedurally embedded" knowledge about a problem domain (e.g. how events relate one to another, what may be conjectured or inferred from a clue, or how one might verify a hypothesis).

When HYPER is in use on a particular task in a domain, hypotheses are created, or instantiated, on the basis of low-level input, high-level beliefs, or any reason in between. The process of establishing the initial hypotheses leads to a propagation of activity (creation, verification, and disconfirmation of hypotheses). Activation patterns will generally vary with the particular task, in heterarchical fashion. A priority mechanism can rank hypotheses in importance depending on the data that contribute to them. Generally, the actions that occur are conditioned by previous assumptions, the data, the success of methods, and other factors. HYPER can be used for planning applications and for multistep vision processing as well as inference (procedures then should generate parallel activity only under tight control). We shall thus allow HYPER to make use of a context-oriented data base (Section 13.1.1). It will use the context mechanism to implement "alternative worlds" in which to reason.

12.5.1 Hypotheses

A HYPER hypothesis is the attribution of a predicate to some arguments; its name is always of the form (PREDICATE ARGUMENTS). Sample hypothesis names could be (HEAD-SHAPED REGION1), (ABOVE A B), (TRIANGLE (X1,Y1) (X2,Y2) (X3,Y3)). A hypothesis is represented as a data structure with four components; the *status, contents, context,* and *links* of the hypothesis.

The *status* represents the state of the HYPER's knowledge of the truth of the hypothesis; it may be T(rue), F(alse), (in either case the hypothesis has been *established*) or P(ending). The *contents* are arbitrary; hypotheses are not just truth-valued assertions. The hypothesis was asserted in the data-base context given in *context.* The *links* of a hypothesis H are pointers to other hypotheses that have asked that H be established because they need H's contents to complete their own computations.

12.5.2 HOW-TO and SO-WHAT Processes

Two processes are associated with every predicate P which appears as the predicate of a hypothesis. Their names are (HOW-TO P) and (SO-WHAT P). In them is embedded the procedural knowledge of the system which remains compiled in from one particular task to another in a problem domain. (HOW-TO P) expresses how to establish the hypothesis (P arguments). It knows what other hypotheses must be established first, the computations needed to establish (P arguments), and so forth. It has a backward-chaining flavor. Similarly, (SO-WHAT P) expresses the consequences of knowing P: what hypotheses could possibly now be established using the contents of (P arguments), what alternative hypotheses should be explored if the status of (P arguments) is *F*, and so on. The feeling here is of forward chaining.

12.5.3 Control Primitives

HYPER hypotheses interact through *primitive control statements*, which affect the investigation of hypotheses and the ramification of their consequences. The primitives are used in HOW-TO and SO-WHAT programs together with other general computations. Most primitives have an argument called priority, which expresses the reliability, urgency, or importance of the action they produce, and is used to schedule processes in a nonparallel computing environment (implemented as a priority job queue [Appendix 2]). The primitives are GET, AFFIRM, DENY, RETRACT, FAIL, WONDERIF, and NUDGE.

GET is to ascertain or establish the status and contents of a hypothesis. It takes a hypothesis H and priority PRI as arguments and returns the status and contents of the hypothesis. If H's status is T or F at the time of execution of the statement, the status and contents are returned immediately. If the status is P (pending), or if H has not been created yet, the current HOW-TO or SO-WHAT program calling GET (call it CURPROG) is exited, the proper HOW-TO job (i.e., the one that deals with H's predicate) is run at priority PRI with argument H, and a link is planted in H back to CURPROG. When H is established, CURPROG will be reactivated through the link mechanism.

AFFIRM is to assert a hypothesis as true with some contents. AFFIRM(H,CONT,PRI) sets H's status to T, its contents to CONT, activates its linked programs and then executes the proper SO-WHAT program on it. The newly activated SO-WHAT programs are performed with priority PRI.

DENY is to assert that a hypothesis with some contents is false. DENY(H,CONT,PRI) is like AFFIRM except that no activation though links occurs, and the status of H is of course set to F.

ASSUME is to assert a hypothesis as true hypothetically. ASSUME(H,CONT,PRI) uses the data base context mechanism to create a new context in which H is AFFIRMED; the original context in which the ASSUME command is given is preserved in the context field of H. H itself is stored into a context-dependent item named LASTASSUMED; this corresponds to remembering a decision point in PLANNER. By using the information in LASTASSUMED and the primitive FAIL (see below), simple backtracking can take place in a tree of contexts.

RETRACT(H) establishes as false a hypothesis that was previously ASSUMEd. RETRACT is always carried out at highest priority, on the principle that it is good to leave the context of a mistaken assumption as quickly as possible. Information (including the name of the context being exited) is transmitted back to the original context in which H was ASSUMEd by passing it back in the fields of H.

FAIL just RETRACTs the hypothesis that is the value of the item LASTASSUMED in the present context.

WONDERIF is to pass suggested contents to HOW-TO processes for verification. It can be useful if verifying a value is easier than computing it from scratch, and is the primitive that passes substantive suggestions. WONDERIF(H1, CONT, H2, PRI) approximates the notion "H2 wonders if H1 has contents CONT."

NUDGE is to wake up HOW-TO programs. NUDGE(H,PRI) runs the HOW-TO program on H with priority PRI. It is used to awaken hypotheses that might be able to use information just computed. Typically it is a SO-WHAT program that NUDGEs others, since the SO-WHAT program is responsible for using the fact that a hypothesis is known.

12.5.4 Aspects of Active Knowledge

The active knowledge style of computation raises a number of questions or problems for its users.

A hypothesis whose contents may attain a large range can be established for some contents and thus express a perfectly good fact (e.g., that a given location of an x-ray does not contain evidence for a tumor) but such a fact is usually of little help when we want to reason about the predicate (about the location of tumors). The SO-WHAT program for a predicate should be written so as to draw conclusions from such negative facts if possible, and from the conclusions endeavor to establish the hypothesis as true for some contents. Usually, therefore, it would set the status of the hypothesis back to P and initiate a new line of attack, or at its discretion abandon the effort and start an entirely new line of reasoning.

Priorities

A major worry with the scheme as described is that priorities are used to schedule running of HOW-TO and SO-WHAT processes, not to express the importance (or supposition value) of the hypotheses. The hypothesis being investigated has no way to communicate how important it is to the program that operates on it, so it is impossible to accumulate importance through time. A very significant fact may lie ignored because it was given to a self-effacing process that had no way of knowing it had been handed something out of the ordinary.

The obvious answer is to make a supposition value a field of the hypothesis, like its status or contents—a hypothesis should be given a measure of its importance. This value may be used to compute execution priorities for jobs involving it. This solution is used in some successful systems [Turner 1974].

Structuring Knowledge

One has a wide choice in how to structure the "theory" of a complex problem in terms of HYPER primitives, predicates, arguments, and HOW-TO and SO-WHAT processes. The set of HOW-TO and SO-WHAT processes specify the complete theory of the tasks to be performed; HYPER encourages one to consider the interrelations between widely separated and distinct-sounding facts and conjectures about a problem, and the structure it imposes on a problem is minimal.

Since HOW-TO and SO-WHAT processes make explicit references to one another via the primitives, they are not "modular" in the sense that they can easily be plugged in and unplugged. If HOW-TO and SO-WHAT processes are invoked by patterns, instead of by names, some of the edge is taken off this criticism. Removing a primitive from a program could modify drastically the avenues of activation, and the consequences of such a modification are sometimes hard to foresee in a program that logically could be running in parallel.

Writing a large and effective program for one domain may not help to write a program for another domain. New problems of segmenting the theory into predicates, and quantifying their interactions via the primitives, setting up a priority structure, and so forth will occur in the new domain, and it seems quite likely that little more than basic utility programs will carry over between domains.

EXERCISES

12.1 In the production system example, write a production that specifies that blue regions are sky using the opponents color notation. How would you now deal with blue regions that are lakes (a) in the existing color-only system; (b) in a system which has surface orientation information?

12.2 This theorem was posed as a challenge for a clausal automatic theorem prover [Henschen et al. 1980]. It is obviously true: what problems does it present?

$$\{[(\exists x)(\forall y)(P(x) \iff P(y))]$$
$$\iff [[(\exists x)Q(x)\} \iff [(\forall y)(P(y))]]\} \iff$$
$$\{[(\exists x)(\forall y)(Q(x) \iff Q(y))]$$
$$\iff [[(\exists x)P(x)] \iff [(\forall y)(Q(y))]]\}$$

12.3 Prove that the operator of Eq. (12.18) takes probability vectors into probability vectors, thus deriving the reason for Eq. (12.19).

12.4 Verify (12.23).

12.5 How do the c_{ij} of (12.18) affect the labeling? What is their semantics?

12.6 If events X and Y always co-occur, then $p(X, Y) = p(X) = p(Y)$. What is the correlation in this case? If X and Y never co-occur, what values of $p(X)$ and $p(Y)$ produce a minimum correlation? If X and Y are independent, how is $p(X, Y)$ related to $p(X)$ and $p(Y)$? What is the value of the correlation of independent X and Y?

12.7 Complete Table 12.3.

12.8 Use only the labels of Fig. 12.9b and c to compute covariances in the manner of Table 12.3. What do you conclude?

12.9 Show that Eq. (12.29) preserves the important properties of the weight vectors.

12.10 Think of some rival normalization schemes to Eq. (12.29) and describe their properties.

12.11 Implement the linear and nonlinear operators of Section 12.4.3 and 12.4.4 and investigate their properties. Include your ideas from Exercise 12.10.

12.12 Show a case that the nonlinear operator of Eq. (12.29) assigns nonzero weights to maximally incompatible labels (those with $r_{ij} = -1$).

12.13 How can a linear programming relaxation such as the one outlined in sec. 12.4.5 cope with faces or edges of the feasible solution solid that are normal to the preference direction, yielding several solutions of equal preference?

12.14 In Fig. 12.11, what (P, Q) solution is optimal if the preference vector is $(1, 4)$? $(4, 1)$? $(-1, 1)$? $(1, -1)$?

REFERENCES

AIKINS, J. S. "Prototypes and production rules: a knowledge representation for computer consultations." Ph.D. dissertation, Computer Science Dept., Stanford Univ., 1980.

BAJCSY, R. and A. K. JOSHI. "A partially ordered world model and natural outdoor scenes." In *CVS*, 1978.

BARROW, H. G. and J. M. TENENBAUM. "MSYS: a system for reasoning about scenes." Technical Note 121, AI Center, SRI International, March 1976.

BRACHMAN, R. J. "On the epistemological status of semantic networks." In *Associative Networks: Representation and Use of Knowledge by Computers*, N. V. Findler (Ed.). New York: Academic Press, 1979, 3–50.

BROWN, C. M. "The HYPER system." DAI Working Paper 9, Dept. of Artificial Intelligence, Univ. Edinburgh, July 1975.

BUCHANAN, B. G. and E. A. FEIGENBAUM. "DENDRAL and meta-DENDRAL: their applications dimensions." *Artificial Intelligence 11*, 2, 1978, 5–24.

BUCHANAN, B. G. and T. M. MITCHELL. "Model-directed learning of production rules." In *Pattern Directed Inference Systems*, D. A. Waterman and F. Hayes-Roth (Eds.). New York: Academic Press, 1978.

COLLINS, A. "Fragments of a theory of human plausible reasoning." *Theoretical Issues in Natural Language Processing-2*, Univ. Illinois at Urbana-Champaign, July 1978, 194–201.

DAVIS, R. and J. KING. "An overview of production systems." AIM-271, Stanford AI Lab, October 1975.

DAVIS, L. S. and A. ROSENFELD. "Applications of relaxation labelling 2. Spring-loaded template matching." Technical Report 440, Computer Science Center, Univ. Maryland, 1976.

DELIYANNI, A. and R. A. KOWALSKI. "Logic and semanatic networks." *Comm. ACM 22*, 3, March 1979, 184–192.

ERMAN, L. D. and V. R. LESSER. "A multi-level organization for problem solving using many, diverse, cooperating sources of knowledge." *Proc.*, 4th IJCAI, September 1975, 483–490.

FELDMAN, J. A. and Y. YAKIMOVSKY. "Decision theory and artificial intelligence: I. A semantics-based region analyser." *Artificial Intelligence 5*, 4, 1974, 349–371.

FIKES, R. E. "Knowledge representation in automatic planning systems." In *Perspectives on Computer Science*, A. Jones (Ed). New York: Academic Press, 1977.

FIKES, R. E. and N. J. NILSSON. "STRIPS: a new approach to the application of theorem proving to problem solving." *Artificial Intelligence 2*, 3/4, 1971, 189–208.

FREUDER, E. C. "A computer system for visual recognition using active knowledge." Ph.D. dissertation, MIT, 1975.

FREUDER, E. C. "Synthesizing constraint expressions." *Comm. ACM 21*, 11, November 1978, 958–965.

GARFINKEL, R. S. and G. L. NEMHAUSER. *Integer Programming.* New York: Wiley, 1972.

GOMORY, R. E. "An algorithm for integer solutions to linear programs." *Bull. American Mathematical Society 64*, 1968, 275–278.

HARALICK, R. M. "The characterization of binary relation homomorphisms." *International J. General Systems 4*, 1978, 113–121.

HARALICK, R. M. and J. S. KARTUS. "Arrangements, homomorphisms, and discrete relaxation." *IEEE Trans. SMC 8*, 8, August 1978, 600–612.

HARALICK, R. M. and L. G. SHAPIRO. "The consistent labeling problem: Part I." *IEEE Trans. PAMI 1*, 2, April 1979, 173–184.

HARALICK, R. M., L. S. DAVIS, and A. ROSENFELD. "Reduction operations for constraint satisfaction." *Information Sciences 14*, 1978, 199–219.

HAYES, P. J. "In defense of logic." *Proc.*, 5th IJCAI, August 1977, 559–565.

HAYES, P. J. "Naive physics: ontology for liquids." Working paper, Institute for Semantic and Cognitive Studies, Geneva, 1978a.

HAYES, P. J. "The naive physics manifesto." Working paper, Institute for Semantic and Cognitive Studies, Geneva, 1978b.

HAYES, P. J. "The logic of frames." *The Frame Reader.* Berlin: DeGruyter, in press, 1981.

HENDRIX, G. G. "Encoding knowledge in partitioned networks." In *Associative Networks: Representation and Use of Knowledge by Computers*, N. V. Findler (Ed.). New York: Academic Press, 1979, 51–92.

HENSCHEN, L., E. LUSK, R. OVERBEEK, B. SMITH, R. VEROFF, S. WINKER, and L. WOS. "Challenge Problem 1." *SIGART Newsletter 72*, July 1980, 30–31.

HERBRAND, J. "Recherches sur la théorie de la démonstration." *Travaux de la Société des Sciences et des Lettres de Varsovie, Classe III, Sciences Mathématiques et Physiques, 33*, 1930.

HEWITT, C. "Description and theoretical analysis (using schemata) of PLANNER" (Ph.D. dissertation). AI-TR-258, AI Lab, MIT, 1972.

HINTON, G. E. "Relaxation and its role in vision." Ph.D. dissertation, Univ. Edinburgh, December 1979.

HUMMEL, R. A. and S. W. ZUCKER. "On the foundations of relaxation labelling processes." TR-80-7, Computer Vision and Graphics Lab, Dept. of Electrical Engineering, McGill Univ., July 1980.

KOWALSKI, R. A. "Predicate logic as a programming language." *Information Processing 74.* Amsterdam: North-Holland, 1974, 569–574.

KOWALSKI, R. A. *Logic for Problem Solving.* New York: ElsevierNorth-Holland (AI Series), 1979.

LINDSAY, R. K., B. G. BUCHANAN, E. A. FEIGENBAUM, and J. LEDERBERG. *Applications of Artificial Intelligence to Chemistry: The DENDRAL Project.* New York: McGraw-Hill, 1980.

LOVELAND, D. "A linear format for resolution." *Proc.*, IRIA 1968 Symp. on Automatic Demonstration, Versailles, France. New York: Springer-Verlag, 1970.

LOVELAND, D. *Automated Theorem Proving: A Logical Basis.* Amsterdam: North-Holland, 1978.

McCARTHY, J. "Circumscription induction—a way of jumping to conclusions." Unpublished report, Stanford AI Lab, 1978.

McCARTHY, J. and P. J. HAYES. "Some philosophical problems from the standpoint of artificial intelligence." In *MI4*, 1969.

McDERMOTT, D. "The PROLOG phenomenon." *SIGART Newsletter 72*, July 1980, 16–20.

MENDELSON, E. *Introduction to Mathematical Logic.* Princeton, NJ: D. Van Nostrand, 1964.

MINSKY, M. L. "A framework for representing knowledge." In *PCV*, 1975.

NEWELL, A., J. SHAW, and H. SIMON. "Empirical explorations of the logic theory machine." In *Computers and Thought*, E. Feigenbaum and J. Feldman (Eds.). New York: McGraw-Hill, 1963.

NILSSON, N. J. *Problem-Solving Methods in Artificial Intelligence.* New York: McGraw-Hill, 1971.

NILSSON, N. J. *Principles of Artificial Intelligence.* Palo Alto, CA: Tioga, 1980.

REITER, R. "On reasoning by default." *Theoretical Issues in Natural Language Processing-2*, Univ. Illinois at Urbana-Champaign, July 1978, 210–218.

ROBINSON, J. A. "A machine-oriented logic based on the resolution principle." *J. ACM 12*, 1, January 1965, 23–41.

ROSENFELD, A., R. A. HUMMEL and S. W. ZUCKER. "Scene labelling by relaxation operations." *IEEE Trans. SMC 6*, 1976, 420.

RYCHNER, M. "An instructable production system: basic design issues." In *Pattern Directed Inference Systems*, D. A. Waterman and F. Hayes-Roth (Eds.). New York: Academic Press, 1978.

SHORTLIFFE, E. H. *Computer-Based Medical Consultations: MYCIN.* New York: American Elsevier, 1976.

SLOAN, K. R. "World model driven recognition of natural scenes." Ph.D. dissertation, Moore School of Electrical Engineering, Univ. Pennsylvania, June 1977.

SLOAN, K. R. and R. BAJCSY. "World model driven recognition of outdoor scenes." TR40, Computer Science Dept., Univ. Rochester, September 1979.

SUSSMAN, G. J. and D. McDERMOTT. "Why conniving is better than planning." AI Memo 255, AI Lab, MIT, 1972.

TURNER, K. J. "Computer perception of curved objects using a television camera." Ph.D. dissertation, School of Artificial Intelligence, Univ. Edinburgh, 1974.

WARREN, H. D., L. PEREIRA, and F. PEREIRA. "PROLOG: The language and its implementation compared with LISP." *Proc.*, Symp. on Artificial Intelligence and Programming Languages, SIGPLAN/SIGART, 1977; *SIGPLAN Notices 12*, 8, August 1977, 109–115.

WATERMAN, D. A. and F. HAYES-ROTH (Eds.). *Pattern-Directed Inference Systems.* New York: Academic Press, 1978.

WINOGRAD, T. "Extended inference modes in reasoning by computer systems." *Proc.*, Conf. on Inductive Logic, Oxford Univ., August 1978.

ZADEH, L. "Fuzzy sets." *Information and Control 8*, 1965, 338–353.

ZUCKER, S. W. "Relaxation labelling and the reduction of local ambiguities." Technical Report 451, Computer Science Dept., Univ. Maryland, 1976.

Goal Achievement   13

Goal Achievement and Vision

Goals and plans are important for visual processing.

- Some skilled vision actually is like problem solving.
- Vision for information gathering can be part of a planned sequence of actions.
- Planning can be a useful and efficient way to guide many visual computations, even those that are not meant to imply "conscious" cognitive activity.

The artificial intelligence activity often called *planning* traditionally has dealt with "robots" (real or modeled) performing actions in the real world. Planning has several aspects.

- Avoid nasty "subgoal interactions" such as getting painted into a corner.
- Find the plan with optimal properties (least risk, least cost, maximized "goodness" of some variety).
- Derive a sequence of steps that will achieve the goal from the starting situation.
- Remember effective action sequences so that they may be applied in new situations.
- Apply planning techniques to giving advice, presumably by simulating the advisee's actions and making the next step from the point they left off.
- Recover from errors or changes in conditions that occur in the middle of a plan.

Traditional planning research has not concentrated on plans with information gathering steps, such as vision. The main interest in planning research has been the expensive and sometimes irrevocable nature of actions in the world. Our goal is to give a flavor of the issues that are pursued in much more detail in the planning literature [Nilsson 1980; Tate 1977; Fahlman 1974; Fikes and Nilsson 1971; Fikes et al. 1972a; 1972b; Warren 1974; Sacerdoti 1974; 1977; Sussman 1975].

Planning concerns an active agent and its interaction with the world. This conception does not fit with the idea of vision as a passive activity. However, one claim of this book is that much of vision is a constructive, active, goal-oriented process, replete with uncertainty. Then a model of vision as a sequence of decisions punctuated by more or less costly information gathering steps becomes more compelling. Vision often is a sequential (recursive, cyclical) process of alternating information gathering and decision making. This paradigm is quite common in computer vision [Shirai 1975; Ballard 1978; Mackworth 1978; Ambler et al. 1975]. However, the formalization of the process in terms of minimizing cost or maximizing utility is not so common [Feldman and Sproull 1977; Ballard 1978; Garvey 1976]. This section examines the paradigms of planning, evaluating plans with costs and utilities, and how plans may be applied to vision processing.

13.1 SYMBOLIC PLANNING

In artificial intelligence, planning is usually a form of problem-solving activity involving a formal "simulation" of a physical world. (Planning, theorem proving, and state-space problem solving are all closely related.) There is an agent (the "robot") who can perform actions that transform the state of the simulated world. The robot planner is confronted with an initial world state and a set of goals to be achieved. Planning explores world states resulting from actions, and tries to find a sequence of actions that achieves the goals. The states can be arranged in a tree with initial state as the root, and branches resulting from applying different actions in a state. Planning is a search through this tree, resulting in a path or sequence of actions, from the root to a state in which the goals are achieved. Usually there is a metric over action sequences; the simplest is that there be as few actions as possible. More generally (Section 13.2), actions may be assigned some cost which the planner should minimize.

13.1.1 Representing the World

This section illustrates planning briefly with a classical example—block stacking. In one simple form there are three blocks initially stacked as shown on the left in Fig. 13.1, to be stacked as shown.

This task may be "formalized" [Bundy 1978] using only the symbolic objects Floor, A, B, and C. (A formalization suitable for a real automated planner must be much more careful about details than we shall be). Assume that only a single block can be picked up at a time. Necessary predicates are CLEAR(X) which is true if a block may be put directly on X and which must be true before X may be picked up, and ON(X, Y), which is true if X is resting directly on Y. Let us stipulate that the Floor is always CLEAR, but otherwise if ON(X, Y) is true, Y is not CLEAR. Then the initial situation in Fig. 13.1 is characterized by the following assertions.

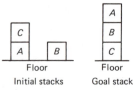

Fig. 13.1 A simple block stacking task.

INITIAL STATE: ON(C,A), ON(A, Floor), ON(B, Floor),
CLEAR(C), CLEAR(B), CLEAR(Floor)

The goal state is one in which the following two assertions are true.

GOAL ASSERTIONS: ON(A,B), ON(B,C)

With only these rules, the formalization of the block stacking world yields a very "loose" semantics. (The task easily translates to sorting integers with some restrictions on operations, or to the "seriation" task of arranging blocks horizontally in order of size, or a host of others.)

Actions transform the set of assertions describing the world. For problems of realistic scale, the representation of the tree of world states is a practical problem. The issue is one of maintaining several coexisting "hypothetical worlds" and reasoning about them. This is another version of the frame problem discussed in sec. 12.1.6. One way to solve this problem is to give each assertion an extra argument, naming the hypothetical world (usually called a situation [Nilsson 1980; McCarthy and Hayes 1969]) in which the assertion holds. Then actions map situations to situations as well as introducing and changing assertions.

An equivalent way to think about (and implement) multiple, dependent, hypothetical worlds is with a tree-structured *context-oriented data base*. This idea is a general one that is useful in many artificial intelligence applications, not just symbolic planning. Such data bases are included in many artificial intelligence languages and appear in other more traditional environments as well. A context-oriented data base *acts* like a tree of data bases; at any node of the tree is a set of assertions that makes up the data base. A new data base (context) may be spawned from any context (data base) in the tree. All assertions that are true in the spawning (ancestor) context are initially true in the spawned (descendant) context. However, new assertions added in any context or deleted from it do not affect its ancestor. Thus by going back to the ancestor, all data base changes performed in the descendent context disappear.

Implementing such a data base is an interesting exercise. Copying all assertions to each new context is possible, but very wasteful if only a few changes are made in each context. The following mechanism is much more efficient. The root or initial context has some set of assertions in it, and each descendant context is merely an *add list* of assertions to add to the data base and a *delete list* of assertions to delete. Then to see if an assertion is true in a context, do the following.

1. If the context is the root context, look up "as usual."
2. Otherwise, if the assertion is on the *add list* of this context, return *true*. If the assertion is on the *delete list* of this context, return *false*.

3. Otherwise, recursively apply this procedure to the ancestor of this context.

In a general programming environment, contexts have names, and there is the facility of executing procedures "in" particular contexts, moving around the context tree, and so forth. However, in what follows, only the ability to look up assertions in contexts is relevant.

13.1.2 Representing Actions

Represent an action as a triple.

ACTION ::= [PATTERN, PRECONDITIONS, POSTCONDITIONS].

Here the pattern gives the name of the action and names for the objects with which it deals—its "formal parameters." Preconditions and postconditions may use the formal variables of the pattern. In a sense, the preconditions and postconditions are the "body" of the action, with subroutine-like "variable bindings" taking place when the action is to be performed. The preconditions give the world states in which the action may be applied. Here the preconditions are assumed simply to be a list of assertions all of which must be true. The postconditions describe the world state that results from performing the action. The context-oriented data base of hypothetical worlds can be used to implement the postconditions.

POSTCONDITIONS ::= [ADD-LIST, DELETE-LIST].

An action is then performed as follows.

1. Bind the pattern variables to entities in the world, thus binding the associated variables in the preconditions and postconditions.
2. If the preconditions are met (the bound assertions exist in the data base), do the next step, else exit reporting failure.
3. Delete the assertions in the delete list, add those in the add list, and exit reporting success.

Here is the *Move* action for our block-stacking example.

	Move Object X from Y to Z		
PATTERN	*PRECONDITIONS*	*DELETE-LIST*	*ADD-LIST*
Move(X,Y,Z)	CLEAR(X)	ON(X,Y)	ON(X,Z)
	CLEAR(Z)	CLEAR(Z)	CLEAR(Y)
	ON(X,Y)		

Here X, Y, and Z are all variables bound to world entities. In the initial state of Fig. 13.1, Move(C,A,Floor) binds X to C, Y to A, Z to Floor, and the preconditions are satisfied; the action may proceed.

However, notice two things.

1. The action given above deletes the CLEAR(Floor) assertion that always should be true. One must fix this somehow; putting CLEAR(Floor) in the add-list does the job, but is a little inelegant.

2. What about an action like Move(C,A,C)? It meets the preconditions, but causes trouble when the add and delete lists are applied. One fix here is to keep in the data base ("world model") a set of assertions such as Different (A,B), Different(A,Floor), . . . , and to add assertions such as Different (X,Z) to the preconditions of Move.

Such housekeeping chores and details of axiomatization are inherent in applying basically syntactic, formal solution methods to problem solving. For now, let us assume that CLEAR(Floor) is never deleted, and that Move(X,Y,Z) is applied only if Z is different from X and Y.

13.1.3 Stacking Blocks

In the block-stacking example, the goal is two simultaneous assertions, ON(A,B) and ON(B,C). One solution method proceeds by repeatedly picking a goal to work on, finding an operator that moves closer to the goal, and applying it. In this case of only one action the question is how to apply it—what to move where. This is answered by looking at the postconditions of the action in the light of the goal. The reasoning might go like this: ON(B,C) can be made true if X is B and Z is C. That is possible in this state if Y is A; all preconditions are satisfied, and the goal ON(B,C) can be achieved with one action.

Part of the world state (or context) tree the planner must search is shown in Fig. 13.2, where states are shown diagrammatically instead of through sets of assertions. Notice the following things in Fig. 13.2.

1. Trying to achieve ON(B,C) first is a mistake (Branch 1).

2. Trying to achieve ON(A,B) first is also a mistake for less obvious reasons (Branch 2).

3. Branches 1 and 2 show "subgoal interaction." The goals as stated are not independent. Branch 3 must be generated somehow, either through backtracking or some intelligent way of coping with interaction. It will never be found by the single-minded approach of (1) and (2). However, if ON(C,Floor) were one of the goal assertions, Branch 3 could be found.

Clearly, representing world and actions is not the whole story in planning. Intelligent search of the context is also necessary. This search involves subgoal selection, action selection, and action argument selection. Bad choices anywhere can mean inefficient or looping action sequences, or the generation of impossible subgoals. "Intelligent" search implies a meta-level capability: the ability of a program to reason about its own plans. "Plan critics" are often a part of sophisticated planners; one of their main jobs is to isolate and rectify unwanted subgoal interaction [Sussman 1975].

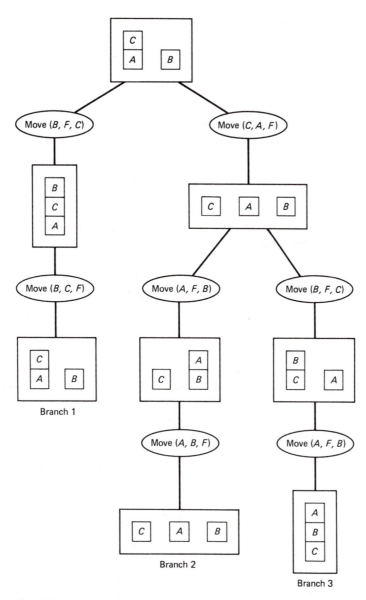

Fig. 13.2 A state tree generated in planning how to stack three blocks.

Intelligent choice of actions is the crux of planning, and is a major research issue. Several avenues have been and are being tried. Perhaps subgoals may be ordered by difficulty and achieved in that order. Perhaps planning should proceed at various levels of detail (like multiresolution image understanding), where the strategic skeleton of a plan is derived without details, then the details are filled in by applying the planner in more detail to the subgoals in the low-resolution plan.

13.1.4 The Frame Problem

All planning is plagued by aspects of the *frame problem* (introduced in Section 12.1.6).

1. It is impractical (and boring) to write down in an action all the things that stay the same when an action is applied.
2. Similarly, it is impractical to reassert in the data base all the things that remain true when an action is applied.
3. Often an action has effects that cannot be represented with simple add and delete lists.

The add and delete list mechanism and the context-oriented data base mechanism addressed the first two problems. The last problem is more troublesome.

Add and delete lists are simple ideas, whereas the world is a complex place. In many interesting cases, the add and delete lists depend on the current state of the world when the action is applied. Think of actions TURNBY(X) and MOVEBY(Z) in a world where orientation and location are important. The orientation and location after an action depend not just on the action but on the state of the world just before the action.

Again, the action may have very complex effects if there are complex dependencies between world objects. Consider the problem of the "monkey and bananas," where the monkey plans to push the box under the bananas and climb on it to reach them (Fig. 13.3). Implementation of realistically powerful add and delete lists may in fact require arbitrary amounts of deduction and computation.

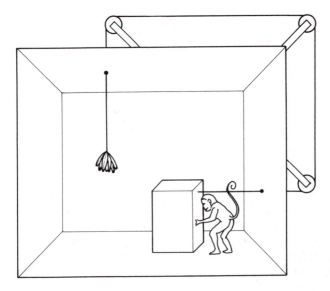

Fig. 13.3 Actions may have complex effects.

This quick précis of symbolic planning does not address many "classical" topics, such as learning or remembering useful plans. Also not discussed are: planning at varying levels of abstraction, plans with uncertain information, or plans with costs. The interested reader should consult the References for more information. The next section addresses plans with costs since they are particularly relevant to vision; some of the other issues appear in the Exercises.

13.2 PLANNING WITH COSTS

Decision making under uncertainty is an important topic in its own right, being of interest to policymakers and managers [Raiffa 1968]. Analytic techniques that can derive the strategy with the "optimal expected outcome" or "maximal expected utility" can be based on Bayesian models of probability.

In [Feldman and Sproull 1977] these techniques are explored in the context of action planning for real-world actions and vision. As an example of the techniques, they are used to model an extended version of the "monkey and bananas" problem of the last section, with multiple boxes but without the maddening pulley arrangement. In the extended problem, there are boxes of different weights which may or may not support the monkey, and he can apply tests (e.g., vision) at some cost to determine whether they are usable. Pushing weighted boxes costs some effort, and the gratification of eating the bananas is "worth" only some finite amount of effort. This extended set of considerations is more like everyday decision making in the number of factors that need balancing, in the uncertainty inherent in the universe, and in the richness of applicable tests. In fact, one might make the claim that human beings always "maximize their expected utility," and if one knew a person's utility functions, his behavior would become predictable. The more intuitive claim that humans beings plan only as far as "sufficient expected utility" can be cast as a maximization operation with nonzero "cost of planning."

The sequential decision-making model of planning with the goal of maximizing the goodness of the expected outcome was used in a travel planner [Sproull 1977]. Knowledge of schedules and costs of various modes of transportation and the attendant risks could be combined with personal prejudices and preferences to produce an itinerary with the maximum expected utility. If unexpected situations (canceled flights, say) arose en route, replanning could be initiated; this incremental plan ramification is a natural extension of sequential decision making.

This section is concerned with measuring the expected performance of plans using a single number. Although one might expect one number to be inadequate, the central theorem of decision theory [DeGroot 1970] shows essentially that one number is enough. Using a numerical measure of goodness allows comparisons between normally incomparable concepts to be made easily. Quite frequently numerical scores are directly relevant to the issues at stake in planning, so they are not obnoxiously reductionistic. Decision theory can also help in the process of applying a plan—the basic plan may be simple, but its application to the world may be complex, in terms of when to declare a result established or an action unsuccessful. The decision-theoretic approach has been used in several artificial intelligence and vision programs [Feldman and Yakimovsky 1974; Bolles 1977; Garvey 1976; Ballard 1978; Sproull 1977].

13.2.1 Planning, Scoring, and Their Interaction

For didactic purposes, the processes of plan generation and plan scoring are considered separately. In fact, these processes may cooperate more or less intimately. The planner produces "sequences" of *actions* for evaluation by the scorer. Each action (computation, information gathering, performing a real-world action) has a *cost*, expressing expenditure of resources, or associated unhappiness. An action has a set of possible *outcomes*, of which only one will really occur when the action is performed. A *goal* is a state of the world with an associated "happiness" or *utility*. For the purposes of uniformity and formal manipulation, goals are treated as (null) actions with no outcomes, and negative utilities are used to express costs. Then the plan has only actions in it; they may be arranged in a strict sequence, or be in loops, be conditional on outcomes of other actions, and so forth.

The *scoring* process evalutes the *expected utility* of a plan. In an uncertain world, a plan prior to execution has only an expected goodness—something might go wrong. Such a scoring process typically is not of interest to those who would use planners to solve puzzles or do proofs; what is interesting is the result, not the effort. But plans that are "optimal" in some sense are decidedly of interest in real-world decision making. In a vision context, plans are usually useful only if they can be evaluated for efficiency and efficacy.

Scoring can take place on "complete" plans, but it can also be used to guide plan generation. The usual artificial intelligence problem-solving techniques of progressive deepening search and branch-and-bound pruning may be applied to planning if scoring happens as the plan is generated [Nilsson 1980]. Scoring can be used to assess the cost of planning and to monitor planning horizons (how far ahead to look and how detailed to make the plan). Scoring will penalize plans that loop without producing results. Plan improvements, such as replanning upon failure, can be assessed with scores, and the contribution of additional steps (say for extra information gathering) can be assessed dynamically by scoring. Scoring can be arbitrarily complex utility functions, thus reflecting such concepts as "risk aversion" and nonlinear value of resources [Raiffa 1968].

13.2.2 Scoring Simple Plans

Scoring and an Example

A *simple plan* is a tree of nodes (there are no loops). The nodes represent actions (and goals). Outcomes are represented by labeled arcs in the tree. A probability of occurrence is associated with each possible outcome; since exactly one outcome actually occurs per action, the probabilities for the possible outcomes of any action sum to unity.

The *score* of a plan is its *expected utility*. The expected utility of any node is recursively defined as its utility times the probability of reaching that node in the plan, plus the expected utilities of the actions at its (possible) outcomes. The probability of reaching any "goal state" in the plan is the product of probabilities of outcomes forming a path from the root of the plan to the goal state.

As an example, consider the plan shown in Fig. 13.4. If the plan of Fig. 13.4

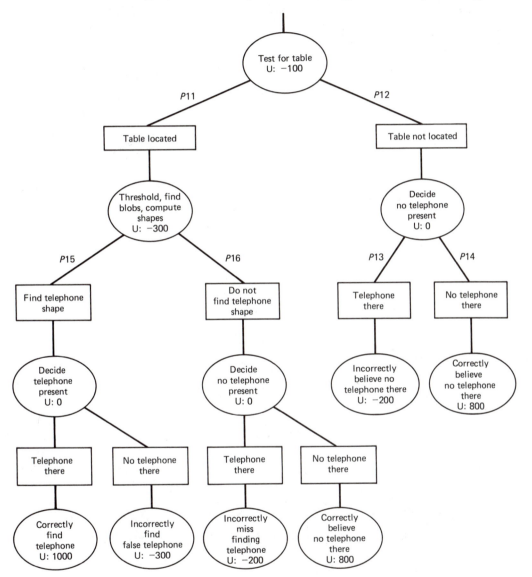

Fig. 13.4 This plan to find a telephone in an office scene involves finding a table first and looking there in more detail. The actions and outcomes are shown. The probabilities of outcomes are assigned symbols (P10, etc.). Utilities (denoted by U:) are given for the individual actions. Note that negative utilities may be considered costs. In this example, decision-making takes no effort, image processing costs vary, and there are various penalties and rewards for correct and incorrect finding of the telephone.

has probabilities assigned to its outcomes, we may compute its expected utility. Figure 13.5 shows the calculation. The probability of correctly finding the telephone is 0.34, and the expected utility of the plan is 433.

Although the generation of a plan may not be easy, scoring a plan is a trivial exercise once the probabilities and utilities are known. In practice, the assignment of probabilities is usually a source of difficulty. The following is an example using

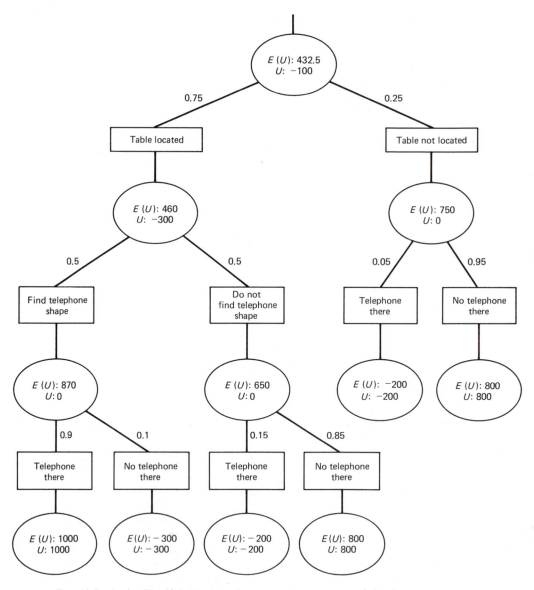

Fig. 13.5 As for Fig. 13.4. U: gives the utility of each action. E(U): gives the expected utillity of the action, which depends on the outcomes below it. Values for outcome probabilities are given on the outcome arcs.

the telephone-finding plan and some assumptions about the tests. Different assumptions yield different scores.

Computing Outcome Probabilities: An Example

This example relies heavily on Bayes' rule:

$$P(B|A)P(A) = P(A \wedge B) = P(A|B)P(B). \qquad (13.1)$$

Let us assume a specific a priori probability that the scene contains a telephone.

$$P_1 = \text{apriori probability of Telephone} \qquad (13.2)$$

Also assume that something is known about the behavior of the various tests in the presence of what they are looking for. This knowledge may accrue from experiments to see how often the table test found tables when telephones (or tables) were and were not present. Let us assume that the following are known probabilities.

$$P_3 = \text{P(table located|telephone in scene)} \qquad (13.3)$$

$$P_5 = \text{P(table located|no telephone in scene)} \qquad (13.4)$$

Either there is a telephone or there is not, and a table is located or it is not, so

$$P_2 = \text{a priori probability of no telephone} = 1 - P_1 \qquad (13.5)$$

$$P_4 = \text{P(no table located|telephone in scene)} = 1 - P_3 \qquad (13.6)$$

$$P_6 = \text{P(no table located|no telephone in scene)} = 1 - P_5 \qquad (13.7)$$

Similarly with the "shape test" for telephones: assume probabilities

$$P_7 = \text{P(telephone shape located|telephone)} \qquad (13.8)$$

$$P_9 = \text{P(telephone shape located|no telephone)} \qquad (13.9)$$

with

$$P_8 = 1 - P_7, \qquad P_{10} = 1 - P_9 \qquad (13.10)$$

as above.

There are a few points to make: First, it is not necessary to know exactly these probabilities in order to score the plan; one could use related probabilities and Bayes' rule. Other useful probabilities are of the form

$$\text{P(telephone|telephone shape located).}$$

In some systems [Garvey 1976] these are assumed to be available directly. This section shows how to derive them from known conditional probabilities that describe the behavior of detectors given certain scene phenomena.

Second, notice the assumption that although both the outcome of the table test and the shape test depend on the presence of telephones, they are taken to be independent of each other. That is, having found a table tells us nothing about the likelihood of finding a telephone shape. Independence assumptions such as this are useful to limit computations and data gathering, but can be somewhat unrealistic. To account for the dependence, one would have to measure such quantities as

$$P(\text{telephone shape found}\,|\,\text{table located}).$$

Now to compute some outcome probabilities: Consider the probability

$$P_{11} = P(\text{table located}) \tag{13.11}$$

Let us write

TL for Table Located
TNL for Table Not Located.

A table may be located whether or not a telephone is in the scene. In terms of known probabilities, Bayes' rule yields

$$P_{11} = P_3 \, P_1 + P_5 \, P_2 \tag{13.12}$$

Then

$$P_{12} = P(\text{TNL}) = 1 - P_{11} \tag{13.13}$$

Calculating P_{13} shows a neat trick using Bayes' Rule:

$$P_{13} = P(\text{telephone}\,|\,\text{TNL}) \tag{13.14}$$

That is, P_{13} is the probability that there is a telephone in the scene given that search for a table was unsuccessful. This probability is not known directly, but

$$
\begin{aligned}
P_{13} &= \frac{P(\text{telephone and TNL})}{P(\text{TNL})} \\
&= \frac{P(\text{TNL and telephone})}{P_{12}} \\
&= \frac{[P(\text{TNL}\,|\,\text{telephone})P(\text{telephone})]}{P_{12}} \\
&= \frac{[P_4 \, P_1]}{P_{12}}
\end{aligned}
\tag{13.15}
$$

Then, of course

$$P_{14} = 1 - P_{13} \tag{13.16}$$

Reasoning in this way using the conditional probabilities and assumptions about their independence allows the completion of the calculation of outcome probabilities (see the Exercises). One possibly confusing point occurs in calculation of P_{15}, which is

$$P_{15} = P(\text{telephone shape found}\,|\,\text{table located}) \tag{13.17}$$

By assumption, these events are only indirectly related. By the simplifying assumptions of independence, the shape operator and the table operator are independent in their operation. (Such assumptions might be false if they used common image processing subroutines, for example.) Of course, the probability of success of each depends on the presence of a telephone in the scene. Therefore their performance is linked in the following way (see the Exercises). (Write TSL for Telephone Shape Located.)

$$P_{15} = P(\text{TSL}|\text{TL})P(\text{TSL}|\text{telephone})P(\text{telephone}|\text{TL}) \qquad (13.18)$$
$$+P(\text{TSL}|\text{no telephone})P(\text{no telephone}|\text{TL})$$

13.2.3 Scoring Enhanced Plans

The plans of Section 13.2.2 were called "simple" because of their tree structure, complete ordering of actions, and the simple actions of their nodes. With a richer output from the symbolic planner, the plans may have different structure. For example, there may be *OR* nodes, any one of whose sons will achieve the action at the node; *AND* nodes, all of which must be satisfied (in any order) for the action to be satisfactorily completed; *SEQUENCE* nodes, which specify a set of actions and a particular order in which to achieve them. The plan may have loops, shared subgoal structure, or goals that depend on each other. How enhanced plans are interpreted and executed depends on the scoring algorithms, the possibilities of parallel execution, whether execution and scoring are interleaved, and so forth. This treatment ignores parallelism and limits discussion to expanding enhanced plans into simple ones.

It should be clear how to go about converting many of these enhanced plans to simple plans. For instance, sequence nodes simply go to a unique path of actions. Alternatively, depending on assumptions about outcomes of such actions (say whether they can fail), they may be coalesced into one action, as was the "threshold, find blobs, and compute shapes" action in the telephone-finding plan.

Rather more interesting are the OR and AND nodes, the order of whose subgoals is unspecified. Each such node yields many simple plans, depending on the order in which the subgoals are attacked. One way to score such a plan is to generate all possible simple plans and score each one, but perhaps it is possible to do better. For example, loops and mutual dependencies in plans can be dealt with in various ways. A loop can be analyzed to make sure that it contains an exit (such as a branch of an OR node that can be executed). One can make ad hoc assumptions that the cost of execution is always more than the cost of planning [Garvey 1976], and score the loop by its executable branch. Another idea is to plan incrementally with a finite horizon, expanding the plan through some progressive deepening, heuristic search, or pruning strategy. The accumulated cost of going around a loop will soon remove it from further consideration.

Recall (Figs. 13.4 and 13.5) that the expected utility of a plan was defined as the sum of the utility of each leaf node times the probability of reaching that node. However, the utilities need not combine linearly in scoring. Different monotonic functions of utility express such different conceptions as "aversion to risk" or "gambling addiction." These considerations are real ones, and nonlinear utilities are the rule rather than the exception. For instance, the value of money is notoriously nonlinear. Many people would pay $5 for an even chance to win $15; not so many people would pay $5,000 for an even chance to win $15,000.

One common way to compute scores based on utilities is the "cost/benefit" ratio. This, in the form "cost/confidence" ratio, is used by Garvey in his planning vision system. This measure is examined in Section 13.2.5; roughly, his "cost" was the effort in machine cycles to achieve goals, and his "confidence" approximated the probability of a goal achieving the correct outcome. The utility of correct outcomes was not explicitly encoded in his planner.

Sequential plan elaboration or partial plan elaboration can be interleaved with execution and scoring. Most practical planning is done in interaction with the world, and the plan scoring approach lends itself well to assessing such interactions. In Section 13.2.5 considers a planning vision system that uses enhanced plans and a limited replanning capability.

A thorny problem for decision making is to assess the cost of planning itself. The planning process is given its own utility (cost), and is carried only out as far as is indicated. Of course, the problem is in general infinitely recursive, since there is also the cost of assessing the cost of planning, etc. If, however, there is a known upper bound on the utility of the best achievable plan, then it is known that infinite planning could not improve it. This sort of reasoning is weaker than that needed to give the expected benefits of planning; it measures only the cost and maximum value of planning.

Another more advanced consideration is that the results of actions can be continuous and multidimensional, and discrete probabilities can be extended to probability distribution functions. Such techniques can reflect the precision of measurements.

An obviously desirable extension to a planner is a "learner," that can abstract rules for action applicability and remember successful plans. One approach would be to derive and remember ranges of planning parameters arising during execution; a range could be associated with a rule specifying appropriate action. This problem is difficult and the subject of current research.

13.2.4 Practical Simplifications

The expected utility calculations allow plans to be evaluated in a more or less "realistic" manner. However, in order to complete the calculations certain probabilities are necessary, and many of these reflect detailed knowledge about the interaction of phenomena in the world. It is thus often impractical to go about a full-blown treatment of scoring in the style of Section 13.2.2. This section presents some possible simplifications.

Of course, in many planning problems, such as those whose costs are nil or irrelevant, or all of whose goals are equally valuable, there is no need to address utility of plans at all. Such plans are typically not concerned with expenditure of real-world or planning resources.

Independence of various probabilities is one of the most helpful and pervasive assumptions in the calculation of probabilities. An example appeared in Section 13.2.2 with the table and telephone shape detectors.

Certain information can be ignored. Garvey [Garvey 1976] ignores failure information. His planning parameters include the "cost" of an action (strictly negative utilities reflecting effort), the probability of the action "succeeding," and the conditional probability that the state of the world is correctly indicated, given success. Related to ignoring some information is the assumption that certain outcomes are more reliable than others. For instance, the decision not to plan past "failure" reports means that they are assumed reliable.

Non-Bayesian rules of inference abound in planners [Shortliffe 1976]; the idea of assigning a single numerical utility score to plans is by no means the only way to make decisions.

13.2.5 A Vision System Based on Planning

Overview

This section outlines some features of a working vision system whose actions are controlled by the planning paradigm [Garvey 1976]. As with all large vision systems, more issues are addressed in this work than planning as a control mechanism. For one thing, the system uses multisensory input, including range and color information. An interactive facility aids in developing and testing low-level operators and "strategies" for object location. The machine-usable representation of knowledge about the objects in the scene domains and how they could be located is, of course, a central component.

The domain is office scenes (Fig. 13.6). For the task of locating different objects in such scenes, a "uniform strategy" is adopted. That is, the vision task is always broken down into a sequence of major goals to be performed in order. Such uniform strategies, if they are imposed on a system at all, tend to vary with different tasks, with different sensors or domain, or with different research goals.

Garvey's uniform strategy consists of the following steps.

1. *Acquire* some pixels thought to be in the desired region (the area of scene making up the image of the desired object).

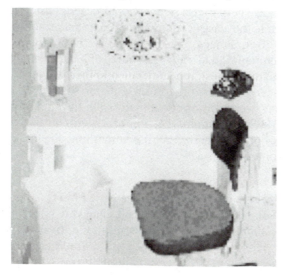

Fig. 13.6 The planning vision system uses input scenes such as these, imaged in different wavelengths and with a rangefinder.

2. *Verify* to some confidence that indeed the region was the desired one.

3. *Bound* the region accurately.

The outline the plan generation, scoring, and execution used in the system are described in the following paragraphs. The plans generated by the system are typically enhanced versions of plans like the telephone finder. Plan scoring proceeds as expected for such plans; allowances are made for the enhanced semantics of plan nodes. A "cost/confidence" scoring function is used, and various practical simplifications are made that do not affect the planning paradigm itself.

An Example Plan and Its Execution

The system's plans are enhanced plans, in the sense of Section 13.2.3. Actions can be *AND*, *OR* or *SEQUENCE* actions, and shared plan structure and loops are permitted. Loops that contain only internal, planning actions would never terminate. However, a loop with an OR node can terminate (has an exit) if one of the subactions of the OR is executable. A plan for locating a chair in an office scene is shown in Fig. 13.7. In Fig. 13.7, the acquire–validate–bound strategy is evident in the two SEQUENCE subgoals of the Find Chair main goal, which is an AND goal. The loop in the plan is evident, and makes sense here because often planning is done for information gathering, not for real world actions.

As noted in Section 13.2.3, an enhanced plan may not be completely specified. If it is to be executed one subgoal at a time (no parallelism is allowed), sequences of subactions must be determined for its AND and OR actions. In Garvey's planner, these sequences are determined initially on the basis of apriori information, but the partial results of actions are "fed back," so that dynamic rescoring and hence dynamic reordering of goal sequences is possible. For example, if one subgoal of an AND action fails, the AND action is abandoned. Thus this planner is to some degree incremental.

In execution, Fig. 13.7 might result in the sequence of actions depicted in Fig. 13.8. The acquisition phase of object location has the most alternatives, so plan generation effort is mainly spent there. Acquisition proceeds either directly or indirectly. Direct acquisition is the classification of input data gathered from a random sampling of a window in the image; the input data are rich enough to allow basic pattern recognition techniques to identify the source of individual pixels.

Indirect acquisition is the use of the location of other "objects" (really identified regions) in the scene to locate the desired region. The desired region might be found by "scanning" vertically or horizontally from the already identified region, for instance. The idea is a planning version of a common one (e.g., the geometric location networks of Section 10.3.2): use something already located to limit and direct search for something else.

Plan Generation

A plan such as Fig. 13.7 is "elaborated" from the basic Find Chair goal by recursively expanding goals. Some goals (such as to find a chair) are not directly executable; they need further elaboration. Elaboration continues until all the subgoals are executable. Executable subgoals are those that analyze the image, run filters and detectors over parts of it, and generate decisions about the presence or absence

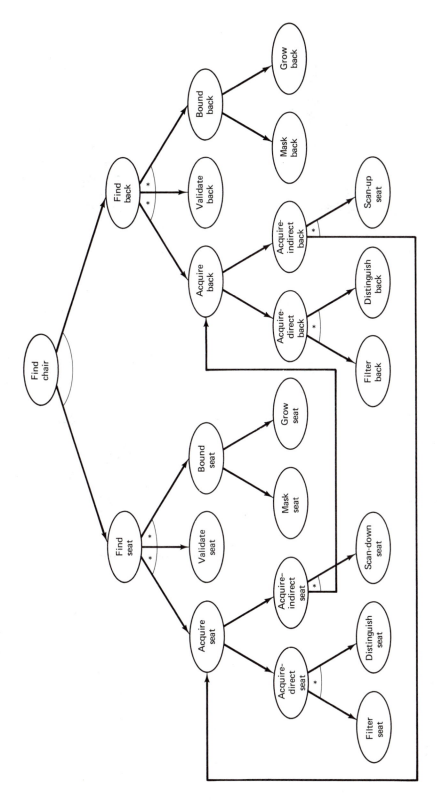

Fig. 13.7 An enhanced plan to locate a chair in an office scene. Untied multiple arcs denote OR actions, arcs tied together denote AND actions, those with *'s denote SE-QUENCE actions. The loop in the plan has executable exits.

(a)

(b)

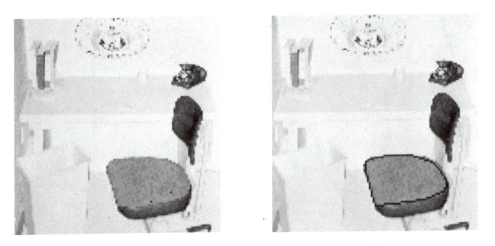

(c)

(d)

Fig. 13.8 The plan of Fig. 13.7 finds the most promising execution sequence for finding the chair in the scene of Fig. 13.6: find the seat first, then scan upwards from the seat looking for the back. Acquisition of the seat proceeds by sampling (a), followed by classification (b). The Validation procedure eliminates non-chair points (c), and the Bounding procedure produces the seat region (d). To find the back, scanning proceeds in the manner indicated by (e) (actually fewer points are examined in each scan). The back is acquired and bounded, leading to the final location of the chair regions (f).

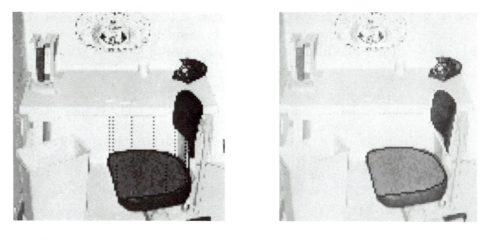

| (e) | (f) |

Fig. 13.8 (cont.)

of image phenomena. This straightforward elaboration is akin to macro expansion, and is not a very sophisticated planning mechanism (the program cannot criticize and manipulate the plan, only score it). A fully elaborated plan is presented for scoring and execution.

The elaboration process, or planner, has at its disposal several sorts of knowledge embodied as modules that can generate subgoals for a goal. Some are general (to find something, find all its parts); some are less general (a chair has a back and a seat); some are quite specific, being perhaps programs arising from an earlier interactive method-generation phase. The elaborator is guided by information stored about objects, for instance this about a tabletop:

OBJECT	PROPERTIES	RELATIONS
Table TOP	Hue:26–58	Supports Telephone 0.6
	Sat.: 0.23–0.32	Supports Book 0.4
	Bright.: 18–26	Occludes Wall 1
	Height: 26–28	
	Orient.: −7–7	

Here the orientation information indicates a vertical surface normal. The planner knows that it has a method of locating horizontal surfaces, and the plan elaborator can thus create a goal of direct acquisition by first locating a horizontal plane. The relational information allows for indirect acquisition plans. The elaborator puts direct and indirect alternatives under an *OR* node in the plan. Information not used for acquisition (height, color) may be used for validation.

Loops may occur in an elaborated plan because each newly generated goal is checked against goals already existing. Should it or an equivalent goal already exist, the existing goal is substituted for the newly generated one. Goals may thus have more than one ancestor, and may depend on one another.

At this stage, the planner does not use any planning parameters (cost, utilities, etc.); it is strictly symbolic. As mentioned above, important information about execution sequences in an enhanced plan is provided by scoring.

Plan Scoring and Execution

The scoring in the vision plan is a version of that explained in Sections 13.2.2 through 13.2.4. Each action in a plan is assumed either to succeed (S) in locating an object or to fail. Each action may report either success ("S") or failure. An action is assumed to report failure correctly, but possibly to be in error in reporting success. Each action has three "planning parameters" associated with it. They are C, its "cost" (in machine cycles), $P("S")$ the probability of it reporting success, and $P(S|"S")$, the probability of success given a report of success.

As shown earlier, the product

$$P(S|"S")P("S") \tag{13.19}$$

is the probability that the action has correctly located an object and reported success. This product is called the "confidence" of the action. An action has structure as shown in Fig. 13.9.

The score of an action is computed as

$$\text{score} = \frac{\text{cost}}{\text{confidence}} \tag{13.20}$$

The planner thus must minimize the score.

The initial planning parameters of an executable action typically are determined by experimentation. The parameters of internal (AND, OR, SEQUENCE) actions by scoring methods alluded to in Sections 13.2.2, 13.2.3, and the Exercises (there are a few idiosyncratic ad hoc adjustments.).

It may bear repeating that planning, scoring, and execution are not separated temporally in this sytem. Scoring is used after the enhanced plan is generated to derive a simple plan (with ordered subgoals). Execution can affect the scores of nodes, and so execution can alternate with "replanning" (really rescoring resulting in a reordering). Recall the example of failure of an *AND* or *SEQUENCE* subgoal, which can immediately fail the entire goal. More generally, the entire goal and ultimately the plan may be rescored. For instance, the parameters of a successful action are modified by setting the cost of the executed action to 0 and its confidence to its second parameter, $P(S|"S")$.

Given a scored plan, execution is then easy; the execution program starts at the top goal of the plan, working its way down the best path as defined by the scores of nodes it encounters. When an executable subgoal is found (e.g. "look for a green region"), it is passed to an evaluation function that "runs" the action associated with the subgoal.

The subgoal is either achieved or not; in either case, information about its outcome is propagated back up the plan. Failure is easy; a failed subgoal of an AND or SEQUENCE goal fails the goal, and this failure is propagated. A failed subgoal of an OR goal is removed from the plan. The use of success information is more complex, involving the adjustment of confidences and planning parameters illustrated above.

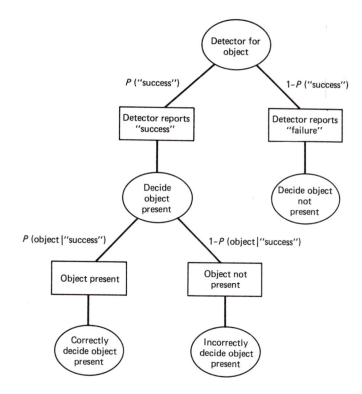

Fig. 13.9 This is the microstructure of a node ("action") of Garvey's planning system in terms of simple plans. Think of actions as being object detectors which announce "Found" or "Not Found." Garvey's planning parameters are P("Found") and P(Object is there|"Found"). Confidence in the action is their product; it is the probability of correctly detecting the object. All other outcomes are lumped together and not used for planning.

After the outcome of a goal is used to adjust the parameters of other goals, the plan is rescored and another cycle of execution performed. The execution can use knowledge about the image picked up along the way by prior execution. This is how results (such as acquired pixels) are passed to later processing stages (such as the validation process). Such a mechanism can even be used to remember successful subplans for later use.

EXERCISES

13.1 Complete the computation of outcome probabilities in the style of Section 13.2.2, using the assumptions given there. Check your work by showing (symbolically) that the probabilities of getting to the terminal actions ("goal states") of the plan sum to 1.

13.2 Assume in Section 13.2.2 that the results of the "table" and "telephone shape" detectors are not independent. Formulate your assumptions and compute the new outcome probabilities for Fig. 13.4.

13.3 Show that

$$P(A|(B \wedge C)) = \frac{P(B|(A \wedge C))P(A|C)}{P(B|C)}$$

13.4 B and C are independent if $P(B \wedge C) = P(B) P(C)$. Assuming that B and C are independent, show that

$$P(B|C) = P(B)$$

$$P((B \wedge C)|A) = P(B|A)P(C|A)$$

$$P(B|(A \wedge C)) = P(B|A)$$

13.5 Starting from the fact that

$$P(A \wedge B) = P(A \wedge B \wedge C) + P(A \wedge B \wedge (-C))$$

show how P_{15} was computed in Section 13.2.2.

13.6 A sequence $D(N)$ of N detectors is used to detect an object; the detectors either succeed or fail. Detector outputs are assumed independent of each other, being conditioned only on the object. Using previous results, show that the probability of an object being detected by applying a sequence of N detectors $D(N)$ is recursively rewritable in terms of the output of the first detector $D1$ and the remaining sequence $D(N-1)$ as

$$P(O|D(N)) = \frac{P(D1|O)P(O|D(N-1))}{P(D1|D(N-1))}$$

13.7 Consider scoring a plan containing an OR node (action). Presumably, each subgoal of the OR has an expected utility. The OR action is achieved as soon as one of the subgoals is achieved. Is it possible to order the subgoals for trial so as to maximize the expected utility of the plan? (This amounts to a unique "best" rewriting of the plan to make it a simple plan.)

13.8 Answer question 13.7 for an AND node; remember that the AND will fail as soon as any of its subgoals fails.

13.9 What can you say about how the cost/confidence ratio of Garvey's planner is related to the expected utility calculations of Section 13.2.2?

13.10 You are at Dandy Dan's used car lot. *Consumer Reports* says that the a priori probability that any car at Dandy Dan's is a lemon is high. You know, though, that to test a car you kick its tire. In fact, with probability:

$P("C"|C)$: a kick correctly announces "creampuff" when the
car actually is a creampuff

$P("C"|L)$: a kick incorrectly announces "creampuff" when
the car is actually a lemon

$P(L)$: the a priori probability that the car is a lemon

Your plan for dealing with Dandy Dan is shown below; give expressions for the probabilities of arriving at the nodes labeled S_1, S_2, F_1, F_2, and F_3. Give numeric answers using the following values

$$P("C"|C) = 0.5, \quad P("C"|L) = 0.5, \quad P(L) = 0.75$$

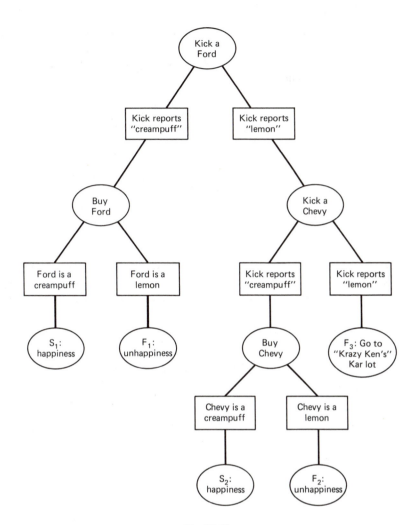

Ex. 13.10

13.11 Two bunches of bananas are in a room with a monkey and a box. One of the bunches is lying on the floor, the other is hanging from the ceiling. One of the bunches is made of wax. The box may be made of flimsy cardboard. Given that:

$$P(WH) \quad = 0.2 : \text{probability that the hanging bananas are wax}$$
$$P(WL) \quad = 0.8 : \text{probability that the lying bananas are wax}$$
$$P(C) \quad = 0.5 : \text{probability that the box is cardboard}$$
$$U(\text{eat}) \quad = 200 : \text{utility of eating a bunch of bananas}$$
$$C(\text{walk}) \quad = -10 : \text{cost of walking a unit distance}$$
$$C(\text{push}) \quad = -20 : \text{cost of pushing the box a unit distance}$$
$$C(\text{climb}) \quad = -20 : \text{cost of climbing up on box}$$

(a) Analyze two different plans for the monkey, showing all paths and calculations. Give criteria (based upon extra information not given here) that would allow the monkey to choose between these plans.

(b) Suppose the monkey knows that the probability that the box will collapse is inversely proportional to the cost of pushing the box a unit distance (and that he can sense this cost after pushing the box 1 unit distance). For example,

$$P(C) = 1.0 - [C(\text{push}) \times 0.01]$$
$$P(C(\text{push}) = 10) = 0.1$$
$$P(C(\text{push}) = 20) = 0.1$$
$$P(C(\text{push}) = 100) = 0.1$$

Repeat part(a) (in detail).

REFERENCES

AMBLER, A. P., H. G. BARROW, C. M. BROWN, R. M. BURSTALL, and R. J. POPPLESTONE. "A versatile system for computer controlled assembly." *Artificial Intelligence 6*, 2, 1975, 129–156.

BALLARD, D. H. "Model-directed detection of ribs in chest radiographs." TR11, Computer Science Dept., U. Rochester, March 1978.

BOLLES, R. C. "Verification vision for programmable assembly." *Proc.*, 5th IJCAI, August 1977, 569–575.

BUNDY, A. *Artificial Intelligence: An introductory course.* New York: North Holland, 1978.

DEGROOT, M. H. *Optimal Statistical Decisions.* New York: McGraw-Hill, 1970.

FAHLMAN, S. E. "A planning system for robot construction tasks." *Artificial Intelligence 5*, 1, Spring 1974, 1–49.

FELDMAN, J. A. and R. F. SPROULL. "Decision theory and artificial intelligence: II. The hungry monkey." *Cognitive Science 1*, 2, 1977, 158–192.

FELDMAN, J. A. and Y. YAKIMOVSKY. "Decision theory and artificial intelligence: I. A semantics-based region analyser." *Artificial Intelligence 5*, 4, 1974, 349–371.

FIKES, R. E. and N. J. NILSSON. "STRIPS: A new approach to the application of theorem proving to problem solving." *Artificial Intelligence 2*, 3/4, 1971, 189–208.

FIKES, R. E., P. E. HART, and N. J. NILSSON. "New Directions in robot problem solving." In *MI7*, 1972a.

FIKES, R. E., P. E. HART, and N. J. NILSSON. "Learning and executing generalized robot plans." *Artificial Intelligence 3*, 4, 1972b, 251–288.

GARVEY, J. D. "Perceptual strategies for purposive vision." Technical Note 117, AI Center, SRI Int'l, 1976.

MACKWORTH, A. K. "Vision research strategy: Black magic, metaphors, mechanisms, mini-worlds, and maps." In *CVS*, 1978.

MCCARTHY, J. and P. J. HAYES. "Some philosophical problems from the standpoint of artificial intelligence." In *MI4*, 1969.

NILSSON, N. J. *Principles of Artificial Intelligence.* Palo Alto, CA: Tioga Publishing Company, 1980.

RAIFFA, H. *Decision Analysis.* Reading, MA: Addison-Wesley, 1968.

SACERDOTI, E. D. "Planning in a hierarchy of abstraction spaces." *Artificial Intelligence 5*, 2, 1974, 115–135.

SACERDOTI, E. D. *A Structure for Plans and Behavior.* New York: Elsevier, 1977.

SHIRAI, Y. "Analyzing intensity arrays using knowledge about scenes." In *PCV*, 1975.

SHORTLIFFE, E. H. *Computer-Based Medical Consultations: MYCIN.* New York: American Elsevier, 1976.

SPROULL, R. F. "Strategy construction using a synthesis of heuristic and decision-theoretic methods." Ph.D. dissertation, Dept. of Computer Science, Stanford U., May 1977.

SUSSMAN, G. J. *A Computer Model of Skill Acquisition.* New York: American Elsevier, 1975.

TATE, A. "Generating project networks." *Proc.,* 5th IJCAI, August 1977, 888–983.

WARREN, D. H. D. "WARPLAN: A system for generating plans." Memo 76, Dept. of Computational Logic, U. Edinburgh, June 1974.

Some
Mathematical Tools

Appendix 1

A1.1 COORDINATE SYSTEMS

A1.1.1 Cartesian

The familiar two- and three-dimensional rectangular (Cartesian) coordinate systems are the most generally useful ones in describing geometry for computer vision. Most common is a right-handed three-dimensional system (Fig. A1.1.). The coordinates of a point are the perpendicular projections of its location onto the coordinate axes. The two-dimensional coordinate system divides two-dimensional space into quadrants, the three-dimensional system divides three-space into octants.

A1.1.2 Polar and Polar Space

Coordinate systems that measure locations partially in terms of angles are in many cases more natural than Cartesian coordinates. For instance, locations with respect

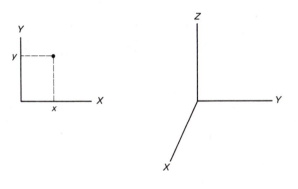

Fig. A1.1 Cartesian coordinate systems.

to the pan-tilt head of a camera or a robot arm may most naturally be described using angles. Two- and three-dimensional polar coordinate systems are shown in Fig. A1.2.

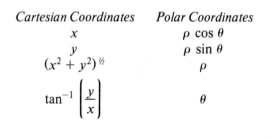

Cartesian Coordinates	Polar Coordinates
x	$\rho \cos \theta$
y	$\rho \sin \theta$
$(x^2 + y^2)^{\frac{1}{2}}$	ρ
$\tan^{-1}\left(\dfrac{y}{x}\right)$	θ

Cartesian Coordinates	Polar Space Coordinates
(x, y, z)	$(\rho \cos \xi, \rho \cos \eta, \rho \cos \zeta)$
$(x^2 + y^2 + z^2)^{\frac{1}{2}}$	ρ
$\cos^{-1}\left(\dfrac{x}{\rho}\right)$	ξ
$\cos^{-1}\left(\dfrac{y}{\rho}\right)$	η
$\cos^{-1}\left(\dfrac{z}{\rho}\right)$	ζ

In these coordinate systems, the Cartesian quadrants or octants in which points fall are often of interest because many trigonometric functions determine only an angle modulo $\pi/2$ or π (one or two quadrants) and more information is necessery to determine the quadrant. Familiar examples are the inverse angle functions (such as arctangent), whose results are ambiguous between two angles.

A1.1.3 Spherical and Cylindrical

The spherical and cylindrical systems are shown in Fig. A1.3.

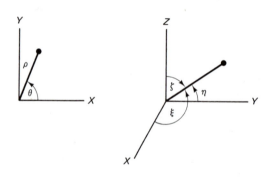

Fig. A1.2 Polar and polar space coordinate systems.

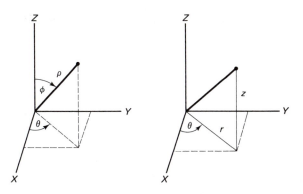

Fig. A1.3 Spherical and cylindrical coordinate systems.

Cartesian Coordinates	Spherical Coordinates
x	$\rho \sin \phi \cos \theta$
y	$\rho \sin \phi \sin \theta = x \tan \theta$
z	$\rho \cos \theta$
$(x^2 + y^2 + z^2)^{1/2}$	ρ
$\tan^{-1}\left[\dfrac{y}{x}\right]$	θ
$\cos^{-1}\left[\dfrac{z}{\rho}\right]$	ϕ

Cartesian Coordinates	Cylindrical Coordinates
x	$r \cos \theta$
y	$r \sin \theta$
z	z
$(x^2 + y^2)^{1/2}$	r
$\tan^{-1}\left[\dfrac{y}{x}\right]$	θ

A1.1.4 Homogeneous Coordinates

Homogeneous coordinates are a very useful tool in computer vision (and computer graphics) because they allow many important geometric transformations to be represented uniformly and elegantly (see Section A1.7). Homogeneous coordinates are redundant: a point in Cartesian n-space is represented by a line in homogeneous $(n + 1)$-space. Thus each (unique) Cartesian coordinate point corresponds to infinitely many homogeneous coordinates.

Cartesian Coordinates	Homogeneous Coordinates
(x, y, z)	(wx, wy, wz, w)
$\left[\dfrac{x}{w}, \dfrac{y}{w}, \dfrac{z}{w}\right]$	(x, y, z, w)

Here x, y, z, and w are real numbers, wx, wy, and wz are the products of the two reals, and x/w and so on are the indicated quotients.

A1.2. TRIGONOMETRY

A1.2.1 Plane Trigonometry

Referring to Fig. A1.4, define

sine: $\sin (A)$ (sometimes sin A) $= \dfrac{a}{c}$

cosine: $\cos (A)$ (or cos A) $= \dfrac{b}{c}$

tangent: $\tan (A)$ (or tan A) $= \dfrac{a}{b}$

The inverse functions arcsin, arccos, and arctan (also written $\sin^{-1}$, $\cos^{-1}$, $\tan^{-1}$) map a value into an angle. There are many useful trigonometric identities; some of the most common are the following.

$$\tan (x) = \frac{\sin (x)}{\cos (x)} = -\tan(-x)$$

$$\sin (x + y) = \sin (x) \cos (y) + \cos (x) \sin (y)$$

$$\cos (x + y) = \cos (x) \cos (y) - \sin (x) \sin (y)$$

$$\tan (x \pm y) = \frac{\tan (x) \mp \tan (y)}{1 \mp \tan (x) \tan(y)}$$

In any triangle with angles A, B, C opposite sides a, b, c, the Law of Sines holds:

$$\frac{a}{\sin A} = \frac{b}{\sin B} = \frac{c}{\sin C}$$

as does the Law of Cosines:

$$a^2 = b^2 + c^2 - 2bc \cos A$$

$$a = b \cos C + c \cos B$$

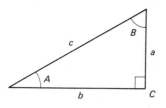

Fig. A1.4   Plane right triangle.

A1.2.2. Spherical Trigonometry

The sides of a spherical triangle (Fig. A1.5) are measured by the angle they subtend at the sphere center; its angles by the angle they subtend on the face of the sphere.

Some useful spherical trigonometric identities are the following.

$$\frac{\sin A}{\sin a} = \frac{\sin B}{\sin b} = \frac{\sin C}{\sin c}$$

$$\cos a = \cos b \cos c + \sin b \sin c \cos A = \frac{\cos b \cos (c \pm \theta)}{\cos \theta}$$

Where $\tan \theta = \tan b \cos A,$

$$\cos A = -\cos B \cos C + \sin B \sin C \cos a$$

A1.3. VECTORS

Vectors are both a notational convenience and a representation of a geometric concept. The familiar interpretation of a vector $\mathbf{v}$ as a directed line segment allows for a geometrical interpretation of many useful vector operations and properties. A more general notion of an n-dimensional vector $\mathbf{v} = (v_1, v_2, \ldots, v_n)$ is that of an n-tuple abiding by mathematical laws of composition and transformation. A vector may be written horizontally (a row vector) or vertically (a column vector).

A point in n-space is characterized by its n coordinates, which are often written as a vector. A point at X, Y, Z coordinates $x, y,$ and z is written as a vector $\mathbf{x}$ whose three components are (x, y, z). Such a vector may be visualized as a directed line segment, or arrow, with its tail at the origin of coordinates and its head at the point at (x, y, z). The same vector may represent instead the direction in which it points—toward the point (x, y, z) starting from the origin. An important type of direction vector is the normal vector, which is a vector in a direction perpendicular to a surface, plane, or line.

Vectors of equal dimension are equal if they are equal componentwise. Vectors may be multiplied by scalars. This corresponds to stretching or shrinking the vector arrow along its original direction.

$$\lambda \mathbf{x} = (\lambda x_1, \lambda x_2, \ldots, \lambda x_n)$$

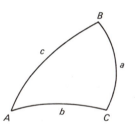

Fig. A1.5 Spherical triangle.

Vector addition and subtraction is defined componentwise, only between vectors of equal dimension. Geometrically, to add two vectors **x** and **y**, put **y**'s tail at **x**'s head and the sum is the vector from **x**'s tail to **y**'s head. To subtract **y** from **x**, put **y**'s head at **x**'s head; the difference is the vector from **x**'s tail to **y**'s tail.

$$\mathbf{x} \pm \mathbf{y} = (x_1 \pm y_1, x_2 \pm y_2, \ldots, x_n \pm y_n)$$

The length (or magnitude) of a vector is computed by an n-dimensional version of Euclidean distance.

$$|\mathbf{x}| = (x_1^2 + x_2^2 + \cdots + x_n^2)^{1/2}$$

A vector of unit length is a unit vector. The unit vectors in the three usual Cartesian coordinate directions have special names.

$$\mathbf{i} = (1, 0, 0)$$
$$\mathbf{j} = (0, 1, 0)$$
$$\mathbf{k} = (0, 0, 1)$$

The inner (or scalar, or dot) product of two vectors is defined as follows.

$$\mathbf{x} \cdot \mathbf{y} = |\mathbf{x}||\mathbf{y}| \cos\theta = x_1 y_1 + x_2 y_2 + \cdots + x_n y_n$$

Here θ is the angle between the two vectors. The dot product of two nonzero numbers is 0 if and only if they are orthogonal (perpendicular). The projection of **x** onto **y** (the component of vector **x** in the direction **y**) is

$$|\mathbf{x}|\cos\theta = \frac{\mathbf{x} \cdot \mathbf{y}}{|\mathbf{y}|}$$

Other identities of interest:

$$\mathbf{x} \cdot \mathbf{y} = \mathbf{y} \cdot \mathbf{x}$$
$$\mathbf{x} \cdot (\mathbf{y} + \mathbf{z}) = \mathbf{x} \cdot \mathbf{y} + \mathbf{x} \cdot \mathbf{z}$$
$$\lambda (\mathbf{x} \cdot \mathbf{y}) = (\lambda\mathbf{x}) \cdot \mathbf{y} = \mathbf{x} \cdot (\lambda\mathbf{y})$$
$$\mathbf{x} \cdot \mathbf{x} = |\mathbf{x}|^2$$

The cross (or vector) product of two three-dimensional vectors is defined as follows.

$$\mathbf{x} \times \mathbf{y} = (x_2 y_3 - x_3 y_2, \ x_3 y_1 - x_1 y_3, \ x_1 y_2 - x_2 y_1)$$

Generally, the cross product of **x** and **y** is a vector perpendicular to both **x** and **y**. The magnitude of the cross product depends on the angle θ between the two vectors.

$$|\mathbf{x} \times \mathbf{y}| = |\mathbf{x}||\mathbf{y}| \sin\theta$$

Thus the magnitude of the product is zero for two nonzero vectors if and only if they are parallel.

Vectors and matrices allow for the short formal expression of many symbolic expressions. One such example is the formal determinant (Section A1.4) which expresses the definition of the cross product given above in a more easily remembered form.

$$\mathbf{x} \times \mathbf{y} = \det \begin{vmatrix} \mathbf{i} & \mathbf{j} & \mathbf{k} \\ x_1 & x_2 & x_3 \\ y_1 & y_2 & y_3 \end{vmatrix}$$

Also,

$$\mathbf{x} \times \mathbf{y} = -\mathbf{y} \times \mathbf{x}$$

$$\mathbf{x} \times (\mathbf{y} \pm \mathbf{z}) = \mathbf{x} \times \mathbf{y} \pm \mathbf{x} \times \mathbf{z}$$

$$\lambda (\mathbf{x} \times \mathbf{y}) = \lambda \mathbf{x} \times \mathbf{y} = \mathbf{x} \times \lambda \mathbf{y}$$

$$\mathbf{i} \times \mathbf{j} = \mathbf{k}$$

$$\mathbf{j} \times \mathbf{k} = \mathbf{i}$$

$$\mathbf{k} \times \mathbf{i} = \mathbf{j}$$

The triple scalar product is $\mathbf{x} \cdot (\mathbf{y} \times \mathbf{z})$, and is equivalent to the value of the determinant

$$\det \begin{bmatrix} x_1 & x_2 & x_3 \\ y_1 & y_2 & y_3 \\ z_1 & z_2 & z_3 \end{bmatrix}$$

The triple vector product is

$$\mathbf{x} \times (\mathbf{y} \times \mathbf{z}) = (\mathbf{x} \cdot \mathbf{z})\mathbf{y} - (\mathbf{x} \cdot \mathbf{y})\mathbf{z}$$

A1.4. MATRICES

A matrix A is a two-dimensional array of elements; if it has m rows and n columns it is of dimension $m \times n$, and the element in the ith row and jth column may be named a_{ij}. If m or $n = 1$, a row matrix or column matrix results, which is often called a vector. There is considerable punning among scalar, vector and matrix representations and operations when the same dimensionality is involved (the 1×1 matrix may sometimes be treated as a scalar, for instance). Usually, this practice is harmless, but occasionally the difference is important.

A matrix is sometimes most naturally treated as a collection of vectors, and sometimes an $m \times n$ matrix M is written as

$$M = [\mathbf{a}_1 \quad \mathbf{a}_2 \quad \cdots \quad \mathbf{a}_n]$$

or

$$M = \begin{bmatrix} \mathbf{b}_1 \\ \mathbf{b}_2 \\ . \\ . \\ . \\ \mathbf{b}_m \end{bmatrix}$$

where the **a**'s are column vectors and the **b**'s are row vectors.

Two matrices A and B are equal if their dimensionality is the same and they are equal elementwise. Like a vector, a matrix may be multiplied (elementwise) by a scalar. Matrix addition and subtraction proceeds elementwise between matrices of like dimensionality. For a scalar k and matrices A, B, and C of like dimensionality the following is true.

$$A = B \pm C \quad \text{if } a_{ij} = b_{ij} \pm c_{ij} \quad 1 \leqslant i \leqslant m, \quad 1 \leqslant j \leqslant n$$

Two matrices A and B are conformable for multiplication if the number of columns of A equals the number of rows of B. The product is defined as

$$C = AB \quad \text{where an element } c_{ij} \text{ is defined by} \quad c_{ij} = \sum_k a_{ik} b_{kj}$$

Thus each element of C is computed as an inner product of a row of A with a column of B. Matrix multiplication is associative but not commutative in general. The multiplicative identity in matrix algebra is called the identity matrix I. I is all zeros except that all elements in its main diagonal have value 1 ($a_{ij} = 1$ if $i = j$, else $a_{ij} = 0$). Sometimes the $n \times n$ identity matrix is written I_n.

The transpose of an $m \times n$ matrix A is the $n \times m$ matrix A^T such that the i,jth element of A is the j,ith element of A^T. If $A^T = A$, A is symmetric.

The inverse matrix of an $n \times n$ matrix A is written A^{-1}. If it exists, then

$$AA^{-1} = A^{-1}A = I$$

If its inverse does not exist, an $n \times n$ matrix is called singular.

With k and p scalars, and A, B, and C $m \times n$ matrices, the following are some laws of matrix algebra (operations are matrix operations):

$$A + B = B + A$$

$$(A + B) + C = A + (B + C)$$

$$k(A + B) = kA + kB$$

$$(k + p)A = kA + pA$$

$$AB \neq BA \quad \text{in general}$$

$$(AB)C = A(BC)$$

$$A(B + C) = AB + AC$$

$$(A + B)C = AC + BC$$

$$A(kB) = k(AB) = (kA)B$$

$$I_m A = AI_n = A$$

$$(A + B^T) = A^T + B^T$$

$$(AB)^T = B^T A^T$$

$$(AB)^{-1} = B^{-1} A^{-1}$$

The determinant of an $n \times n$ matrix is an important quantity; among other things, a matrix with zero determinant is singular. Let A_{ij} be the $(n-1) \times (n-1)$ matrix resulting from deleting the ith row and jth column from an $n \times n$ matrix A. The determinant of a 1×1 matrix is the value of its single element. For $n > 1$,

$$\det A = \sum_{i=1}^{n} a_{ij} (-1)^{i+j} \det A_{ij}$$

for any j between 1 and n. Given the definition of determinant, the inverse of a matrix may be defined as

$$(a^{-1})_{ij} = \frac{(-1)^{i+j} \det A_{ji}}{\det A}$$

In practice, matrix inversion may be a difficult computational problem, but this important algorithm has received much attention, and robust and efficient methods exist in the literature, many of which may also be used to compute the determinant. Many of the matrices arising in computer vision have to do with geometric transformations, and have well-behaved inverses corresponding to the inverse transformations. Matrices of small dimensionality are usually quite computationally tractable.

Matrices are often used to denote linear transformations; if a row (column) matrix X of dimension n is post (pre)multiplied by an $n \times n$ matrix A, the result $X' = XA$ ($X' = AX$) is another row (column) matrix, each of whose elements is a linear combination of the elements of X, the weights being supplied by the values of A. By employing the common pun between row matrices and vectors, $\mathbf{x}' = \mathbf{x}A$ ($\mathbf{x}' = A\mathbf{x}$) is often written for a linear transformation of a vector $\mathbf{x}$.

An eigenvector of an $n \times n$ matrix A is a vector $\mathbf{v}$ such that for some scalar λ (called an eigenvalue),

$$\mathbf{v}A = \lambda\mathbf{v}$$

That is, the linear transformation A operates on $\mathbf{v}$ just as a scaling operation. A matrix has n eigenvalues, but in general they may be complex and of repeated values. The computation of eigenvalues and eigenvectors of matrices is another computational problem of major importance, with good algorithms for general matrices being complicated. The n eigenvalues are roots of the so-called characteristic polynomial resulting from setting a formal determinant to zero:

$$\det (A - \lambda I) = 0.$$

Eigenvalues of matrices up to 4×4 may be found in closed form by solving the characteristic equation exactly. Often, the matrices whose eigenvalues are of interest are symmetric, and luckily in this case the eigenvalues are all real. Many algorithms exist in the literature which compute eigenvalues and eigenvectors both for symmetric and general matrices.

A1.5. LINES

An infinite line may be represented by several methods, each with its own advantages and limitations. An example of a representation which is not often very useful is two planes that intersect to form the line. The representations below have proven generally useful.

A1.5.1 Two Points

A two-dimensional or three-dimensional line (throughout Appendix 1 this shorthand is used for "line in two-space" and "line in three-space"; similarly for "two (three) dimensional point") is determined by two points on it, **x1** and **x2**. This representation can serve as well for a half-line or a line segment. The two points can be kept as the rows of a $(2 \times n)$ matrix.

A1.5.2 Point and Direction

A two-dimensional or three-dimensional line (or half-line) is determined by a point **x** on it (its endpoint) and a direction vector **v** along it. This representation is essentially the same as that of Section A1.5.1, but the interpretation of the vectors is different.

A1.5.3 Slope and Intercept

A two-dimensional line can often be represented by the Y value b where the line intersects the Y axis, and the slope m of the line (the tangent of its inclination with the x axis). This representation fails for vertical lines (those with infinite slope). The representation is in the form of an equation making explicit the dependence of y on x:

$$y = mx + b$$

A similar representation may of course be based on the X intercept.

A1.5.4 Ratios

A two-dimensional or three-dimensional line may be represented as an equation of ratios arising from two points $\mathbf{x1} = (x_1, y_1, z_1)$ and $\mathbf{x2} = (x_2, y_2, z_2)$ on the line.

$$\frac{x - x_1}{x_2 - x_1} = \frac{y - y_1}{y_2 - y_1} = \frac{z - z_1}{z_2 - z_1}$$

A1.5.5 Normal and Distance from Origin (Line Equation)

This representation for two-dimensional lines is elegant in that its parts have useful geometric significance which extends to planes (not to three-dimensional lines). The coefficients of the general two-dimensional linear equation represent a two-dimensional line and incidentally give its normal (perpendicular) vector and its (perpendicular) distance from the origin (Fig. A1.6).

From the ratio representation above, it is easy to derive (in two dimensions) that

$$(x - x_1) \sin \theta \; - (y - y_1) \cos \theta = 0$$

so for

$$d = -(x_1 \sin \theta - y_1 \cos \theta),$$

$$x \sin \theta - y \cos \theta + d = 0$$

This equation has the form of a dot product with a formal homogeneous vector $(x, y, 1)$:

$$(x, y, 1) \cdot (\sin \theta, -\cos \theta, d) = 0$$

Here the two-dimensional vector $(\sin \theta, -\cos \theta)$ is perpendicular to the line (it is a unit normal vector, in fact), and d is the signed distance in the direction of the normal vector from the line to the origin. Multiplying both sides of the equation by a constant leaves the line invariant, but destroys the interpretation of d as the distance to the origin.

This form of line representation has several advantages besides the interpretations of its parameters. The parameters never go to infinity (this is useful in the Hough algorithm described in Chapter 4). The representation extends naturally to representing n-dimensional planes. Least squared error line fitting (Section A1.9) with this form of line equation (as opposed to slope-intercept) minimizes errors perpendicular to the line (as opposed to those perpendicular to one of the coordinate axes).

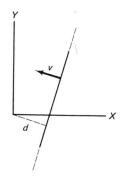

Fig. A1.6 Two-dimensional line with normal vector and distance to origin.

A1.5.6 Parametric

It is sometimes useful to be able mathematically to "walk along" a line by varying some parameter t. The basic parametric representation here follows from the two-point representation. If $\mathbf{x1}$ and $\mathbf{x2}$ are two particular points on the line, a general point on the line may be written as

$$\mathbf{x} = \mathbf{x1} + t(\mathbf{x2} - \mathbf{x1})$$

In matrix terms this is

$$\mathbf{x} = [t \quad 1]L$$

where L is the $2 \times n$ matrix whose first row is $(\mathbf{x2} - \mathbf{x1})$ and whose second is $\mathbf{x1}$. Parametric representations based on points on the lines may be transformed by the geometric point transformations (Section A1.7).

A1.6. PLANES

The most common representation of planes is to use the coordinates of the plane equation. This representation is an extension of the line-equation representation of Section A1.5.5. The plane equation may be written

$$ax + by + cz + d = 0$$

which is in the form of a dot product $\mathbf{x} \cdot \mathbf{p} = 0$. Four numbers given by $\mathbf{p} = (a, b, c, d)$ characterize a plane, and any homogeneous point $\mathbf{x} = (x, y, z, w)$ satisfying the foregoing equation lies in the plane. In $\mathbf{p}$, the first three numbers (a, b, c) form a normal vector to the plane. If this normal vector is made to be a unit vector by scaling $\mathbf{p}$, then d is the signed distance to the origin from the plane. Thus the dot product of the plane coefficient vector and any point (in homogeneous coordinates) gives the distance of the point to the plane (Fig. A1.7).

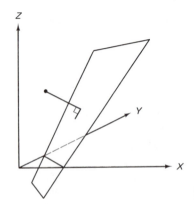

Fig. A1.7 Distance from a point to a plane.

App. 1   Some Mathematical Tools

Three noncollinear points **x1**, **x2**, **x3** determine a plane **p**. To find it, write

$$\begin{vmatrix} \mathbf{x1} & & & \\ \mathbf{x2} & & & \\ \mathbf{x3} & & & \\ 0 & 0 & 0 & 1 \end{vmatrix} \mathbf{p} = \begin{bmatrix} 0 \\ 0 \\ 0 \\ 1 \end{bmatrix}$$

If the matrix containing the point vectors can be inverted, the desired vector **p** is thus proportional to the fourth column of the inverse.

Three planes **p1**, **p2**, **p3** may intersect in a point **x**. To find it, write

$$\mathbf{x} \begin{vmatrix} \mathbf{p1} & \mathbf{p2} & \mathbf{p3} & 0 \\ & & & 0 \\ & & & 0 \\ & & & 1 \end{vmatrix} = [0 \quad 0 \quad 0 \quad 1]$$

If the matrix containing the plane vectors can be inverted, the desired point **p** is given by the fourth row of the inverse. If the planes do not intersect in a point, the inverse does not exist.

A1.7 GEOMETRIC TRANSFORMATIONS

This section contains some results that are well known through their central place in the computer graphics literature, and illustrated in greater detail there. The idea is to use homogeneous coordinates to allow the writing of important transformations (including affine and projective) as linear transformations. The transformations of interest here map points or point sets onto other points or point sets. They include rotation, scaling, skewing, translation, and perspective distortion (point projection) (Fig. A1.8).

A point x in three-space is written as the homogeneous row four-vector (x, y, z, w), and postmultiplication by the following transformation matrices accomplishes point transformation. A set of m points may be represented as an $m \times 4$ matrix of row point vectors, and the matrix multiplication transforms all points at once.

A1.7.1 Rotation

Rotation is measured clockwise about the named axis while looking along the axis toward the origin.

Rotation by θ about the X axis:

$$\begin{bmatrix} 1 & 0 & 0 & 0 \\ 0 & \cos\theta & -\sin\theta & 0 \\ 0 & \sin\theta & \cos\theta & 0 \\ 0 & 0 & 0 & 1 \end{bmatrix}$$

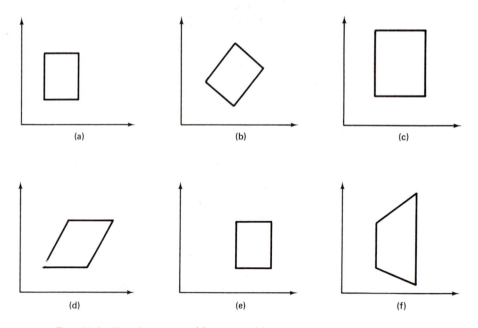

Fig. A1.8 Transformations: (a) original, (b) rotation, (c) scaling, (d) skewing, (e) translation, and (f) perspective.

Rotation by θ about the Y axis:

$$\begin{bmatrix} \cos\theta & 0 & \sin\theta & 0 \\ 0 & 1 & 0 & 0 \\ -\sin\theta & 0 & \cos\theta & 0 \\ 0 & 0 & 0 & 1 \end{bmatrix}$$

Rotation by θ about the Z axis:

$$\begin{bmatrix} \cos\theta & -\sin\theta & 0 & 0 \\ \sin\theta & \cos\theta & 0 & 0 \\ 0 & 0 & 1 & 0 \\ 0 & 0 & 0 & 1 \end{bmatrix}$$

A1.7.2 Scaling

Scaling is stretching points out along the coordinate directions. Scaling can transform a cube to an arbitrary rectangular parallelepiped.

Scale by S_x, S_y, and S_z in the X, Y, and Z directions:

$$\begin{bmatrix} S_x & 0 & 0 & 0 \\ 0 & S_y & 0 & 0 \\ 0 & 0 & S_z & 0 \\ 0 & 0 & 0 & 1 \end{bmatrix}$$

A1.7.3 Skewing

Skewing is a linear change in the coordinates of a point based on certain of its other coordinates. Skewing can transform a square into a parallelogram in a simple case:

$$\begin{bmatrix} 1 & 0 & 0 & 0 \\ d & 1 & 0 & 0 \\ 0 & 0 & 1 & 0 \\ 0 & 0 & 0 & 1 \end{bmatrix}$$

In general, skewing is quite powerful:

$$\begin{bmatrix} 1 & k & n & 0 \\ d & 1 & p & 0 \\ e & m & 1 & 0 \\ 0 & 0 & 0 & 1 \end{bmatrix}$$

Rotation is a composition of scaling and skewing (Section A1.7.7).

A1.7.4 Translation

Translate a point by (t, u, v):

$$\begin{bmatrix} 1 & 0 & 0 & 0 \\ 0 & 1 & 0 & 0 \\ 0 & 0 & 1 & 0 \\ t & u & v & 1 \end{bmatrix}$$

With a three-dimensional Cartesian point representation, this transformation is accomplished through vector addition, not matrix multiplication.

A1.7.5 Perspective

The properties of point projection, which model perspective distortion, were derived in Chapter 2. In this formulation the viewpoint is on the positive Z axis at $(0, 0, f, 1)$ looking toward the origin: f acts like a "focal length". The visible world is projected through the viewpoint onto the $Z = 0$ image plane (Fig. A1.9).

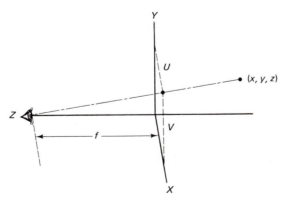

Fig. A1.9 Geometry of image formation.

Similar triangles arguments show that the image plane point for any world point (x, y, z) is given by

$$(U, V) = \left(\frac{fx}{f-z}, \frac{fy}{f-z} \right)$$

Using homogeneous coordinates, a "perspective distortion" transformation can be written which distorts three-dimensional space so that after orthographic projection onto the image plane, the result looks like that required above for perspective distortion. Roughly, the transformation shrinks the size of things as they get more distant in Z. Although the transformation is of course linear in homogeneous coordinates, the final step of changing to Cartesian coordinates by dividing through by the fourth vector element accomplishes the nonlinear shrinking necessary.

Perspective distortion (situation of Fig. A1.9):

$$\begin{bmatrix} 1 & 0 & 0 & 0 \\ 0 & 1 & 0 & 0 \\ 0 & 0 & 1 & \frac{-1}{f} \\ 0 & 0 & 0 & 1 \end{bmatrix}$$

Perspective from a general viewpoint has nonzero elements in the entire fourth column, but this is just equivalent to a rotated coordinate system and the perspective distortion above (Section A1.7).

A1.7.6 Transforming Lines and Planes

Line and plane equations may be operated on by linear transformations, just as points can. Point-based parametric representations of lines and planes transform as do points, but the line and plane equation representations act differently. They have an elegant relation to the point transformation. If T is a transformation matrix (3×3 for two dimensions, 4×4 for three dimensions) as defined in Sections A1.7.1 to A1.7.5, then a point represented as a row vector is transformed as

$$\mathbf{x}' = \mathbf{x} T$$

and the linear equation (line or plane) when represented as a column vector $\mathbf{v}$ is transformed by

$$\mathbf{v}' = T^{-1} \mathbf{v}$$

A1.7.7 Summary

The 4×4 matrix formulation is a way to unify the representation and calculation of useful geometric transformations, rigid (rotation and translation), and nonrigid (scaling and skewing), including the projective. The semantics of the matrix are summarized in Fig. A1.10.

Since the results of applying a transformation to a row vector is another row vector, transformations may be concatenated by repeated matrix multiplication. Such composition of transformations follows the rules of matrix algebra (it is associative but not commutative, for instance). The semantics of

$$\mathbf{x}' = \mathbf{x}ABC$$

is that $\mathbf{x}'$ is the vector resulting from applying transformation A to $\mathbf{x}$, then B to the transformed $\mathbf{x}$, then C to the twice-transformed $\mathbf{x}$. The single 4×4 matrix $D = ABC$ would do the same job. The inverses of geometric transformation matrices are just the matrices expressing the inverse transformations, and are easy to derive.

A1.8. CAMERA CALIBRATION AND INVERSE PERSPECTIVE

The aim of this section is to explore the correspondence between world and image points. A (half) line of sight in the world corresponds to each image point. Camera calibration permits prediction of where in the image a world point will appear. Inverse perspective transformation determines the line of sight corresponding to an image point. Given an inverse perspective transform and the knowledge that a visible point lies on a particular world plane (say the floor, or in a planar beam of light), then its precise three-dimensional coordinates may be found, since the line of sight generally intersects the world plane in just one point.

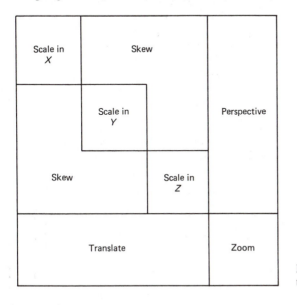

Fig. A1.10   The 4×4 homogeneous transformation matrix.

A1.8.1 Camera Calibration

This section is concerned with the "camera model"; the model takes the form of a 4×3 matrix mapping three-dimensional world points to two-dimensional image points. There are many ways to derive a camera model. The one given here is easy to state mathematically; in practice, a more general optimization technique such as hill climbing can be most effective in finding the camera parameters, since it can take advantage of any that are already known and can reflect dependencies between them.

Let the image plane coordinates be U and V; in homogeneous coordinates an image plane point is (u, v, t). Thus

$$U = \frac{u}{t}$$

$$V = \frac{v}{t}$$

Call the desired camera model matrix C, with elements C_{ij} and column four-vectors C_j. Then for any world point (x, y, z) a C is needed such that

$$(x, y, z, 1)\,C = (u, v, t)$$

So

$$u = (x, y, z, 1)\,C_1$$
$$v = (x, y, z, 1)\,C_2$$
$$t = (x, y, z, 1)\,C_3$$

Expanding the inner products and rewriting $u - Ut = 0$ and $v - Vt = 0$,

$$xC_{11} + yC_{21} + zC_{31} + C_{41} - UxC_{13} - UyC_{23} - UzC_{33} - UC_{43} = 0$$
$$xC_{12} + yC_{22} + zC_{32} + C_{42} - VxC_{13} - VyC_{23} - VzC_{33} - VC_{43} = 0$$

The overall scaling of C is irrelevant, thanks to the homogeneous formulation, so C_{43} may be arbitrarily set to 1. Then equations such as those above can be written in matrix form:

$$
\begin{bmatrix}
x^1 & y^1 & z^1 & 1 & 0 & 0 & 0 & 0 & -U^1x^1 & -U^1y^1 & -U^1z^1 \\
0 & 0 & 0 & 0 & x^1 & y^1 & z^1 & 1 & -V^1x^1 & -V^1y^1 & -V^1z^1 \\
x^2 & y^2 & z^2 & 1 & . & & . & & . & & \\
. & & & & & & & & & & \\
. & & & & & & & & & & \\
. & & & & & & & & & & \\
0 & 0 & 0 & 0 & x^n & y^n & z^n & 1 & -V^nx^n & -V^ny^n & -V^nz^n
\end{bmatrix}
\begin{bmatrix}
C_{11} \\
C_{21} \\
. \\
. \\
. \\
. \\
C_{34}
\end{bmatrix}
=
\begin{bmatrix}
U^1 \\
V^1 \\
. \\
. \\
. \\
U^n \\
V^n
\end{bmatrix}
$$

Eleven such equations allow a solution for C. Two equations result for every association of an (x, y, z) point with a (U, V) point. Such an association must be established using visible objects of known location (often placed for the purpose). If more than $5\frac{1}{2}$ such observations are used, a least-squared-error solution to the overdetermined system may be obtained by using a pseudo-inverse to solve the resulting matrix equation (Section A1.9).

A1.8.2 Inverse Perspective

Finding the world line corresponding to an image point relies on the fact that the perspective transformation matrix also affects the z component of a world point. This information is lost when the z component is projected away orthographically, but it encodes the relation between the focal point and the z position of the point. Varying this third component references points whose world positions vary in z but which project onto the same position in the image. The line can be parameterized by a variable p that formally occupies the position of that z coordinate in three-space that has no physical meaning in imaging.

Write the inverse perspective transform P^{-1} as

$$(x', y', p, 1)P^{-1} = (x', y', p, 1 + \frac{p}{f})$$

Rewriting this in the usual way gives these relations between the (x, y, z) points on the line.

$$(x, y, z, 1) = \left| \frac{fx'}{f + p}, \frac{fy'}{f + p}, \frac{fp'}{f + p}, 1 \right|$$

Eliminating the parameter p between the expressions for z and x and those for z and y leaves

$$x = \frac{x'}{y'}, \quad y = \frac{-x'}{f}(z - f)$$

Thus x, y, and z are linearly related; as expected, all points on the inverse perspective transform of an image point lie in a line, and unsurprisingly both the viewpoint $(0, 0, f)$ and the image point $(x', y', 0)$ lie on it.

A camera matrix C determines the three-dimensional line that is the inverse perspective transform of any image point. Scale C so that $C_{43} = 1$, and let world points be written $\mathbf{x} = (x, y, z, 1)$ and image points $\mathbf{u} = (u, v, t)$. The actual image points are then

$$U = \frac{u}{t}, \quad V + \frac{v}{t}, \quad \text{so } u = Ut, \quad v + Vt$$

Since

$$\mathbf{u} = \mathbf{x}C,$$

$$u = Ut = \mathbf{x}C_1$$

$$v = Vt = \mathbf{x}C_2$$

$$t = \mathbf{x}C_3$$

Substituting the expression for t into that for u and v gives

$$U \mathbf{x} C_3 = \mathbf{x} C_1$$

$$V \mathbf{x} C_3 = \mathbf{x} C_2$$

which may be written

$$\mathbf{x}(C_1 - U C_3) = 0$$

$$\mathbf{x}(C_2 - V C_3) = 0$$

These two equations are in the form of plane equations. For any U, V in the image and camera model C, there are determined two planes whose intersection gives the desired line. Writing the plane equations as

$$a_1 x + b_1 y + c_1 z + d_1 = 0$$

$$a_2 x + b_2 y + c_2 z + d_2 = 0$$

then

$$a_1 = C_{11} - C_{13} U \qquad a_2 = C_{12} - C_{13} V$$

and so on. The direction (λ, μ, ν) of the intersection of two planes is given by the cross product of their normal vectors, which may now be written as

$$(\lambda, \mu, \nu) = (a_1, b_1, c_1) \times (a_2, b_2, c_2)$$

$$= (b_1 c_2 - b_2 c_1, \; c_1 a_2 - c_2 a_1, \; a_1 b_2 - a_2 b_1)$$

Then if $\nu \neq 0$, for any particular z_0,

$$x_0 = \frac{b_1 (c_2 z_0 + d_2) - b_2 (c_1 z_0 - d_1)}{a_1 b_2 - b_1 a_2}$$

$$y_0 = \frac{a_2 (c_1 z_0 + d_1) - a_1 (c_2 z_0 - d_2)}{a_1 b_2 - b_1 a_2}$$

and the line may be written

$$\frac{x - x_0}{\lambda} = \frac{y - y_0}{\mu} = \frac{z - z_0}{\nu}$$

A1.9. LEAST-SQUARED-ERROR FITTING

The problem of fitting a simple functional model to a set of data points is a common one, and is the concern of this section. The subproblem of fitting a straight line to a set of (x, y) points ("linear regression") is the first topic. In computer vision, this line-fitting problem is encountered relatively often. Model-fitting methods try to find the "best" fit; that is, they minimize some error. Methods which yield closed-form, analytical solutions for such best fits are at issue here.

The relevant "error" to minimize is determined partly by assumptions of dependence between variables. If x is independent, the line may be represented as $y = mx + b$ and the error defined as the vertical displacement of a point from the line. Symmetrically, if x is dependent, horizontal error should be minimized. If neither variable is dependent, a reasonable error to minimize is the perpendicular distance from points to the line. In this case the line equation $ax + by + 1 = 0$ can be used with the method shown here, or the eigenvector approach of Section A1.9.2 may be used.

A1.9.1 Pseudo-Inverse Method

In fitting an $n \times 1$ observations matrix y by some linear model of p parameters, the prediction is that the linear model will approximate the actual data. Then

$$Y = XB + E$$

where X is an $n \times p$ formal independent variable matrix, B is a $p \times 1$ parameter matrix whose values are to be determined, and E represents the difference between the prediction and the actuality: it is an $n \times 1$ error matrix.

For example, to fit a straight line $y = mx + b$ to some data (x_i, y_i) points, form Y as a column matrix of the y_i.

$$X = \begin{bmatrix} 1 & x_1 \\ 1 & x_2 \\ 1 & x_3 \\ & \cdot \\ & \cdot \\ & \cdot \end{bmatrix}$$

$$B = \begin{bmatrix} b \\ m \end{bmatrix}$$

Now the task is to find the parameter B (above, the b and m that determine the straight line) that minimizes the error. The error is the sum of squared difference from the prediction, or the sum of the elements of E squared, or $E^T E$ (if we do not mind conflating the one-element matrix with a scalar). The mathematically attractive properties of the squared-error definition are almost universally taken to compensate for whatever disadvantages it has over what is really meant by error (the absolute value is much harder to calculate with, for example).

To minimize the error, simply differentiate it with respect to the elements of B and set the derivative to 0. The second derivative is positive: this is indeed a minimum. These elementwise derivatives are written tersely in matrix form. First rewrite the error terms:

$$E^T E = (Y - XB)^T (Y - XB)$$
$$= Y^T Y - B^T X^T Y - Y^T XB + B^T X^T XB$$
$$= Y^T Y - 2B^T X^T Y + B^T X^T XB$$

(here, the combined terms were 1×1 matrices.) Now differentiate: setting the derivative to 0 yields

$$0 = X^T X B - X^T Y$$

and thus

$$B = (X^T X)^{-1} X^T Y = X^{\dagger} Y$$

where $X^{\dagger}$ is called the pseudo-inverse of X.

The pseudo-inverse method generalizes to fitting any parametrized model to data (Section A1.9.3). The model should be chosen with some care. For example, Fig. A1.11 shows a disturbing case in which the model above (minimize vertical errors) is used to fit a relatively vertical swarm of points. The "best fit" line in this case is not the intuitive one.

A1.9.2 Principal Axis Method

The principal axes and moments of a swarm of points determine the direction and amount of its dispersion in space. These concepts are familiar in physics as the principal axes and moments of inertia. If a swarm of (possibly weighted) points is translated so that its center of mass (average location) is at the origin, a symmetric matrix M may be easily calculated whose eigenvectors determine the best-fit line or plane in a least-squared-perpendicular-error sense, and whose eigenvalues tell how good the resulting fit is.

Given a set $\{\mathbf{x}^i\}$ row of vectors with weights w^i, define their "scatter matrix" to be the symmetric matrix M, where $\mathbf{x}^i = (x_1^i, x_2^i, x_3^i)$:

$$M = \sum_i \mathbf{x}^{i^T} \mathbf{x}^i$$

$$M_{kp} = \sum_i x_k^i x_p^i \qquad 1 \leqslant k, \, p \leqslant 3$$

Define the dispersion of the $\mathbf{x}^i$ in a direction $\mathbf{v}$ (i.e., "dispersion around the plane whose normal is $\mathbf{v}$") to be the sum of weighted squared lengths of the $\mathbf{x}^i$ in the direction $\mathbf{v}$. This squared error E^2 is

$$E^2 = \sum_i w^i (\mathbf{x}^i \cdot \mathbf{v})^2 = \mathbf{v} \left(\sum_i w^i \mathbf{x}^{i^T} \mathbf{x}^i \right) \mathbf{v}^T = \mathbf{v} M \mathbf{v}^T$$

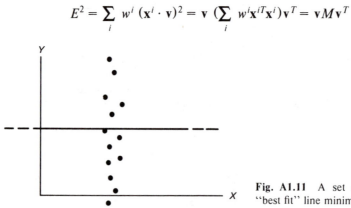

Fig. A1.11 A set of points and the "best fit" line minimizing error in Y.

To find the direction of minimum dispersion (the normal to the best-fit line or plane), note that the minimum of $\mathbf{v}M\mathbf{v}^T$ over all unit vectors $\mathbf{v}$ is the minimum eigenvalue λ_1 of M. If $\mathbf{v}_1$ is the corresponding eigenvector, the minimum dispersion is attained at $\mathbf{v} = \mathbf{v}_1$. The best fit line or plane of the points goes through the center of mass, which is at the origin; inverting the translation that brought the centroid to the origin yields the best fit line or plane for the original point swarm.

The eigenvectors correspond to dispersions in orthogonal directions, and the eigenvalues tell how much dispersion there is. Thus with a three-dimensional point swarm, two large eigenvalues and one small one indicate a planar swarm whose normal is the smallest eigenvector. Two small eigenvalues and one large one indicate a line in the direction of the normal to the "worst fit plane", or eigenvector of largest eigenvalue. (It can be proved that in fact this is the best-fit line in a least squared perpendicular error sense). Three equal eigenvalues indicate a "spherical" swarm.

A1.9.3 Fitting Curves by the Pseudo-Inverse Method

Given a function $f(\mathbf{x})$ whose value is known on n points $\mathbf{x}_1, \ldots, \mathbf{x}_n$, it may be useful is to fit it with a function $g(\mathbf{x})$ of m parameters $(b_1, \ldots, b_m)$. If the squared error at a point x_i is defined as

$$(e_i)^2 = [f(\mathbf{x}_i) - g(\mathbf{x}_i)]^2$$

a sequence of steps similar to that of Section A1.9.1 leads to setting a derivative to zero and obtaining

$$0 = G^T G \mathbf{b} - G^T \mathbf{f}$$

where $\mathbf{b}$ is the vector of parameters, $\mathbf{f}$ the vector of n values of $f(\mathbf{x})$, and

$$G = \frac{\partial \mathbf{g}}{\partial \mathbf{b}} = \begin{bmatrix} \dfrac{\partial g(x_1)}{\partial b_1} & \dfrac{\partial g(x_2)}{\partial b_2} & \cdots \\ \cdot & & \\ \cdot & & \\ \cdot & \cdots & \dfrac{\partial g(x_n)}{\partial b_m} \end{bmatrix}$$

As before, this yields

$$\mathbf{b} = (G^T G)^{-1} G^T \mathbf{f}$$

Explicit least-squares solutions for curves can have nonintuitive behavior. In particular, say that a general circle is represented

$$f(x, y) = x^2 + y^2 + 2Dx + 2Ey + F$$

this yields values of D, E, and F which minimize

$$e^2 = \sum_{i=1}^{n} f(x_i, y_i)^2$$

for n input points. The error term being minimized does not turn out to accord with our intuitive one. It gives the intuitive distance of a point to the curve, but weighted by a factor roughly proportional to the radius of the curve (probably not desirable). The best fit criterion thus favors curves with high average curvature, resulting in smaller circles than expected. In fitting ellipses, this error criterion favors more eccentric ones.

The most successful conic fitters abandon the luxury of a closed-form solution and go to iterative minimization techniques, in which the error measure is adjusted to compensate for the unwanted weighting, as follows.

$$e^2 = \sum_{i=1}^{n} \left[\frac{f(x_i, y_i)}{|\nabla f(x_i, y_i)|} \right]^2$$

A1.10 CONICS

The conic sections are useful because they provide closed two-dimensional curves, they occur in many images, and they are well-behaved and familiar polynomials of low degree. This section gives their equations in standard form, illustrates how the general conic equation may be put into standard form, and presents some sample specific results for characterizing ellipses.

All the standard form conics may be subjected to rotation, translation, and scaling to move them around on the plane. These operations on points affect the conic equation in a predictable way.

Circle: r = radius $\qquad x^2 + y^2 = r^2$

Ellipse: a, b = major, minor axes $\qquad \dfrac{x^2}{a^2} + \dfrac{y^2}{b^2} = 1$

Parabola: $(p, 0)$ = focus, p = directrix $\qquad y^2 = 4px$

Hyperbola: vertices $(\pm a, 0)$, asymptotes $y = \pm \left(\dfrac{b}{a}\right) x \qquad \dfrac{x^2}{a^2} - \dfrac{y^2}{b^2} = 1$

The general conic equation is

$$Ax^2 + 2Bxy + Cy^2 + 2Dx + 2Ey + F = 0$$

This equation may be written formally as

$$(x \quad y \quad 1) \begin{bmatrix} A & B & D \\ B & C & E \\ D & E & F \end{bmatrix} \begin{pmatrix} x \\ y \\ 1 \end{pmatrix} = \mathbf{x} M \mathbf{x}^T = 0$$

Putting the general conic equation into one of the standard forms is a common analytic geometry exercise. The symmetric 3×3 matrix M may be diagonalized, thus eliminating the coefficients B, D, and E from the equation and reducing it to be close to standard form. The diagonalization amounts to a rigid motion that puts the conic in a symmetric position at the origin. The transformation is in fact the 3×3 matrix E whose rows are eigenvectors of M. Recall that if $\mathbf{v}$ is an eigenvector of M,

$$\mathbf{v}M = \lambda \mathbf{v}$$

Then if D is a diagonal matrix of the three eigenvalues, $\lambda_1, \lambda_2, \lambda_3$,

$$EM = DE$$

but then

$$EME^{-1} = DEE^{-1} = D$$

and M has been transformed by a similarity transformation into a diagonal matrix such that

$$\mathbf{x}D\mathbf{x}^T = 0$$

This general idea is of course related to the principal axis calculation given in Section A1.9.2, and extends to three-dimensional quadric surfaces such as the ellipsoid, cone, hyperbolic paraboloid, and so forth. The general result given above has particular consequences illustrated by the following facts about the ellipse. Given a general conic equation representing an ellipse, its center (x_c, y_c) is given by

$$x_c = \frac{CD - BE}{B^2 - AC}$$

$$y_c = \frac{EA - BD}{B^2 - AC}$$

The orientation is

$$\theta = \tfrac{1}{2}\tan^{-1}\left|\frac{B}{A - C}\right|$$

The major and minor axes are

$$\frac{-2G}{(A + C) \pm [B^2 + (A - C)^2]^{\frac{1}{2}}}$$

where

$$G = F - (Ax_c^2 + Bx_c y_c + Cy_c^2)$$

A1.11 INTERPOLATION

Interpolation fits data by giving values between known data points. Usually, the interpolating function passes through each given data point. Many interpolation methods are known; one of the simplest is Lagrangean interpolation.

A1.11.1 One-Dimensional

Given $n + 1$ points (x_j, y_j), $x_0 < x_1 < \cdots < x_n$, the idea is to produce an nth-degree polynomial involving $n + 1$ so-called Lagrangean coefficients. It is

$$f(x) = \sum_{j=0}^{n} L_j(x)y_j$$

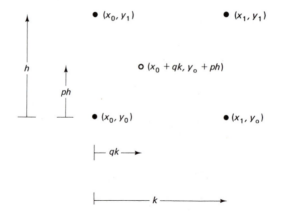

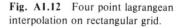

Fig. A1.12 Four point lagrangean interpolation on rectangular grid.

where $L_j(x)$ is the jth coefficient;

$$L_j(x) = \frac{(x - x_0)\ (x - x_1)\ \cdots\ (x - x_{j-1})\ (x - x_{j+1})\ \cdots\ (x - x_n)}{(x_j - x_0)\ (x_j - x_i)\ \cdots\ (x_j - x_{j-1})\ (x_j - x_{j+1})\ \cdots\ (x_j - x_n)}$$

Other interpolative schemes include divided differences, Hermite interpolation for use when function derivatives are also known, and splines. The use of a polynomial interpolation rule can always produce surprising results if the function being interpolated does not behave locally like a polynomial.

A1.11.2 Two-Dimensional

The four-point Lagrangean method is for the situation shown in Fig. A1.12. Let $f_{ij} = f(x_i, y_j)$. Then

$$f(x_0 + qk, y_0 + ph) = (1 - p)\ (1 - q)\ f_{00} + q\ (1 - p)\ f_{10} + p\ (1 - q)\ f_{01} + pqf_{11}$$

A1.12 THE FAST FOURIER TRANSFORM

The following routine computes the discrete Fourier transform of a one-dimensional complex array XIn of length $N = 2^{\log N}$ and produces the one-dimensional complex array XOut. It uses an array W of the N complex Nth roots of unity, computed as shown, and an array Bits containing a bit-reversal table of length N. N, LogN, W, and Bits are all global to the subroutine as written. If the logical variable Forward is TRUE, the FFT is performed; if Forward is FALSE, the inverse FFT is performed.

```
SUBROUTINE FFT(XIn, KOut, Forward)
GLOBAL W, Bits, N, LogN
LOGICAL Forward
COMPLEX XIn, Xout, W, A, B
INTEGER Bits
ARRAY(0:N) W, Bits, XIn, XOut
```

```
DO (I = 0, N − 1) XOut(I) = XIn(Bits(I))
JOff = N/2
JPnt = N/2
JBk = 2
IOFF = 1
DO  (I = 1, LogN)
.      DO  (IStart = 0, N − 1, JBk)
.      .      JWPnt = 0
.      .      DO  (K = IStart, IStart + IOff − 1)
.      .      .      WHEN (Forward)
.      .      .      .      A = XOut(K + IOff) * W(JWPnt) + XOut(K)
.      .      .      .      B = XOut(K + IOff) * W(JWPnt + JOff) + XOut(K)
.      .      .      ...  FIN
.      .      .      ELSE
.      .      .      .      A = XOut (K + IOff) * CONJG(W(JWPnt)) + XOut(K)
.      .      .      .      B = XOut(K + IOff) * CONJG(W(JWPnt + JOff)) + XOut(K)
.      .      .      ...  FIN
.      .      .      XOut(K) = A
.      .      .      XOut(K + IOff) = B
.      .      .      JWPnt = JWPnt + JPnt
.      .      ...  FIN
.      ...  FIN
.      JPnt = JPnt/2
.      IOff = JBk
.      JBk = JBk * 2
...  FIN
UNLESS (Forward)
.      DO  (I = 0, N − 1) XOut(I) = XOut(I)/N
...  FIN
END


────────────────────────────────────────


TO  INIT-W
.      Pi = 3.14159265
.      DO  (K = 0, N − 1)
.      .      Theta = 2 * Pi/N
.      .      W(K) = CMPLX(COS(Theta * K), SIN(Theta * K))
.      ...  FIN
...  FIN


────────────────────────────────────────


TO  BIT-REV
.      Bits(0) = 0
.      M = 1
.      DO  (I = 0, LogN − 1)
.      .      DO  (J = 0, M − 1)
.      .      .      Bits(J) = Bits(J) * 2
```

```
.     .     .        Bits(J + M ) = Bits(J) + 1
.     .     ...   FIN
.     .     M = M * 2
.     ...  FIN
...   FIN
```

A1.13 THE ICOSAHEDRON

Geodesic dome constructions provide a useful way to partition the sphere (hence the three-dimensional directions) into relatively uniform patches. The resulting polyhedra look like those of Fig. A1.13.

   The icosahedron has 12 vertices, 20 faces, and 30 edges. Let its center be at the origin of Cartesian coordinates and let each vertex be a unit distance from the center. Define

$$t, \text{ the golden ratio} = \frac{1 + \sqrt{5}}{2}$$

$$a = \frac{\sqrt{t}}{5^{1/4}}$$

$$b = \frac{1}{(\sqrt{t} \; 5^{1/4})}$$

$$c = a + 2b = \frac{1}{b}$$

$$d = a + b = \frac{t^{3/2}}{5^{1/4}}$$

$$A = \text{angle subtended by edge at origin} = \arccos(\frac{\sqrt{5}}{5})$$

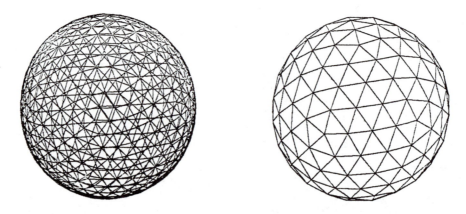

Fig. A1.13   Multifaceted polyhedra from the icosahedron.

Then

> angle between radius and an edge $= b = \arccos(b)$
> edge length $= 2b$
> distance from origin to center of edge $= a$
> distance from origin to center of face $= \dfrac{ta}{\sqrt{3}}$

The 12 vertices may be placed at

$$(\ 0, \ \pm a, \ \pm b)$$
$$(\pm b, \ \ 0, \ \pm a)$$
$$(\pm a, \ \pm b, \ \ 0)$$

Then midpoints of the 20 faces are given by

$$\tfrac{1}{3}(\pm d, \ \pm d, \ \pm d)$$
$$\tfrac{1}{3}(\ \ 0, \ \pm a, \ \pm c)$$
$$\tfrac{1}{3}(\pm c, \ \ \ 0, \ \pm a)$$
$$\tfrac{1}{3}(\pm a, \ \pm c, \ \ \ 0)$$

To subdivide icosahedral faces further, several methods suggest themselves, the simplest being to divide each edge into n equal lengths and then construct n^2 congruent equilateral triangles on each face, pushing them out to the radius of the sphere for their final position. (There are better methods than this if more uniform face sizes are desired.)

A1.14 ROOT FINDING

Since polynomials of fifth and higher degree are not soluble in closed form, numerical (approximate) solutions are useful for them as well as for nonpolynomial functions. The Newton–Raphson method produces successive approximations to a real root of a differentiable function of one variable.

$$x^{i+1} = x^i - \frac{f(x^i)}{f'(x^i)}$$

Here x^i is the ith approximation to the root, and $f(x^i)$ and $f'(x^i)$ are the function and its derivative evaluated at x^i. The new approximation to the root is x^{i+1}. The successive generation of approximations can stop when they converge to a single value. The convergence to a root is governed by the choice of initial approximation to the root and by the behavior of the function in the vicinity of the root. For instance, several roots close together can cause problems.

The one-dimensional form of this method extends in a natural way to solving systems of simultaneous nonlinear equations. Given n functions F_i, each of n parameters, the problem is to find the set of parameters that drives all the functions to zero. Write the parameter vector **x**.

$$\mathbf{x} = \begin{bmatrix} x_1 \\ x_2 \\ \cdot \\ \cdot \\ \cdot \\ x_n \end{bmatrix}$$

Form the function column vector $\mathbf{F}$ such that

$$\mathbf{F}(\mathbf{x}) = \begin{bmatrix} F_1(\mathbf{x}) \\ F_2(\mathbf{x}) \\ \cdot \\ \cdot \\ \cdot \\ F_n(\mathbf{x}) \end{bmatrix}$$

The Jacobean matrix J is defined as

$$J = \begin{bmatrix} \dfrac{\partial F_1}{\partial x_1} & \dfrac{\partial F_1}{\partial x_2} & \cdots & \dfrac{\partial F_1}{\partial x_n} \\ \cdot & & & \\ \cdot & & & \\ \cdot & & & \\ \dfrac{\partial F_n}{\partial x_1} & & \cdots & \dfrac{\partial F_n}{\partial x_n} \end{bmatrix}$$

Then the extension of the Newton–Raphson formula is

$$\mathbf{x}^{i+1} = \mathbf{x}^i - J^{-1}(\mathbf{x}^i)F(\mathbf{x}^i)$$

which requires one matrix inversion per iteration.

EXERCISES

A1.1 $\mathbf{x}$ and $\mathbf{y}$ are two two-dimensional vectors placed tail to tail. Prove that the area of the triangle they define is $|\mathbf{x} \times \mathbf{y}|/2$.

A1.2 Show that points $\mathbf{q}$ in a plane defined by the three points $\mathbf{x}$, $\mathbf{y}$, and $\mathbf{z}$ are given by
$$q \cdot \left[(\mathbf{y} - \mathbf{x}) \times (\mathbf{z} - \mathbf{x}) \right] = \mathbf{x} \cdot (\mathbf{y} \times \mathbf{z})$$

A1.3 Verify that the vector triple product may be written as claimed in its definition.

A1.4 Given an arctangent routine, write an arcsine routine.

A1.5 Show that the closed form for the inverse of a 2×2 A matrix is
$$\frac{1}{\det A} \begin{bmatrix} a_{22} & -a_{12} \\ -a_{21} & a_{11} \end{bmatrix}$$

A1.6 Prove by trigonometry that the matrix transformations for rotation are correct.

A1.7 What geometric transformation is accomplished when a_{44} of a geometric transformation matrix A varies from unity?

A1.8 Establish conversions between the given line representations.

A1.9 Write a geometric transform to mirror points about a given plane.

A1.10 What is the line-equation representation of a line $L1$ through a point $\mathbf{x}$ and perpendicular to a line $L2$ (similarly represented)? Parallel to $L2$?

A1.11 Derive the ellipse results given in Section A1.10.

A1.12 Explicitly derive the values of D, E, and F minimizing the error term

$$\sum_{i=1}^{n} [f(x_i, y_i)]^2$$

in the general equation for a circle

$$x^2 + y^2 + 2Dx + 2Ey + F = 0$$

A1.13 Show that if points and lines are transformed as shown in Section A1.7.6, the transformed points indeed lie on the transformed lines.

A1.14 Explicitly derive the least-squared-error solution for lines represented as $ax + by + 1 = 0$.

A1.15 If three planes intersect in a point, is the inverse of

$$\begin{bmatrix} p1 & p2 & p3 & 0 \\ & & & 0 \\ & & & 0 \\ & & & 1 \end{bmatrix}$$

guaranteed to exist?

A1.16 What is the angle between two three-space lines?

A1.17 In two dimensions, show that two lines $\mathbf{u}$ and $\mathbf{v}$ intersect at a point $\mathbf{x}$ given by $\mathbf{x} = \mathbf{u} \times \mathbf{v}$.

A1.18 How can you tell if two line segments (defined by their end points) intersect in the plane?

A1.19 Find a 4×4 matrix that transforms an arbitrary direction (or point) to lie on the Z axis.

A1.20 Derive a parametric representation for planes based on three points lying in the plane.

A1.21 Devise a scheme for interpolation on a triangular grid.

A1.22 What does the homogeneous point $(x, y, z, 0)$ represent?

REFERENCES AND FURTHER READING

Computer Graphics

1. NEWMAN, W. M., and R. F. SPROULL. *Principles of Interactive Computer Graphics*, 2nd Ed. New York: McGraw-Hill, 1979.

2. CHASEN, S. H. *Geometric Principles and Procedures for Computer Graphic Applications*. Englewood Cliffs, NJ: Prentice-Hall, 1978.

3. FAUX, I. D., and M. J. PRATT. *Computational Geometry for Design and Manufacture*, Chichester, UK: Ellis Horwood Ltd, 1979.

4. ROGERS, D. F., and J. A. ADAMS. *Mathematical Elements for Computer Graphics.* New York: McGraw-Hill 1976.

Computer Vision

5. HORN, B. K. P. "VISMEM: Vision Flash 34." AI Lab, MIT, December 1972.

6. SOBEL, I. "Camera models and machine perception." AIM-21, Stanford AI Lab, May 1970.

7. DUDA, R. O. and P. E. HART. *Pattern Classification and Scene Analysis.* New York: Wiley, 1973.

8. ROSENFELD, A., and A. C. KAK. *Digital Picture Processing.* New York: Academic Press, 1976.

9. PAVLIDIS, T. *Structural Pattern Recognition.* New York: Springer-Verlag, 1977.

Geometry, Calculus, Numerical Analysis

10. WEXLER, C. *Analytic Geometry: A Vector Approach.* Reading, MA: Addison Wesley, 1962.

11. APOSTOL, T. M. *Calculus.*, Vol. 2, Waltham, MA: Blaisdell, 1962.

12. CONTE, S. D. and C. DEBOOR. *Elementary Numerical Analysis: An Algorithmic Approach.* New York: McGraw-Hill, 1972.

13. RALSTON, A. *A First Course in Numerical Analysis.* New York: McGraw-Hill, 1965.

14. ABRAMOWITZ, M, and I. A. STEGUN. *Handbook of Mathematical Functions.* New York: Dover, 1964.

15. HODGEMAN, C. D. (Ed.). *CRC Standard Mathematical Tables.* West Palm Beach, FL: CRC Press.

Geodesic Tesselations

16. BROWN, C. "Fast display of well-tesselated surfaces." *Computers and Graphics 4*, 1979, 77–85.

17. CLINTON, J. D. "Advanced structural geometry studies, part I: polyhedral subdivision concepts for structural applications," NASA CR-1734/35, September 1971.

Fast Transformations

18. PRATT, W. K. *Digital Image Processing.* New York: Wiley-Interscience, 1978.

19. ANDREWS, H. C., and J. KANE. "Kronecker matrices, computer implementation, and generalized spectra." *J. ACM*, April 1970.

Advanced
Control Mechanisms
Appendix 2

This appendix is concerned with specific control mechanisms that are provided by programming languages or that may be implemented on top of existing languages as aids to doing computer vision. The treatment here is brief; our aim is to expose the reader to several ideas for control of computer programs that have been developed in the artificial intelligence context, and to indicate how they relate to the main computational goals of computer vision.

A2.1 STANDARD CONTROL STRUCTURES

For completeness, we mention the control mechanisms that are provided as a matter of course by conventional research programming languages, such as Pascal, Algol, POP-2, SAIL, and PL/1. The influential language LISP, which provides a base language for many of the most advanced control mechanisms in computer vision, ironically is itself missing (in its pure form) a substantial number of these more standard constructs. Another common language missing some standard control mechanisms is SNOBOL. These standard constructions are so basic to the current conception of a serial von Neumann computer that they are often realized in the instruction set of the machine. In this sense we are almost talking here of computer hardware.

The standard mechanisms are the following:

1. *Sequence.* Advance the program counter to the next intruction.
2. *Branch instruction.* Go to a specific address.
3. *Conditional branch.* Go to a specific address if a condition is true, otherwise, go to the next instruction.
4. *Iteration.* Repeat a sequence of instructions until a condition is met.

5. *Subroutines.* Go to a certain location; execute a set of instructions using a set of supplied parameters; then return to the next instruction after the subroutine call.

All the standard control structures should be in the toolkit of a programmer. They will be used, together with the data structures and data types supplied in the working language, to implement other control mechanisms. The remainder of this appendix deals with "nonstandard" control mechanisms; those not typically provided in commercial programming languages and which have no close correlates in primitive machine instructions. Nonstandard control mechanisms, although not at all domain-specific, have developed to meet needs that are not the "lowest common denominator" of computer programming. They impose their own view of problem decomposition just as do the standard structures.

Less standard mechanisms are *recursion* and *co-routining*. Co-routining can be thought of as a form of recursion.

A2.1.1 Recursion

Recursion obeys all the constraints of subroutining, except that a routine may call upon "itself." The user sees no difference between recursive and nonrecursive subroutines, but internally recursion requires slightly more bookkeeping to be performed in the language software, since typically the hardware of a computer does not extend to managing recursion (although some machines have instructions that are quite useful here).

A typical use of a recursive control paradigm in computer vision might be:

To Understand-Scene (X);
(
If Immediately-Apparent(X)
then Report-Understanding-Of(X);
else
   (SimplerParts ← Decompose(X);
   ForEach Part *in* SimplerParts
     Understand-Scene(Part);
)
[ )

Recursion is an elegant way to specify many important algorithms (such as tree traversals), but in a way it has no conceptual differences from subroutining. A routine is broken up into subroutines (some of which may involve smaller versions of the original task); these are attacked sequentially, and they must finish before they return control to the routine that invokes them.

A2.1.2 Co-Routining

Co-routines are simply programs that can call (invoke) each other. Most high-level languages do not directly provide co-routines, and thus they are a nonstandard control structure. However, co-routining is a fundamental concept [Knuth 1973] and serves here as a bridge between standard and nonstandard control mechanisms.

Subroutines and their calling programs have a "slave–master" aspect: control is always returned to the master calling program after the subroutine has carried out its job. This mechanism not only leads to efficiencies by reducing the amount of executable code, but is considered to be so useful that it is built into the instruction set of most computers. The pervasiveness of subroutining has subtle effects on the approach to problem decomposition, encouraging a hierarchical subproblem structure. The co-routine relationship is more egalitarian than the subroutine relationship. If co-routine *A* needs the services of co-routine *B*, it can call *B*, and (here is the difference) conversely, *B* can call *A* if *B* needs *A*'s services.

Here is a simple (sounding) problem [Floyd 1979]: "Read lines of text, until a completely blank line is found. Eliminate redundant blanks between the words. Print the text, 30 characters to a line, without breaking words between lines." This problem is hard to program elegantly in most languages because the iterations involved do not nest well (try it!). However, an elegant solution exists if the job is decomposed into three co-routines, calling each other to perform input, formatting, and output of a character stream.

A useful paradigm for problem solving, besides the strictly hierarchical, is that of a "heterarchical" community of experts, each performing a job and when necessary calling on other experts. A heterarchy can be implemented by co-routines. Many of the nonstandard mechanisms discussed below are in the spirit of co-routines.

A2.2 INHERENTLY SEQUENTIAL MECHANISMS

A2.2.1 Automatic Backtracking

The PLANNER language [Hewitt 1972] implicitly implemented the feature of "automatic backtracking." The advisability of uniformly using this technique, which is equivalent to depth-first search, was questioned by those who wished to give the programmer greater freedom to choose which task to activate next [Sussman and McDermott 1972].

A basic backtracking discipline may be provided by recursive calls, in which a return to a higher level is a "backtrack." The features of automatic backtracking are predicated on an ability to save and reinstate the computational state of a process automatically, without explicit specification by the programmer.

Automatic backtracking has its problems. One basic problem occurs in systems that perform inferences while following a particular line of reasoning which may ultimately be unsuccessful. The problem is that along the way, perhaps many perfectly valid and useful computations were performed and many facts were added to the internal model. Mixed in with these, of course, are wrong deductions which ultimately cause the line of reasoning to fail. The problem: After having restored control to a higher decision point after a failure is noticed, how is the system to know which deductions were valid and which invalid? One expensive way suggested by automatic backtracking is to keep track of all hypotheses that contributed to deriving each fact. Then one can remove all results of failed deduction paths. This is generally the wrong thing to do; modern trends have abandoned the automatic backtracking idea and allow the programmer some control over what is restored upon failure-driven backtracking. Typically, a compromise is implemented in which the programmer may mark certain hypotheses for deletion upon backtracking.

A2.2.2 Context Switching

Context switching is a general term that is used to mean switching of general process state (a control primitive) or switching a data base context (a data access primitive). The two ideas are not independent, because it could be confusing for a process to put itself to sleep and be reawakened in a totally different data context.

Backtracking is one use of general control context switching. The most general capability is a "general GO TO." A regular GO TO allows one to go only to a particular location defined in a static program. After the GO TO, all bindings and returnpoints are still determined by the current state of processing. In contrast, a general GO TO allows a transfer not only across program "space," but through program "time" as well. Just as a regular GO TO can go to a predefined program label, a general GO TO can go to a "tag" which is created to save the entire state of a process. To GO TO such a tag is to go back in time and recreate the local binding, access, control, and process state of the process that made the tag.

A good example of the use of such power is given in a problem-solving program that constructs complex structures of blocks [Fahlman 1974].

A2.3 SEQUENTIAL OR PARALLEL MECHANISMS

Some language constructs explicity designate parallel computing. They may actually reflect a parallel computing environment, but more often they control a simulated version in which several control paths are maintained and multi-processed under system control. Examples here are module and message primitives given below and statements such as the CO-BEGIN, CO-END pairs which can bracket notionally parallel blocks of code in some Algol-like language extensions.

A2.3.1 Modules and Messages

Modules and messages form a useful, versatile control paradigm that is relatively noncommittal. That is, it forces no particular problem decomposition or methodological style on its user, as does a pure subroutine paradigm, for example. Message passing is a general and elegant model of control which can be used to subsume others, such as subroutining, recursion, co-routining, and parallelism [Feldman 1979].

There are many antecedents to the mechanism of modules communicating by messages described here. They include [Feldman and Sproull 1971; Hewitt and

Smith 1975; Goldberg and Kay 1976; Birtwhistle et al. 1973]. In the formulation presented by Hewitt, the message-passing paradigm can be extended down into the lowest level of machine architecture. The construction outlined here [Feldman 1979] is more moderate, since in it the base programming language may be used with its full power, and itself is not module and message based.

A program is made up of *modules.* A module is a piece of code with associated local data. The crucial point is that the internal state of a module (e.g. its data) is not accessible to other modules. Within a module, the base programming language, such as Algol, may be used to its full power (subroutine calls, recursion, iteration, and so forth are allowed). However, modules may not in any sense "call upon" each other. Modules communicate only by means of *messages.* A module may send a message to another module; the message may be a request for service, an informational message, a signal, or whatever. The module to whom the message is sent may, when it is ready, receive the message and process it, and may then itself send messages either to the original module, or indeed to any combination of other modules.

The module–message paradigm has several advantages over subroutine (or co-routine) calls.

1. If subroutines are in different languages, the subroutine call mechanisms must be made compatible.

2. Any sophisticated lockout mechanism for resource access requires the internal coding of queues equivalent to that which a message switcher provides.

3. A subroutine that tries to execute a locked subroutine is unable to proceed with other computation.

4. Having a resource always allocated by a single controlling module greatly simplifies all the common exclusion problems.

5. For inherently distributed resources, message communication is natural. Module-valued slots provide a very flexible but safe discipline for control transfers.

Another view of messages is as a generalization of parameter lists in subroutine or coroutine calls. The idea of explicitly naming parameters is common in assembly languages, where the total number of parameters to a routine may be very large. More important, the message discipline presents to a module a collection of suggested parameters rather than automatically filling in the values of parameters. This leads naturally to the use of semantic checks on the consistency of parameters and to the use of default values for unspecified ones, which can be a substantial improvement on type checking. The use of return messages allows multiple-valued functions; an answer message may have several slots. Messages solve the so-called "uniform reference problem"—one need not be concerned with whether an answer (say an array element) is computed by a procedure or a table.

There is yet another useful view of messages. One can view a message as a partially specified relation (or pattern), with some slot values filled in and some unbound. This is common in relational data bases [Astrahan et al. 1976] and artificial intelligence languages [Bobrow and Raphael 1974]. In this view, a message is a task specification with some recipient and some complaint departments to talk to about it. Various modules can attempt to satisfy or contract out parts of the task of filling in the remaining slots. A module may handle messages containing slots unknown to it. This allows several modules to work together on a task while maintaining locality. For example, an executive module could route messages (on the basis of a few slots that it understands) to modules that deal with special aspects of a problem using different slots in the message.

There is no apparent conflict between these varying views of messages. It is too early in their development to be sure, but the combined power of these paradigms seems to provide a qualitative improvement in our ability to develop vision programs.

A2.3.2 Priority Job Queue

In any system of independent processes on a serial computer, there must be a mechanism for scheduling activation. One general mechanism for accomplishing scheduling is the priority job queue. Priority queues are a well-known abstraction [Aho et al. 1974]. Informally, a priority job queue is just an ordered list of processes to be activated. A monitor program is responsible for dequeueing processes and executing them; processes do not give control directly to other processes, but only to the monitor. The only way for a process to initiate another is to enqueue it in the job queue. It is easiest to implement a priority job queue if processes are definable entities in the programming language being used; in other words, programs should be manipulable datatypes. This is possible in LISP and POP-2, for example.

If a process needs another job performed by another process, it enqueues the sub job on the job queue and *suspends* itself (it is *deactivated*, or put to sleep). The sub job, when it is dequeued and executed by the monitor, must explicitly enqueue the "calling" process if a subroutine effect is desired. Thus along with usual arguments telling a job what data to work on, a job queue discipline implies passing of control information.

Job queues are a general implementational technique useful for simulating other types of control mechanisms, such as active knowledge (Chapter 12). Also, a job queue can be used to switch between jobs which are notionally executing in parallel, as is common in multiprocessing systems. In this case sufficient information must be maintained to start the job at arbitrary points in its execution.

An example of a priority job queue is a program [Ballard 1978] that locates ribs in chest radiographs. The program maintains a relational model of the ribcage including geometric and procedural knowledge. Uninstantiated model nodes corresponding to ribs might be called hypotheses that those ribs exist. Associated with each hypothesis is a set of procedures that may, under various conditions, be used to verify it (i.e., to find a rib). Procedures carry information about preconditions that must be true in order that they may be executed, and about how to compute estimates of their utility once executed. These descriptive components allow an executive program to rank the procedures by expected usefulness at a given time.

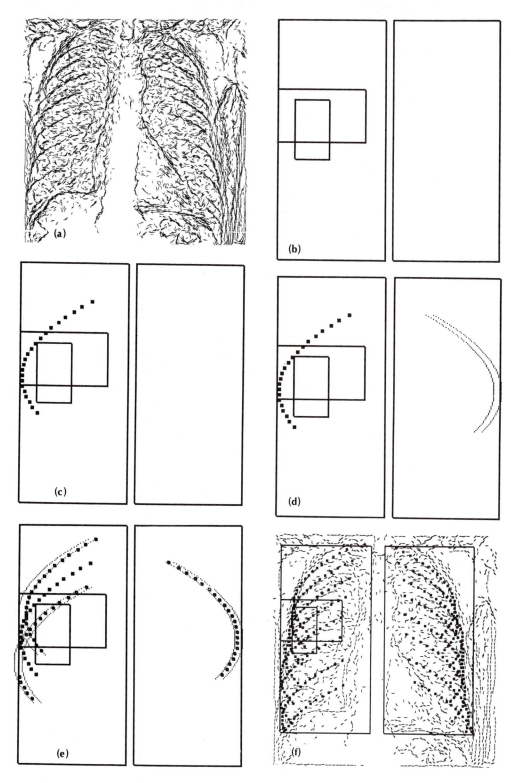

Fig. A2.1 The rib-finding process in action (see text).

There is an initial action that is likely to succeed (locating a particular rib that is usually obvious in the x-ray). In heterarchical fashion, further actions use the results of previous actions. Once the initial rib has been found, its neighbors (both above and below and directly across the body midline) become eligible for consideration.

Eligible rib-finding procedures correspond to short-term plans; they are all put on a job queue to be considered by an *executive* program that must compute the expected utility of expending computational energy on verifying one of the hypotheses by running one of the jobs. The executive computes a priority on the jobs based on how likely they are to succeed, using the utility functions and parameters associated with the individual nodes in the rib model (the individual hypotheses) and the current state of knowledge. The executive not only picks a hypothesis but also the procedure that should be able to verify it with least effort.

The hypothesis is either "verified," "not-verified," or "some evidence is found." Verifying a hypothesis results in related hypotheses (about the neighboring ribs) becoming eligible for consideration. The information found during the verification process is used in several ways that can affect the utility of other procedures.

The position of the rib with respect to instantiated neighbors is used to adjust horizontal and vertical scale factors governing the predicted size of the ribcage. The position of the rib affects the predicted range of locations for other unfound ribs. The shape of the rib also affects the search region for uninstantiated rib neighbors.

If some evidence is found for the rib, but not enough to warrant an instantiation, the rib hypothesis is left on the active list and the rib model node is not instantiated. Rib hypotheses left on the active list will be reconsidered by the executive, which may try them again on the basis of new evidence.

The sequence of figures (Fig. A2.1, p. 503) shows a few steps in the finding of ribs using this program. Figure A2.1a shows the input data. A2.1b shows rectangles enclosing the lung field and the initial area to be searched for a particular rib which is usually findable. Only one rib-finding procedure is applicable for ribs with no neighbors found, so it is invoked and the rib shown by dark boxes in Fig. A2.1b is found. Predicted locations for neighboring ribs are generated and are used in order by the executive which invokes the rib-finding procedures in order of expected utility (A2.1c-e). Predicted locations are shown by dots, actual locations by crosses; in Fig. A2.1f, all modelled ribs are found. The type of procedure that found the rib is denoted by the symbol used to draw in the rib. Figure A2.1f shows the final rib borders superimposed on the data.

A2.3.3 Pattern Directed Invocation

Considerable attention has been focused recently on pattern directed systems (see, e.g., [Waterman and Hayes-Roth 1978]). Another common example of a pattern directed system is the production system, discussed in Section 12.3. The idea behind a pattern directed system is that a procedure will be activated not when its name is invoked, but when a key situation occurs. These systems have in common that their activity is guided by the appearance of "patterns" of data in either input or memory. Broadly construed, all data forms patterns, and hence patterns guide any computation. This section is concerned with a definition of patterns as something very much smaller than the entire data set, together with the specification of control mechanisms that make use of them.

Pattern directed systems have three components.

1. A data structure or data base containing modifiable items whose structure may be defined in terms of patterns
2. Pattern-directed modules that match patterns in the data structure
3. A controlling executive that selects modules that match patterns and activates them

A popular name for a pattern-directed procedure is a *demon*. Demons were named originally by Selfridge [Selfridge 1959]. They are used successfully in many AI programs, notably in a natural language understanding system [Charniak 1972]. Generally, a demon is a program which is associated with a *pattern* that describes part of the knowledge base (usually the pattern is closely related to the form of "items" in a data base). When a part of the knowledge base matching the pattern is added, modified, or deleted, the demon runs "automatically." It is as if the demon were constantly watching the data base waiting for information associated with certain patterns to change. Of course, in most implementations on conventional computers, demons are not always actively watching. Equivalent behavior is simulated by having the demons register their interests with the system routines that access the data base. Then upon access, the system can check for demon activation conditions and arrange for the interested demons to be run when the data base changes.

Advanced languages that support a sophisticated data base often provide demon facilities, which are variously known as if-added and if-removed procedures, antecedent theorems, traps, or triggers.

A2.3.4 Blackboard Systems

In artificial intelligence literature, a "blackboard" is a special kind of globally accessible data base. The term first became prominent in the context of a large pattern directed system to understand human speech [Erman and Lesser 1975; Erman et al. 1980]. More recently, blackboards have been used as a vision control system [Hanson and Riseman 1978]. Blackboards often have mechanisms associated with them for invoking demons and synchronizing their activities. One can appreciate that programming with demons can be difficult. Since general patterns are being used, one can never be sure exactly when a pattern directed procedure will be activated; often they can be activated in incorrect or bizarre sequences not anticipated by their designer. Blackboards attempt to alleviate this uncertainly by controlling the matching process in two ways:

1. Blackboards represent the current part of the model that is being associated with image data;

2. Blackboards incorporate rules that determine which specialized subsystems of demons are likely to be needed for the current job. This structuring of the data base of procedures increases efficiency and loosely corresponds to a "mental set."

These two ideas are illustrated by Figs. A2.2 and A2.3 [Hanson and Riseman 1978]. Figure A2.2 shows the concept of a blackboard as a repository for only model-image bindings. Figure A2.3 shows transformations between model entities that are used to select appropriate groups of demons.

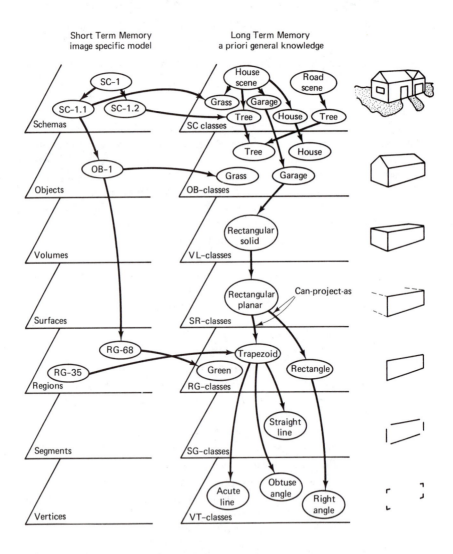

Fig. A2.2 An implementation of the blackboard concept. Here the blackboard is called Short Term Memory; it holds a partial interpretation of a specific image.

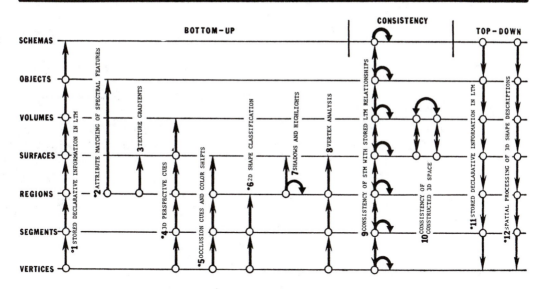

Fig. A2.3 Paths for hypothesis flow, showing transformations between model entities and the sorts of knowledge needed for the transformations.

REFERENCES

AHO, A. V., J. E. HOPCROFT and J. D. ULLMAN. *The Design and Analysis of Computer Algorithms.* Reading, MA: Addison-Wesley, 1974.

ASTRAHAN, M. M. et al. "System R: A relational approach to data base management." IBM Research Lab, February 1976.

BALLARD, D. H. "Model-directed detection of ribs in chest radiographs." Proceedings, Fourth IJCPR, Kyoto, Japan, 1978.

BIRTWHISTLE, G. et al. *Simula Begin.* Philadelphia: Auerbach, 1973.

BOBROW, D. G. and B. RAPHAEL. "New programming languages for artificial intelligence." *Computing Surveys 6,* 3, September 1974, 155–174.

CHARNIAK, E. "Towards a model of children's story comprehension." AI-TR-266, AI Lab, MIT, 1972.

ERMAN, L. D. and V. R. LESSER. "A multi-level organization for problem solving using many diverse cooperating sources of knowledge." *Proc.,* 4th IJCAI, September 1975, 483–490.

ERMAN, L. D., F. HAYES-ROTH, V. R. LESSER, and D. R. REDDY. "The HEARSAY-II speech-understanding system: Integrating knowledge to resolve uncertainty." *Computing Surveys 12,* 2, June 1980, 213–253.

FAHLMAN, S. E. "A planning system for robot construction tasks." *Artificial Intelligence 5,* 1, Spring 1974, 1–49.

FELDMAN, J. A. "High-level programming for distributed computing." *Comm. ACM 22,* 6, July 1979, 363–368.

FELDMAN, J. A. and R. F. SPROULL. "System support for the Stanford hand-eye system." *Proc.,* 2nd IJCAI, September 1971, 183–189.

FLOYD, R. W. "The paradigms of programming." *Comm. ACM 22,* 8, August 1979, 455–460.

GOLDBERG, A. and A. KAY (Eds). "SMALLTALK-72 Instruction Manual." SSL 76-6, Xerox PARC, Palo Alto, CA, 1976.

HANSON, A. R. and E. M. RISEMAN. "Visions: A computer system for interpreting scenes." In *CVS*, 1978.

HEWITT, C. "Description and theoretical analysis (using schemata) of PLANNER" (Ph.D. dissertation). AI-TR-258, AI Lab, MIT, 1972.

HEWITT, C. and B. SMITH. "Towards a programming apprentice." *IEEE Trans. Software Engineering. 1*, 1, March 1975, 26–45.

KNUTH, D. E. *The Art of Computer Programming*, Vol. 1. Reading, MA: Addison-Wesley, 1973.

SELFRIDGE, O. "Pandemonium, a paradigm for learning." In *Proc.*, Symp. on the Mechanisation of Thought Processes, National Physical Laboratory, Teddington, England, 1959.

SUSSMAN, G. J. and D. MCDERMOTT. "Why conniving is better than planning." AI Memo 255A, AI Lab, MIT, 1972.

WATERMAN, D. A. and F. HAYES-ROTH (Eds.). *Pattern-Directed Inference Systems*. New York: Academic Press, 1978.

Author Index

Subject Index

Atomic formula in logic, 384
Attention, control of, 340
Automated inference systems, 396

B-spline, 239–243
Background subtraction:
  low-pass filtering, 72
  spline surface, 72
Backtrack search, 363–365, 372–375
  automatic, 499
  variations and improvements, 364–365
Backward chaining, 342, 399
Bandlimited signal, 41
Basis for color space, 33–35
Bayes' rule, 449 (*See also* Decision, theory and planning)
Bayesian decisions and region growing, 162
Belief maintenance, 319, 346
Bending energy of curve, 256
Binary search correlation, 108
Binary tree, 244
Binocular imaging, 20–22 (*See also* Stereo vision)
Binormal of space curve, 276
Blackboard, 505
Blob finding, 143–146, 151
Block stacking, 322, 438–443
Blocks world:
  vision, 291
  structure matching, 370–372
Bottom-up (*See* Control; Inference)
Boundary, 75, 265
  conditions for B-splines, 241
  detection, 119–148
    in binary images, 143
    divide and conquer, 122
    dynamic programming, 137–143
    Hough algorithm, 123–131
  evaluation, 288–291
  as graph, 131
  representations, 232–247, 265–274
Branch and bound search:
  backtracking improvement, 364
  for boundaries, 136
Breakpoints in linear segmentation, 232

Calculus, predicate (*See* Predicate logic)
Camera model and calibration, 481–484
Cartesian coordinate system, 465
CAT imagery, 1, 56–59
Cell decomposition volume representation, 281
Centroid of volume, 285
Chain code, 235–237, 256, 258
  area calculation, 236
  derivative, 236
  merging, 236

normalized, 236
Chamfer matching, 354
Charge transfer devices, 49
Chessboard metric, 39
Chest radiograph understanding, 321, 344–346
Chromaticity diagram, 37
Chunks of knowledge, 334
Circular arcs, 237
City block metric, 38
Classification:
  in pattern recognition, 181–184
  set membership, 284
  tree for regions, 163
Clause form of predicate logic, 384
Clique, and use in matching, 358, 366–369, 375
Closed curves, 246
Closure operator for sets, 282
Clustering:
  motion detection, 217
  parametric and non-parametric for pattern
    recognition, 181–183
Co-routining, 498 (*See also* Control)
Coherence:
  of knowledge representation, 320
  rule for line-drawing interpretation, 297
Collision detection with optic flow, 201
Color, 31–35
  bases, 33–34
  -space histograms, 153–155
Comb, dirac, 19, 40
Combining operators for volumes, 282
Compactness of region, 256
Completeness of inference system, 389
Complexity of graph algorithms, 359
Component, r-connected, 369, 380
Computer as research tool, 9
Concave line label, 296
Concavity tree of region, 258
Cone, generalized (*See* Generalized, cone)
Confidence:
  planning, 415 (*See also* Supposition value in relaxation)
  region growing, 164
Conic, 239, 488–489
Conjunctive normal form for logic, 388
Connect line label, 303
Connected:
  component of graph, 369, 380
  region, 150, 255
Connectives of logic, 385
Connectivity:
  difference, 375
  image, 36
  matching, 372–375
CONNIVER, 322
Consolidation 102, (*See also* Pyramid)
Constraint (*See also* Relaxation)

inconsistency, 427
as inequality in linear programming, 423
labeling, 408–410
n-ary, 410
propagation, 299, 413–415
relaxation, 408–430
satisfaction for belief maintenance, 347
semantic, on region-growing, 160–164
Constructive solid geometry volume
    representation, 282
Context:
    data base, 440
    switching, 500
Continuity of knowledge representation, 320
Contour:
    following, 143–146
    occluding, 101
Contrast enhancement, 71
Control, 315, 340–350, 497–502
    bottom-up or data-driven, 341, 344–346
    hierarchical and heterarchical, 341–346
    in knowledge representations, 317
    message passing, 501
    mixed top-down and bottom-up, 344–346
    structures, standard and nonstandard, 497–500
    top-down, 342, 343–346
Convergence of relaxation algorithms, 414, 418
Conversions, logic to semantic nets, 332
Convex:
    decomposition of region, 253
    line label, 296
    region, 258
Convolution, 25, 68
    theorem, 30
Cooperative algorithms, 408–430
Coordinate systems, definitions and conversions,
    465–468
Correctness of inference system, 389
Correlation, 25, 30, 66–70
    binary search, 108
    coefficient, 419
    metrics, 362
    non-linear for edge linking, 121
    normalized, 68–70
    periodic and aperiodic, 67
    texture, 187
Correspondence problem, 89
Cost:
    of planning, 452
    in plans, 445–459
Crack edges, 78
Curvature:
    boundary, 256
    in evaluation function, 133
    space curve, 276
Curve 231:

detection, Hough algorithm, 126
fitting, 487
intersection, 247
segmentation techniques, 233–234
Cutting planes in linear programming, 428
Cylinder, generalized, 274–280
Cylindrical coordinate system, 466

Data:
    base, 398, 431, 440
    -driven control, 341–346
    fitting, 239, 484–488
    nodes in location networks, 336, 338
    structure for boundaries, 158
Decision:
    theory and planning, 446–453
    trees for matching, 370–377
Decomposition:
    region, 253
    solid, 287
Default values in knowledge representations, 330,
    334–335
Delete list, 440
Delta function, 18–19, 40
Demon, 412, 429, 505
DeMorgan's laws, 387
Densitometer, 46
Density of image, 44, 74
Dependence, gray-level, 186–188
Depth:
    -first search and variations, 136, 363–365, 372, 412
    from optic flow, 201
Determinant, 473
Difference measurement in motion, 221
Digital images, 35–42
Digitizers, image, 45
Dirac Comb, 19, 40
Direction-magnitude sets, 270
Discrete:
    images, 35–42
    knowledge representation, 320
    labeling algorithms, 410–415
Disparity, 21, 89, 208
Dispersion of knowledge representation, 320
Distance:
    on discrete raster, 36
    image (*See* Image, range)
Distortion, perspective (*See* Projection, perspective)
Divergence theorem for mass properties, 288
Divide and conquer:
    algorithms for CSG, 285
    method for boundary detection, 122
Domain-dependent and -independent motion
    understanding, 196–199, 214–219
Drum scanner, 46

Dual graph, 159
Dynamic programming and search, 137–143

Early processing, 63–65
Eccentricity of region, 255
Edge, 75
  detection
    in binary images, 143–146
    from optic flow, 202–206
    in pyramids, 109
  following, 131–146
    as blob finding, 143–146
    as dynamic programming, 137–143
    as graph search, 131–143
  labels, 296–297
  linking, 119–131
    known approximate location, 121–122
    problems with, 119–120
    Hough algorithm, 123–131
  operator, 64, 75–88
    gradient, 76–80
    Kirsch, 79
    Laplacian, 76–79
    performance, 77, 83–84
    relaxation, 85–88
    templates, 79
    3-D performance, 81–83
  profiles, 75
  representation for surfaces, 266
  strength in evaluation function, 133
  thresholding, 80
Eigenvalues and eigenvectors, 473, 486–487
Element, texture, 166
Elongation of region, 255
Enclosing surface, 265
Energy, texture, 187
Engineering:
  drawings, 291
  knowledge, 407
Entropy, texture, 187
ERTS imagery, 46
Euclidean metric, 38
Euler number of region, 255, 266
Evaluation:
  function for heuristic search, 133
  mechanism in semantic networks, 337
Existential quantifiers, 385
Extended inference, 315, 319, 322, 383, 395–396
Extensional concepts in knowledge representation, 328

Faces, 271
  for surface representation, 265, 271
Feature
  classification and matching, 376–378

texture, 184–186
  vectors and space, 181
Field, television, 46
Figure-ground distinction, 4
Filtering, 25, 64–75
First order predicate logic (*See* Predicate logic)
Fitting data (*See* Data, fitting)
Flat-bed scanner, 46
Flying spot scanner, 45
Focal length, 19, 479
Focus of expansion in optical flow, 199
Formal inference system, 390
Forward chaining, 342, 399
Fourier
  descriptors, 238
  filtering, 65
  transform, 24–30, 490–492
Frame:
  problem, 395, 444
  system theory, 334–335
Frenet frame and formulae, 276
Function:
  image, 18–19
  logic, 385
  Skolem, 387

G-Hough algorithm, 128
Gamma, film, 45
Gaussian sphere, 101, 270
Generalized:
  clipping, 284
  cone, 274–280
    matching to data, 278, 372–375
  cylinder (*See* Generalized, cone)
  image, 6, 14, 320
Geodesic tesselation, 271, 493
Geometric:
  matching, 354
  operations in location networks, 336
  relations and propositions, 332
  representations, 8, 227–311
  structures and matching, 354
  transformations, 477–481
Geometry theorem prover, 322
Gestalt psychology, 116
Goal achievement, 319, 346–347, 438–439
Goodness of fit, 273
Gradient:
  edge operator, 76–80
  space, 95, 301
  techniques, 355
  texture, 168, 189–193
  use in Hough algorithm, 124
Grammar:
  ambiguous, 172

array, 178–181
  on pyramid, 179
  shape, 173–174
  stochastic, 172
  texture, 172–181
  tree, 175–178
Graph 131:
  adjacency for regions, 159
  algorithms, complexity, 359
  association, 358, 365–369
  dual, 159
  isomorphism, 357–359, 364
  matching, 355
  r-connected component, 369, 380
Gray level, 18, 23, 35
  dependence matrices for texture, 186–188
Grazing incidence, 111

H&D curve, 44
Heart volume, 273
Heterarchical control, 341–346, 499
Heuristic search:
  boundaries, 131–133
  dynamic programming, 143
  region growing, 157
Hierarchical (*See also* Pyramid)
  abstractions, 505
  control, 341–346
  textures, 170
High-level:
  models, 317
  motion detection, 196–199
  vision, 2–6
Hill-climbing and matching, 355
Histogram, 70
  equalization and transformation, 70
  splitting for thresholds, 152
  color space, 153–155
Homogeneous:
  coordinate system, 467
  regions, 150
  texture, 188
Hough algorithm, 123–131
  generalized, 128–131
  refinements, 124
  vanishing points, 191
Human body for motion understanding,
  214–219
Hungry monkey planning problem, 445
Hypotheses, 343, 384, 422
  active knowledge, 431
Hypothetical worlds, 432, 440

Iconic structures and matching, 353
Icosahedron, 492
IHS color basis, 33
Image (*See also* Imaging)
  aerial, 1, 335
  CAT, 1, 56–59
  connectivity, 36

digital, 35–42
digitizers, 45
distance on raster, 36
edges, 75–88
ERTS or LANDSAT, 46
formation, 17
function, 18–19
generalized, 6, 14, 320
histogram, 70
intrinsic, 7, 14, 63
irradiance, 23, 73
orthicon, 47
plane, 19
processing, 2, 17, 25
range, 52–56, 64, 88
sampling, 18, 35
segmented, 7
sequence understanding, 207–222
ultrasound, 54
variance, 69
Imaging:
  active, 14
  devices, 42–59
  geometry, 19–21
  light stripe, 22
  model, 17–42
  monocular and binocular, 19–22
  stereo, 20–22, 52–54, 88–93, 98
Inconsistent labeling, 410 (*See also* Labeling)
Indexing property of semantic nets, 324
Inequalities in linear programming, 422, 427
Inference, 314, 319–321, 383
  bottom-up and top-down, 392
  extended, 315, 319, 322, 383, 395
  rules of, 388
  in semantic nets, 327
  systems, formal and informal, 390
  syllogistic, 321
Infinity, point at, 20
Informal inference system, 390
Inheritance of properties in knowledge
    representation, 330, 335
Inhibitory local evidence in line drawings, 295
Intensional concepts in knowledge
    representation, 328
Interaction graph for dynamic programming, 143
Interest operator, 69, 208
Interior operator, 282
Interpolation, 489–490
Interpretation:
  matching, 352
  region-growing, 160–164
Interpreter
  production system, 398–399
  semantic net, 326, 339
Intersection of strip trees, 244
Interval, sampling, 35
Intrinsic:
  image, 7, 14, 63
  parameters, 63

human body for motion, 217–219
in knowledge representation, 9, 317
Modules and messages, 500–502
Modus Ponens and Modus Tollens, 388
Moment of inertia, 255, 286, 473, 486
Monocular imaging (*See* Imaging)
Motion, 195 (*See also* Optic flow)
adjacency and collision detection, 201
body model, 214–219
common in sequence, 199, 208
consistent match, 199
continuity, 197
depth, 201
human body, 214–219
image sequences, 207–222
maximum velocity, 198
moving light displays, 214–217
observer, 206
rigid bodies, 197, 210–214
surface orientation and edge detection, 202
Multi-
dimensional histograms, 153–155
modal sensor, 453–459
resolution images, 100–110 (*See also* Pyramid)

Nearest-neighbor clustering, 183
Network:
interpreter, 326
representation, 391 (*See also* Semantic nets)
Newton-Raphson, 493
Node types in semantic nets, 324–329
Noise, 65
Nonclausal form, 385–387 (*See also* Predicate logic)
Nondeterministic algorithms, 359
Nonrigid:
body motion understanding, 214–217
solids, 264
Nonstandard:
control structures, 499–507 (*See also* Control)
inference (*See* Extended inference)
Normalized correlation, 68–70
NP-completeness, 359
NTSC, 34

Object identification in line drawings, 294
Occluding:
contour, 101
line label, 296
Oct-tree, 281, 287
Office scene understanding, 453–459
Operator (*See* Edge, Interest operator, Interior operator, Closure operator for sets, Planning Relaxation, etc.)
Opponent processes, 33

Optic flow, 65, 102–105, 196, 199–206
Optical system analysis, 23
Optimal labeling, 410
Optimization:
linear programming, 424–425
matching, 354
Orientation of surface (*See* Surface, orientation calculation)
Origami world, 300
Orthicon, image, 47
Orthographic projection (*See* Projection)

PANDEMONIUM, 345
Parallel:
computation, 64, 341, 360
-iterative refinement in graph matching, 358, 378
-iterative relaxation, 64, 412
Parameter:
optimization as matching, 354
space, 123
Parametric:
clustering, 183
edge models, 80–81
line representation, 476
Parseval's theorem, 256
Partial:
knowledge in location networks, 339
matches, 360–362
Partition:
feature space in pattern recognition, 181
Fourier space, 185
semantic nets, 331, 391
space, 150
Pattern:
-directed invocation, 321, 504
matching data base (*See* Data, base)
matching in production systems, 399–400
recognition, 2, 181–184
texture, 166
Performance, edge operators, 77, 83–84
Periodic:
correlation, 67
function, 237
Perspective (*See* Projection)
Photography, 44–45
Photometric stereo vision, 98
Picture element (pixel), 36
Piecewise polynomial, 240
Plane:
curves and regions, 231
cutting in linear programming, 428
representation, 476
transformation, 480
PLANNER, 322